This book reviews and interrelates a large number of theoretical and experimental contributions to research on finite plastic deformation of single crystals and polycrystalline metals made during the past quarter century. An overall theoretical framework for investigation of large strains in crystalline materials is presented that enables the blending of contemporary and earlier experimental research with modern concepts in solid mechanics. Professor Havner has provided an historical perspective throughout, including accurate attribution of ideas and emphasis on pioneering studies, beginning with G. I. Taylor and the German researchers in the 1920s.

Early chapters deal with single crystals, including extensive analyses of particular experimental configurations. Then connections between single crystals and polycrystalline material behavior are presented. Necessary background to the most general theoretical sections is given in an appendix, and approximately 200 references to the primary scientific literature are included.

Engineering scientists, applied mathematicians, and physical metallurgists will find this volume an invaluable guide to the development of finite plastic deformation theory.

CAMBRIDGE MONOGRAPHS ON
MECHANICS AND APPLIED MATHEMATICS

FINITE PLASTIC DEFORMATION OF CRYSTALLINE SOLIDS

Finite Plastic Deformation of Crystalline Solids

K. S. HAVNER

Departments of Civil Engineering and Materials Science and Engineering
North Carolina State University

CAMBRIDGE UNIVERSITY PRESS
Cambridge, New York, Melbourne, Madrid, Cape Town, Singapore, São Paulo

Cambridge University Press
The Edinburgh Building, Cambridge CB2 8RU, UK

Published in the United States of America by Cambridge University Press, New York

www.cambridge.org
Information on this title: www.cambridge.org/9780521392457

First published 1992
This digitally printed version 2008

A catalogue record for this publication is available from the British Library

Library of Congress Cataloguing in Publication data

Havner, K. S.

Finite plastic deformation of crystalline solids / K. S. Havner.

p. cm. – (Cambridge monographs on mechanics and applied mathematics)

Includes index.

ISBN 0-521-39245-4
1. Plasticity. 2. Crystals – Mechanical properties.
3. Deformations (Mechanics) I. Title. II. Series.
QD933.H38 1992
531′.38–dc20 91–24032
CIP

ISBN 978-0-521-39245-7 hardback
ISBN 978-0-521-05420-1 paperback

To Roberta, the "R. H." longest in my life

CONTENTS

PREFACE

Rodney Hill wrote his preface to *The Mathematical Theory of Plasticity* 41 years ago this month. As a reader of the present monograph likely knows, that classic work dealt with the macroscopic theory of metal plasticity and its applications as the subject stood at mid-century; and Hill only briefly (albeit superbly) discussed in his introductory chapter the physical background of the plastic properties of crystals and polycrystalline aggregates. The same year, however, saw publication of the English translation of an earlier (1935) classic specifically concerned with that background, *Kristallplastizität* by E. Schmid and W. Boas. Not entirely coincidentally, both Rodney Hill and this translation were associated with the Cavendish Laboratory, Cambridge, during the period immediately following World War II.

Each of these books when first published was in many respects a treatise on its respective subject, but there was no contemporary work which integrated these fields. Today, I doubt a comprehensive treatise could be written on all that has transpired both in the development of mathematical theory and in the experimental study of plastic behavior of crystalline materials during this century (or even since 1950). Accordingly, in planning and carrying out the writing of the present work, I decided to restrict its scope to those aspects of the broad subject of crystalline plasticity that have particularly interested me and that I have contributed to or at least seriously studied during the years since 1968. Consequently, such important topics as shear band formation and plastic wave propagation are not treated; and numerical methodologies for large computer solution of boundary value problems are mentioned only in passing.

The greater part of the book, Chapters 1 through 5, deals primarily with uniform deformation of single crystals of cubic lattice structure: their finite-strain experimental behavior, and theories and analyses directed toward predicting that behavior. Chapters 6 and 7 focus upon polycrystalline aggregates, the former presenting a rigorous investigation of theoretical connections between individual crystal and macroscopic material response, and the latter chapter providing a survey of approximate polycrystal models from 1938 to 1990.

In setting level and style, I have assumed readers to be familiar with the basic elements of continuum kinematics and the theory of stress at finite strain (including an understanding of rotating reference frames and objective tensors) and also to be acquainted with stereographic projections, Miller index notation, and both face-centered cubic (f.c.c.) and body-centered cubic (b.c.c.) crystal structure. Familiarity with standard concepts from macroscopic plasticity theory also is anticipated but is, I believe, of lesser importance. A rather more sophisticated background, related to general conjugate measures of stress and strain, is called upon from time to time; and a concise but self-contained account of that background is presented in the Appendix.

In organizing the overall work, I chose (primarily for variety and change of pace) to move back and forth between (a) analyses related to specific experimental configurations of cubic crystals or polycrystals and (b) more general and abstract theoretical developments applicable, in principle, to any crystal class finitely deforming by lattice distortion and crystalline slip. Correspondingly, although my intent has been to tell an orderly and coherent story from first page to last, a reader wishing to avoid the most mathematical material, at least on a first reading, may pass over Chapters 3 and 6 (and the Appendix) and direct attention to the balance of the monograph.

With respect to references, I have consistently gone to original sources rather than rely on accounts of them by others; and throughout the book I have sought to provide historical perspective, accurate attribution of original ideas, and recognition of pioneering studies (both experimental and theoretical) as I see them. There is also the matter of selection; and as I have cited only approximately 200 references from a considerably greater literature (particularly on the metallurgical side), some qualifying statements are in order. First, there likely are important and germane contributions of which I am unaware, and I apologize to the authors for having overlooked their work. Second, there are a number of papers that I reviewed and should have liked to cite but did not because reference to them did not arise naturally in the book as it finally evolved. Third, there may be papers whose merits I failed to appreciate on initial reading and so did not cite. I apologize to these authors as well and trust that I shall be wiser at some future occasion and have the opportunity to correct my omissions. (The manuscript, save for minor corrections, was completed on the last day of last year. With the exception of a paper then in press that I co-authored, no work published after fall 1990 is cited.)

There are a number of people whose contributions to this book, direct or indirect, I wish to acknowledge, beginning with T. H. Lin. My reading in October 1967 of his then most recent papers on polycrystalline aggregate analysis inspired and initiated my own study of crystalline plasticity; and I have benefited from his friendship, encouragement, and advice over the years. The papers on metal plasticity of the late Sir Geoffrey Taylor (1886–1975) have been an inspiration since I first began to study them in 1969; and the remarkable and seminal contributions he made to crystal mechanics are in evidence throughout this book. The debt I owe Rodney Hill is greatest of all, but I shall defer expressing it until the end of the preface.

I thank the series editors, George Batchelor and Ben Freund, for their strong support of this project from the outset, and of course for reading and accepting the completed work. Additional thanks are due George Batchelor for helping me recognize while on sabbatical at Cambridge that the book I had originally conceived, and began there in January 1989, was actually two books: one a graduate text I still hope to complete, the other the proposal which became this monograph. Approximately 30 percent of this book was then written in Cambridge during March–June 1989.

I thank Keith Moffatt, Head of the Department of Applied Mathematics and Theoretical Physics, for providing office space and the services and hospitality of the department during my six-month stay in Cambridge, and North Carolina State University for granting my second Cambridge sabbatical. I also acknowledge the privilege and pleasure of membership in Clare Hall on both occasions (where first I was a Visiting Fellow in 1981).

John Hudson's positive reaction to Chapter 1 provided further encouragement before I left Cambridge, and I thank him as well for his later reading of other parts of the book. I also thank my colleague, Bob Douglas, for reading Chapter 1 and several sections of Chapter 4. The contributions of my students whose names appear prominently in Chapters 4 and 5 were essential to completion of the analyses and numerical results presented there, and I thank them all.

It has been a pleasure to work with Alan Harvey and Peter-John Leone, officers of Cambridge University Press, the former throughout this project, first at Cambridge, then in New York; and I appreciate their efforts on my behalf. I thank Annette Maynard for her extraordinarily accurate typing of my handwritten manuscript, and for her unfailing good humor throughout more than ten years of frequently having to face my equations at eight in the morning. I also acknowledge the assistance of Huyen-Tran Ton-nu, Science and Engineering Reference Librarian, University of Virginia, in tracking down the original publication details of a paper by George Green. Finally, thanks are due the National Science Foundation, Solid Mechanics Program, for almost continuous support of my research from 1971 to 1990, and in particular to Cliff Astill, Program Director, 1971–85.

This preface ends as it began, with Rodney Hill. I am grateful for his advice and support through an extensive correspondence spanning 22 years, for the opportunity to have collaborated in 1981, for his early encouragement of my book-writing plans and his positive comments on the finished work, and for the exemplars (in both substance and style) of his own papers on finite deformation, many of which are utilized in this book. Without his continued innovative contributions since publication of his classic monograph in 1950, only a shadow of this work could have been written.

Raleigh, North Carolina
May 9, 1991

1

AN HISTORICAL INTRODUCTION

The scientific study of finite distortion of cubic metal crystals was formally inaugurated by G. I. Taylor's 1923 Bakerian Lecture to the Royal Society (Taylor & Elam 1923), and the development of a rational mechanics of finite plastic deformation of crystalline solids may be said to have begun. The remarkable feature of this pioneering work (and of all Taylor's subsequent experimental investigations of f.c.c. and b.c.c. crystals) is that, in addition to the use of X-ray analysis to determine the changing orientation of the crystal atomic lattice, external measurements sufficient to completely define the uniform distortion of the crystal specimen were made at each stage of the test. All material directions that remained unchanged in length were then established by exact geometric analysis.

Taylor's approach is to be distinguished from that of Mark, Polanyi & Schmid (1923), whose experimental study of crystals of hexagonal structure is of comparable historical importance. They pulled single crystal wires of zinc and assumed that slip lines (or bands) on the specimen surface were traces of a family of planes of single slip, which were then shown to coincide with a crystal plane. Taylor (1926), discussing such methods in general, rather amusingly remarked: "They depend in fact on knowing the form which the answer will take before starting to solve the problem."

Taylor & Elam (1923) firmly established two experimental laws of fundamental and lasting value for the foundations of crystal mechanics that seven decades of experimentation on f.c.c. crystals have only served to reinforce. As concisely restated by Taylor & Elam (1925 p. 28), in reference to their initial work, "when a 'single crystal' bar of aluminium is stretched the whole distortion during a large part of the stretching[1] is

[1] Up to at least 40% extension of the test specimen (Taylor & Elam 1923).

due to a simple shear parallel to an octahedral (111) plane, and in the direction of one of the three diad (110) axes lying in that plane. Of the twelve[2] crystallographically similar possible modes of shearing, the one for which the component of shear stress in the direction of shear was greatest was the one which actually occurred."

In addition to further experimental confirmation of these laws, Taylor & Elam (1925) contains the first identification on a (100) stereographic projection of the 24 slip-system domains for axially loaded f.c.c. crystals that are their consequence. Moreover, this second paper explicitly introduces a deformation-dependent physical property of each slip system (implied in their earlier work) which they call the "resistance to slipping" per unit area of the slip plane and which is, in essence, the *critical strength* of a slip system subsequently adopted in this monograph.

The general and geometrically rigorous method of Taylor, followed in all his experimental investigations of f.c.c. and b.c.c. crystals (Taylor & Elam 1923, 1925, 1926; Taylor 1925, 1926, 1927a, b, 1928; Taylor & Farren 1926), so far as I can find has not been equaled elsewhere in the experimental literature on crystal deformation. Commonly, the rotation of the crystal axes is determined by X-ray analysis, but no distortion measurements are made[3] (apart from extension or compression in the loading direction), and only occasionally is there evidence of efforts comparable to those of Taylor and his collaborators to ensure uniformity of deformation. In particular, one seldom finds the use of scribed grids of parallel lines on crystal faces (as were always employed by Taylor) whose observation and measurement can confirm that uniformity. Taylor's classic experimental papers on crystal plasticity (1923–8) are models nonpareil, and his geometric method is reviewed in some detail in the following.

A section of a uniformly distorted crystal bar with two sets of scribed parallel lines on each longitudinal face is depicted in Fig. 1.1. (Although the bar is shown as a parallelepiped, the analysis applies equally well to crystal specimens triangular in cross section, as in the experiments of Franciosi & Zaoui (1982b), with the third longitudinal face a plane through points A, B, D, E.) This system of marking was introduced by Taylor & Elam (1923) and used in all their tensile experiments on aluminum and iron crystals. The choice of Cartesian reference frame in Fig. 1.1, with

[2] Here they add the footnote: "Or 24 if shears in opposite senses parallel to the same line be regarded as distinct." In this monograph we shall invariably make that distinction.

[3] Notable exceptions to this and the immediately following statement will be analyzed and discussed later in this monograph.

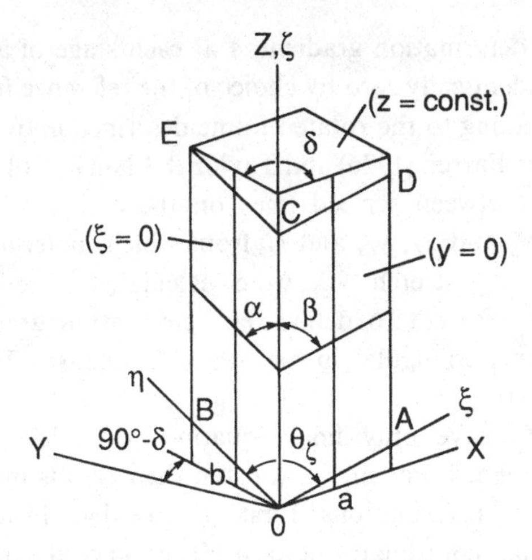

Fig. 1.1. Deformed crystal bar showing scribed lines.

plane $y = 0$ a crystal face and axis Z a longitudinal edge, was first presented in Taylor & Elam (1926). I have adopted it here both for historical reasons and because it also is useful in analysis of the channel die compression test treated later in this monograph (where X, Y are the loading and channel constraint directions, Z is the channel axis, and δ is a constant right angle as the cross section is constrained to remain rectangular). For uniaxial compression experiments on aluminum and iron crystals, Taylor & Farren (1926) and Taylor & Elam (1926) used crystal specimens that were initially right cylindrical discs of small thickness, with grids scribed only on the end faces and the reference frame chosen so that Z was normal to these faces and X was coincident with a scribed line through the center of the original circular cross section. (In other words, the frame in Fig. 1.1 would be rotated such that X coincided with scribed line OA and the xy plane contained scribed line OB. Z would then be normal to material planes AOB and DCE.)

With reference to Fig. 1.1, Taylor & Elam (1923, 1925, 1926) measured angles α, β, δ, axial stretch λ_z, and either transverse dimensions a, b or transverse thicknesses $a \sin\delta$, $b \sin\delta$ at each stage of every tensile experiment (with Z the loading direction) and at several sections along the length of a crystal specimen to assure uniformity of the deformation. As there usually were small differences in measurements (see Taylor & Elam 1923, table 1), they averaged readings at the different sections and on opposite faces to establish the "mean parallelepiped" for a given extension. These six (averaged) measurements are sufficient to completely define the nonsymmetric deformation gradient A at each stage of a test, since a_{xz}, a_{yx} and a_{yz} are identically zero by choice of the reference frame. In compression, corresponding to the rotated frame described above, the measurements of Taylor & Farren (1926) and Taylor & Elam (1926) were, in essence, the angle θ_ζ (between scribed lines on the end faces) and coordinates x_A, x_B, and y_B and x_C, y_C, and z_C from which material line stretches $\lambda_\xi, \lambda_\eta$ and spacing stretch $\lambda_{(z)}$ were calculated. These six measurements again are sufficient to define the deformation gradient, whose matrix array is upper triangular in the rotated frame (see Taylor & Farren 1926, eqs. (5)–(7)).

Taylor & Elam (1926) gave only final equations of deformation corresponding to the reference frame of Fig. 1.1, but their results may be obtained by two simple transformations. First, it is evident from the geometry that the Cartesian coordinates of an arbitrary crystal point at a given stage of an experiment (i.e., in the current state) are

$$\begin{aligned} x &= \lambda_\xi \xi_0 \sin\beta + \lambda_\eta \eta_0 \sin\alpha\cos\delta, \quad y = \lambda_\eta \eta_0 \sin\alpha \sin\delta, \\ z &= \lambda_\xi \xi_0 \cos\beta + \lambda_\eta \eta_0 \cos\alpha + \lambda_\zeta \zeta_0, \end{aligned} \tag{1.1}$$

where ξ_0, η_0, ζ_0 are the initial coordinates of the point on the embedded skew axes and (as previously mentioned) λ_ξ, λ_η, and $\lambda_\zeta = \lambda_z$ are the material stretches after deformation. Then, inverting these equations evaluated in the reference state, one obtains

$$\begin{aligned} \xi_0 &= \frac{x_0}{\sin\beta_0} - \frac{y_0 \cot\delta_0}{\sin\beta_0}, \quad \eta_0 = \frac{y_0}{\sin\alpha_0 \sin\delta_0}, \\ \zeta_0 &= -x_0 \cot\beta_0 + \frac{y_0}{\sin\delta_0}(\cot\beta_0 \cos\delta_0 - \cot\alpha_0) + z_0 \end{aligned} \tag{1.2}$$

(as $\lambda_\xi^0 = \lambda_\eta^0 = \lambda_z^0 = 1$). Upon substituting equations (1.2) and (from the geometry)

$$\lambda_\xi = \frac{a \sin\beta_0}{a_0 \sin\beta}, \quad \lambda_\eta = \frac{b \sin\alpha_0}{b_0 \sin\alpha}$$

into equations (1.1), one obtains Taylor & Elam's (1926) equations

$$\begin{aligned} x &= \frac{a}{a_0} x_0 + \left(\frac{b\cos\delta}{b_0 \sin\delta_0} - \frac{a}{a_0}\cot\delta_0\right) y_0, \quad y = \frac{b \sin\delta}{b_0 \sin\delta_0} y_0, \\ z &= \left(\frac{a}{a_0}\cot\beta - \lambda_z \cot\beta_0\right) x_0 + \left\{\frac{b \cot\alpha}{b_0 \sin\delta_0} - \frac{a}{a_0}\cot\beta\cot\delta_0\right. \\ &\quad \left. + \frac{\lambda_z}{\sin\delta_0}(\cot\beta_0 \cos\delta_0 - \cot\alpha_0)\right\} y_0 + \lambda_z z_0, \end{aligned} \tag{1.3}$$

and the six nonzero components $a_{xx}, a_{xy}, a_{yy}, a_{zx}, a_{zy}$, and a_{zz} of the general deformation gradient $A = \partial\mathbf{x}/\partial\mathbf{x}_0$ are defined. (It may be noted that a counterclockwise rotation of the reference frame of Fig. 1.1 about the ζ-axis until the plane $x = 0$ coincided with crystal face *OBCE* would make the deformation gradient lower triangular.) In the reference frame adopted for uniaxial compression measurements by Taylor & Farren (1926), their equations (5)–(7) for the upper-triangular deformation gradient may be similarly derived.

We now turn to the second part of Taylor's general method, which is the determination of the current and reference positions of the cone of unextended material lines emanating from an arbitrary point of the uniformly strained crystal. Taylor's constructions and arguments in the Bakerian Lecture for finding the "unstretched cone" (Taylor & Elam 1923) are in the geometric spirit of Newton's *Principia* (1687). Here we shall follow the analytical procedure in Taylor & Elam (1926), but in a direct vector/tensor notation.

Let $\boldsymbol{\iota}_0$ and $\boldsymbol{\iota}$ denote unit vector elements in the reference and current deformed crystal configurations of a material line *OL* that is unchanged in length; whence $A\boldsymbol{\iota}_0 = \boldsymbol{\iota}$, and equations

$$\boldsymbol{\iota}_0(A^{\mathrm{T}}A - I)\boldsymbol{\iota}_0 = 0, \quad \boldsymbol{\iota}(A^{-\mathrm{T}}A^{-1} - I)\boldsymbol{\iota} = 0 \tag{1.4}$$

define the unstretched cone in the respective configurations. For the reference frame of Fig. 1.1, the Cartesian components of A are given by equations (1.3), and the elements of the inverse deformation gradient $A^{-1} = \partial\mathbf{x}_0/\partial\mathbf{x}$ may be read from those of A by replacing λ_z by $1/\lambda_z$ and deleting the subscript "0" where it appears in equations (1.3) and adding it where it does not (e.g., $a_{zx}^{-1} = (a_0/a)\cot\beta_0 - (1/\lambda_z)\cot\beta$). Denoting the respective spherical angles of direction *OL* in the reference and current states by θ_0, ϕ_0 and θ, ϕ, we have

$$\begin{aligned} \boldsymbol{\iota}_0 &= (\sin\theta_0\cos\phi_0, \sin\theta_0\sin\phi_0, \cos\theta_0), \\ \boldsymbol{\iota} &= (\sin\theta\cos\phi, \sin\theta\sin\phi, \cos\theta). \end{aligned} \tag{1.5}$$

The equation of the reference position of the unstretched cone then follows upon substitution of equation $(1.5)_1$ and the elements of A from equations (1.3) into equation $(1.4)_1$. The result is

$$\begin{aligned} &\{(a_{xx}\cos\phi_0 + a_{xy}\sin\phi_0)^2 + (a_{yy}\sin\phi_0)^2\}\sin^2\theta_0 \\ &\quad + \{(a_{zx}\cos\phi_0 + a_{zy}\sin\phi_0)\sin\theta_0 + a_{zz}\cos\theta_0\}^2 = 1, \end{aligned} \tag{1.6}$$

which is equivalent to equation (6) of Taylor & Elam (1926). The equation

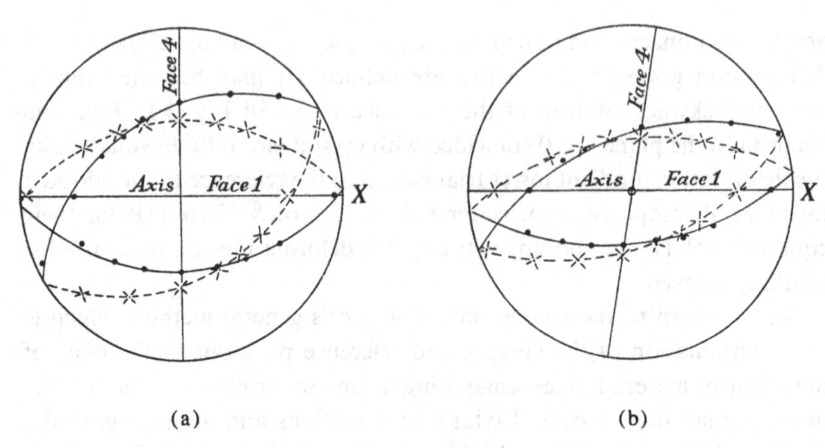

Fig. 1.2. Stereographic projections of (a) reference (unstrained) and (b) current positions of the cone of unstretched material lines in an aluminum crystal after an axial extension of 30% (from Taylor & Elam (1923, figs. 7 and 8), as presented in *The Scientific Papers of Sir Geoffrey Ingram Taylor*, Vol. I).

of the unstretched cone in the deformed crystal may be obtained simply by dropping the subscript 0 in equation (1.6) and replacing each $a_{ij}(i, j = x, y, z)$ by the corresponding element a_{ij}^{-1}. However, this equation is not needed, since a preferred procedure of calculation would be to determine θ, ϕ (for any θ_0, ϕ_0 satisfying equation (1.6)) from

$$\mathbf{\iota} = A\mathbf{\iota}_0, \quad \theta = \arccos \iota_z, \quad \phi = \arctan(\iota_y/\iota_x). \tag{1.7}$$

Varying ϕ_0 between $-90°$ and $90°$ encompasses all families of planes in a crystal specimen that are parallel to longitudinal edge OZ (Fig. 1.1). For a particular plane ϕ_0 through OZ at any stage of the deformation, equation (1.6) gives two angles θ_0, which define the initial positions of the unstretched lines in the chosen plane. These lines may be plotted as points on a stereographic projection (as yet unrelated to the underlying crystal lattice) with Z as the pole and X to the right (Fig. 1.2). Thus, to define a line in the referential unstretched cone, draw a radial line at the chosen angle ϕ_0 counterclockwise from the zx plane ($\phi_0 = 0$) on the stereographic projection and plot each angle θ_0 satisfying equation (1.6) as a radial distance $r = r_0 \tan(\frac{1}{2}\theta_0)$ along line ϕ_0, where r_0 is the radius of the projection. A direction lying in the current configuration of the unstretched cone is similarly plotted on another projection after ϕ and θ are determined for that deformation from equations (1.7).

In Figs. 1.2a, b are shown stereographic projections and plotted points in the reference and deformed configurations from the original experiment of Taylor & Elam (1923) corresponding to a 30 percent extension of an aluminum crystal that was nearly square-cut ($\delta_0 = 90.6°$). Faces 1 and 4 are equivalent to faces $OACD$ and $OBCE$ of the crystal depicted in Fig. 1.1. The dots in Figs. 1.2a, b correspond to $\theta_0 < 90°$, and the crosses to $\theta_0 > 90°$. It is seen that all points calculated from the measurements lie almost exactly on two great circles (each a lens-shaped figure consisting of one continuous curve and one dashed curve). This same result was found for each of 10, 15, 20, and 40 percent extensions. Thus, *the unextended cone is a degenerate form consisting of two planes.* Consistent with this result, the volume change was negligible (only a few tenths of a percent), and Taylor and Elam concluded, "The strain brought about by the extension of 30 percent can therefore be regarded as being due to a simple shear parallel to either of the two unstretched planes" (p. 655). From their measurements, a similar statement applied for all (nominal) axial strains through 40 percent for their aluminum crystal specimen.

To establish that in a simple shear there always are two families of planes on either of which a shear (or continuous slip) of the same amount could have taken place to produce the given deformation, consider the geometric constructions of Fig. 1.3. Here a unit cube of material in the reference state is chosen such that two parallel faces (seen as lines OL and MN in the figure) are planes of unextended directions after a primary shear (defined by angle χ_1) in direction m_1 on the family of planes with normal direction n_1. To construct an alternative simple shear of equal amount on a secondary family of planes that produces exactly the same deformation, proceed as follows. First locate P_1 midway between N and N_1. The original position P of this point is then equidistant from N on the other side so that $PN = NP_1 = P_1N_1 = \tan\alpha = \frac{1}{2}\tan\chi_1$, and line LP represents the second family of unstretched planes. (Angles α, β here are unrelated to α, β in Fig. 1.1.) Construct lines n_2 and OO_2, respectively normal and parallel to LP, choosing point O_2 to be equidistant with O from normal n_2 (as shown) and draw line O_2L. Then construct line M_2N_2 through P parallel to O_2L and of equal length, locating point N_2 to make $PN_2 = P_1N_1$, and draw the parallel lines O_2M_2 and LN_2. (The angle of shear in this secondary system is defined by χ_2 as shown.) From $\angle LPN_2 = 90° + \alpha$, $PN_2 = \tan\alpha$, $LP = \sec\alpha$, and the law of cosines there follows $LN_2 = \sec\chi_1$. Thus, as $LN_1 = \sec\chi_1$, triangle LPN_2 equals triangle LP_1N_1 and $\beta = \chi_1 - \alpha$. But $\chi_2 = \alpha + \beta$ (from $\angle O_2LN_2 = 90° + \alpha + \beta$). Therefore $\chi_2 = \chi_1$ and the two configurations OLM_1N_1 and $O_2LM_2N_2$ are in exact coincidence after a clockwise rigid rotation of the latter by 2α about L.

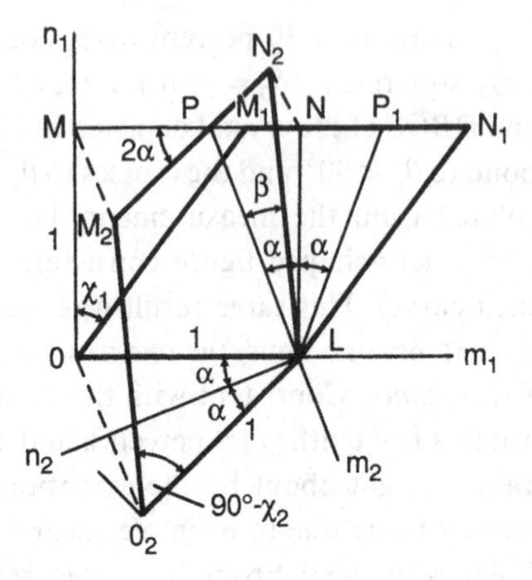

Fig. 1.3. Alternative geometric representations of a simple shear.

In a continuous simple shear one of the two planes represented by lines *OL* and *LP* in Fig. 1.3 is of course only momentarily unstretched, and in the next increment of deformation either *OL* or *LP* will change its length, depending upon whether m_1, n_1 or m_2, n_2 is the actual slip system. From their measurements at each of 10, 15, 20, 30, and 40 percent extensions, Taylor & Elam (1923) were able to readily identify the actual plane of slip as the uniquely unstretched material plane (*OL*, say) common to all levels of distortion. The direction of the finite shear at any stage of the deformation is normal to the line of intersection of the continuously unstretched plane (*OL*) with the momentarily unstretched one (*LP*), and Taylor and Elam demonstrated that, within experimental error, this was a fixed material line of the crystal specimen for all strains. Only after the material plane and direction of the simple shear were unambiguously established from external measurements were the changing orientations of the lattice planes and directions (as determined by X-ray diffraction) compared with the crystal deformations.

The changes in relative lattice angles were found to be very small, and were considered as due primarily to measurement errors[4] whose effects were minimized by an averaging process, so that the gross lattice essentially underwent rigid body motion without distortion. Finally, comparison of the positions of lattice planes and directions with both the plane and direction of simple shear and the relative rotation of the loading axis led to Taylor and Elam's basic experimental laws already given. The carefully

[4] In Taylor & Elam (1925) markedly more accurate measurements were achieved, indicating negligible changes in lattice angles.

worded original statement from the 1923 Bakerian Lecture of the first of their conclusions is worth quoting in full. "The fact that the position of the slip-plane determined by external measurements is so close to the (111) plane determined by X-ray analysis makes one suspect that they really coincide, and that the difference between their measured positions is due to experimental error. This idea is confirmed by the fact that the direction of slip is so exactly parallel to one of the three principal lines of atoms in the (111) plane" (p. 661).

During the period 1926–31 the experimental laws of Taylor and Elam were shown to apply equally to many other f.c.c. metal (and alloy) crystals in tension by Elam in England and by various investigators in Germany, as summarized in Elam (1935, pp. 26–7). The applicability of the laws to f.c.c. crystals (specifically aluminum) in compression was established by Taylor & Farren (1926), adopting the alternative reference frame and measurements described previously and once again applying Taylor's general method of geometric analysis to find the unstretched cone. They concluded, "so far as these experiments go, the distortion of a crystal of aluminium under compression is of the same nature as the distortion which occurs when a uniform single-crystal bar is stretched" (p. 551), restating the basic laws but adding that the choice of slip system depends "not at all on whether the stress normal to the slip plane is a pressure or a tension."

The tangent to the angle of shear, $\gamma = \tan \chi_1$ (fig. 1.3), was explicitly introduced as the kinematic measure of single slip in Taylor & Elam (1923), but when they first displayed their experimental results for the increasing resistance to slipping (critical strength) τ of four aluminum tensile specimens in Taylor & Elam (1925) they plotted τ versus the axial stretch λ_z. Apparently it had not yet occurred to them that a better correlation might be obtained by choosing the work-conjugate variables τ and γ. (In single slip, neglecting lattice straining, $\tau d\gamma$ is differential work per unit lattice volume.) That same year of publication, however, Taylor (1925), in a little-known paper,[5] presented in a single figure (included here as Fig. 1.4) τ versus γ plots for one of the tension specimens (the circles) of Taylor & Elam (1925) and a compression specimen (the crosses) from the series of experiments on aluminum discs subsequently reported at length in Taylor & Farren (1926) and Taylor (1927a, b). As Taylor (1925) comments, "it is very remarkable that the curves for compression and tension are almost identical." Thus, Taylor had demonstrated for the first

[5] I have not seen a previous citation of this work.

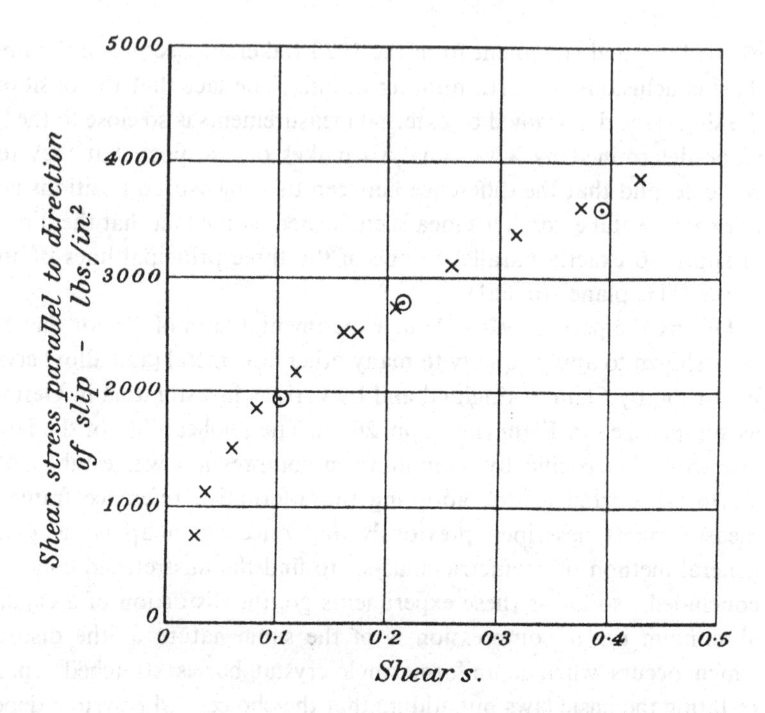

Fig. 1.4. Resolved shear stress vs. slip data ($s = \gamma$) for tension and compression specimens (circle and cross, respectively) of aluminum crystals (from Taylor 1925, fig. 1, as presented in *The Scientific Papers of Sir Geoffrey Ingram Taylor*, Vol. I).

time the likelihood of a general hardening law $\tau = f(\gamma)$ for crystals of a given material in single slip that would be essentially independent of initial lattice orientation and whether the axial loading was tensile or compressive.

E. Schmid (1926) independently recognized that τ, γ are the most suitable crystal parameters in single slip and presented plots of these variables from experiments covering a wide range of lattice orientations of zinc crystals, which slipped easily on the basal plane of the hexagonal lattice to very large shears in the range 3.5–4.5. He proposed a hardening law for zinc of the form $\tau = \tau_0 + k\gamma$, with τ_0 and k constants averaged from the data band (see his fig. 2). The specific idea of an initial critical strength τ_0 that is constant for a given crystalline material and temperature, commonly called *Schmid's critical shear stress law*, had previously been introduced in Schmid (1924) where it was given a reasonable degree of experimental support from tension-test results reported for zinc and rhombohedral bismuth crystals. (See Schmid & Boas (1935, figs. 81(a)–(c)) for a compilation of additional experimental results on initial slipping in tension of various hexagonal metal crystals that generally confirm the critical shear stress law.) Schmid's law of constant τ_0 is almost implied by Taylor & Elam's (1923) second experimental law, although not fully, as they did not attempt to precisely define an initial yield stress in their original experiment nor speculate about its possible variation (or constancy) within the preferred one of 24 stereographic triangles of f.c.c. crystals. Lastly it may be noted that the most impressive demonstration, as acknowledged by Schmid & Boas (1935, p. 124), of what I shall call the general *Taylor–Schmid law* $\tau = f(\gamma)$ in single slip is found in Taylor (1927b, fig. 4). There τ versus γ plots from very carefully performed compression tests of lubricated aluminum discs (of different lattice orientations) are shown to very nearly coincide with each other and with τ versus γ data for an aluminum crystal in tension, to slips of order 1.0.

Single slip on the basal plane apparently was the dominant mode of finite deformation until necking and fracture in the early experimental investigations of hexagonal metal crystals (see Schmid & Boas (1935) for a comprehensive review and extensive bibliography). For f.c.c. and b.c.c crystals, however, single slip generally did not continue throughout a finite distortion test. In the original investigation of an aluminum crystal in tension, Taylor & Elam (1923) found that at 78 percent extension the unstretched cone corresponding to the additional deformation from 40 percent strain was indeed a cone (rather than the degenerate case of two planes which had applied at each strain level from 0 to 40 percent); whence this deformation could not have been a simple shear. They conjectured equal double slip for the last stage of the crystal distortion and showed by X-ray analysis that the unstretched cone of extension from 40 to 78 percent contained the line of intersection $[1\bar{1}0]$ of the original (111) slip plane with another octahedral plane $(\bar{1}\bar{1}1)$.[6] This is beautifully illustrated in their figure 13, included here as Fig. 1.5. The curve "0–40" is the stereographic projection (with the crystal axis as the pole) of the plane of unextended directions at 40 percent strain that was the actual (111) slip plane, determined from measurements at 10, 15, 20, and 30 percent strains. The curve "40–78" (and its dashed counterpart for $\theta > 90°$) represents the locus on the stereographic projection of the unstretched-cone position in the crystal at 40 percent extension corresponding to the additional deformation to 78 percent. (For a given ϕ_0 the pair of θ_0s again may be

[6] Taylor & Elam (1923, fig. 12) used a left-handed lattice frame. Consequently, directions [010] and [001] were interchanged and their second plane and line of intersection were labeled $(\bar{1}1\bar{1})$ (or $1\bar{1}1$)) and $[10\bar{1}]$.

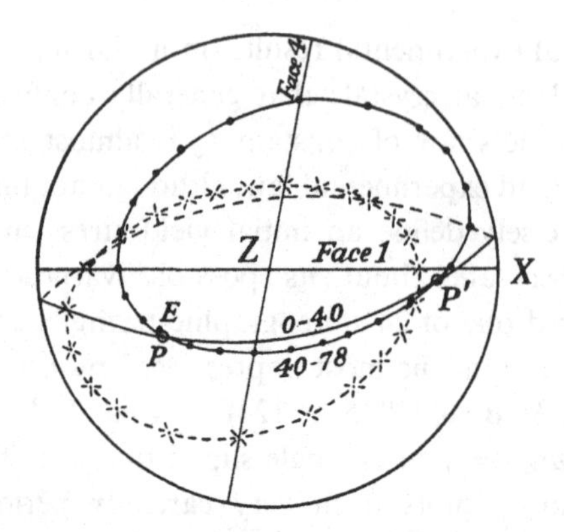

Fig. 1.5. Z-axis stereographic projections, in an aluminum crystal extended 40%, of the positions of the slip plane for extension from 0 to 40% and the unstretched cone for extension from 40 to 78% (from Taylor & Elam (1923, fig. 13), as presented in *The Scientific Papers of Sir Geoffrey Ingram Taylor*, Vol. I).

found from eqs. (1.3) and (1.6), with now $a_0, b_0, \alpha_0, \beta_0, \delta_0$ the measurements at 40 percent strain and λ_z the relative additional stretch 1.27, equal to 1.78 ÷ 1.40.) Point E is the [1$\bar{1}$0] "diad axis," and as seen it is almost exactly coincident with one of the intersections P of the two loci. However, Taylor and Elam could not prove that the [1$\bar{1}$0] intersection was an unextended material line throughout the deformation, with point P' merely representing a momentarily unstretched direction, because they had no reliable distortion measurements for stages between 40 and 78 percent. Nevertheless, they quite reasonably considered their results a verification of double (but not necessarily equal) slip on crystallographic planes (111) and ($\bar{1}\bar{1}$1). Moreover, from the position of the unstretched cone in the crystal extended 78 percent (see their fig. 14) and from X-ray measurements at each of 60 and 78 percent strains, the slips apparently were of comparable magnitudes but not strictly equal as conjectured.

Consider the [001] stereographic projection of Fig. 1.6, in which the labeling of slip planes and directions introduced in Taylor (1938b) and defined in Table 1 has been adopted. The original slip system in Taylor & Elam (1923) may be considered to be $a\bar{2}$, or (111) [$\bar{1}$01] in standard notation, with the initial position of the tensile axis very close to the great circle (not shown) through [$\bar{1}$01] and [011] and approximately 10° from the latter (see their fig. 12, taking into account the left-handedness of their lattice frame). In the Bakerian Lecture, Taylor in effect postulated equal hardening of all 24 slip systems, which will be called *Taylor hardening* in this monograph. He hypothesized that when (after about 50 percent

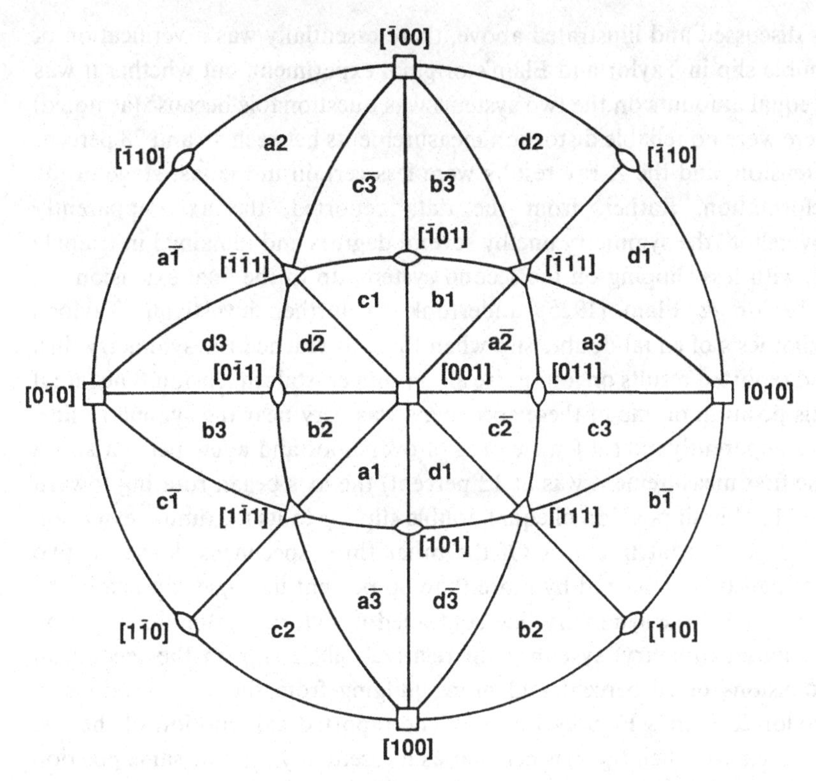

Fig. 1.6. [001] stereographic projection showing f.c.c. crystal slip-system regions in uniaxial tension corresponding to Taylor & Elam's (1923) experimental laws.

Table 1. *Designation of slip systems in f.c.c. crystals*

Plane	(111)			($\bar{1}\bar{1}$1)			($\bar{1}$11)			(1$\bar{1}$1)		
Direction	[0$\bar{1}$1]	[10$\bar{1}$]	[$\bar{1}$10]	[011]	[$\bar{1}$0$\bar{1}$]	[1$\bar{1}$0]	[0$\bar{1}$1]	[$\bar{1}$0$\bar{1}$]	[110]	[011]	[10$\bar{1}$]	[$\bar{1}\bar{1}$0]
System	*a*1	*a*2	*a*3	*b*1	*b*2	*b*3	*c*1	*c*2	*c*3	*d*1	*d*2	*d*3

extension in Taylor and Elam's case) the tensile axis reached the symmetry line between systems $a\bar{2}$ and $b1$, as it rotated on a great circle toward slip direction [$\bar{1}$01] from its initial position within (spherical) triangle $a\bar{2}$,[7] double slip would be initiated and the axis thenceforward would rotate toward the limit position [$\bar{1}$12] (the vector sum of the two slip directions).

[7] Taylor & Elam (1923) had established that the axis of their aluminum crystal so rotated from X-ray analysis at each of 10, 15, 20, 30, and 40% extensions (see their fig. 12, in which the successive positions are plotted).

As discussed and illustrated above, there essentially was a verification of double slip in Taylor and Elam's original experiment, but whether it was of equal amounts on the two systems was questionable because (as noted) there were no reliable distortion measurements between 40 and 78 percent extension, and the X-ray results were less certain in the last stage of the deformation. Rather, from the data reported, the axis apparently "overshot" the symmetry line by several degrees and remained in triangle *b*1, with less slipping on the second system, up to the final extension.

Taylor & Elam (1925) undertook to further investigate Taylor's hypothesis of equal double slip when the axis reached the symmetry line, and reported results on four more aluminum crystals in tension. The initial axis position of one of these specimens was very near the symmetry line, and apparently after a few degrees of overshoot and a few percent strain (the first measurement was at 12 percent) the axis began rotating toward the $[\bar{1}12]$ limit position of equal double slip, up to its maximum extension of 42 percent (at fracture). Of the other three specimens, however, two that had been extended by more than 50 percent in single slip before the axis reached the symmetry line continued to deform in dominant slip on the initial (primary) system, with relatively little slip on the second, to extensions of 70 percent and more, judging from the axis positions in Taylor & Elam's Figures 3 and 5. The reported axis motion of the last specimen (see their fig. 4) is peculiar as it essentially is in the same position (and not a limiting one) from 51 to 70 percent extension, indicating unreliable X-ray measurements (as they acknowledged).

Taylor and Elam also sought to determine the relative amounts of double slip from external measurements, again using Taylor's geometric method and investigating one specimen between 74 and 93 percent extension and another between 51 and 69 percent. For comparisons with the actual unstretched cones in the respective initial configurations (i.e., new reference states) of each range, they derived an analytical cone corresponding to *incremental* double slip of arbitrary relative amounts in the two systems, expressed in terms of the parameter

$$\mu = \frac{d\gamma_1 - d\gamma_2}{d\gamma_1 + d\gamma_2},$$

which theoretically can take on any value from -1 to 1. That is, they established the equation (Taylor & Elam 1925, eq. (5)) of the cone of directions of zero *rates* of extension (or stretch rates) in arbitrary double slip.

From the standard equation for Eulerian strain rate D in terms of the deformation gradient,

$$D = \tfrac{1}{2}(\dot{A}A^{-1} + A^{-\mathrm{T}}\dot{A}^{\mathrm{T}}), \tag{1.8}$$

the general equation (whatever the deformation) of the cone of zero stretch rates in the reference state is determined by

$$\boldsymbol{\iota}_0(\dot{A}_0 + \dot{A}_0^{\mathrm{T}})\boldsymbol{\iota}_0 = 0, \tag{1.9}$$

where $\boldsymbol{\iota}_0$ now represents a unit vector whose range defines this cone. A given $\boldsymbol{\iota}_0$ satisfying equation (1.9) will not ordinarily also satisfy equation $(1.4)_1$, which gives the reference position of the cone of unstretched material lines. Consider the rate of change of that cone, recognizing that $\boldsymbol{\iota}_0$ in equation $(1.4)_1$ is generally not a fixed unit vector as it may only be a momentary reference direction of unstretched length for a particular deformation gradient A:

$$\boldsymbol{\iota}_0(A^{\mathrm{T}}\dot{A} + \dot{A}^{\mathrm{T}}A)\boldsymbol{\iota}_0 + 2\boldsymbol{\iota}_0 A^{\mathrm{T}}A\dot{\boldsymbol{\iota}}_0 = 0.$$

Only for infinitesimal deformations (upon setting $A = I$ and $\dot{A} = \dot{A}_0$ in the limit of zero strain and using $\boldsymbol{\iota}_0 \cdot \dot{\boldsymbol{\iota}}_0 = 0$) does this equation, satisfied by directions in the unstretched cone, reduce to equation (1.9) satisfied by the cone of zero stretch rates.

A simple example illustrating the difference between these cones is the degenerate case of two planes corresponding to single slip. In Fig. 1.3, one of these planes, represented by line OL (the slip plane), is common to both cones. However, the reference position of the second of these planes is represented by line LP as constructed, whereas the second plane of the cone of zero stretch rates in the reference state obviously is LN (equivalently, OM). These planes are exactly coincident only for $\chi_1 = 0$.

Taylor & Elam (1925, p. 42) recognized that these cones were not identical for the additional relative extensions of 11 and 12 percent of their two aluminum crystals (measured after initial extensions of 74 and 51 percent, respectively). Nonetheless, they compared the cones in order to approximately determine their parameter μ during double slip. As the errors in such comparisons probably were not large for these moderate additional strains, their statement that for the two specimens "the amount of the slip on the original plane was nearly three times as great as that on the second slip-plane" (p. 43) may be considered essentially correct. What was needed, of course, was a general equation for the deformation gradient of *finite* double slip for arbitrary μ and reference axis position; but such an equation was not given until 41 years later in the work of

Chin, Thurston & Nesbitt (1966) (with further equations developed by Shalaby & Havner (1978) and Havner (1979)).[8]

For assumed equal double slip in f.c.c. crystals, equations of finite axial deformation and rotation were derived by v. Göler & Sachs (1927) in tension and by Taylor (1927a)[9] in compression. Karnop & Sachs (1927) applied v. Göler and Sach's equations to determine τ versus γ curves from load–extension data for aluminum crystals in (presumed) single and double slip. However, as Basinski & Basinski (1979, p. 335) note in their extensive review of the literature on crystal deformation and hardening, Karnop and Sachs "measured only the initial orientations of their crystals and calculated their curves from measured extensions *assuming* single glide followed by double glide at the boundary." That procedure in fact commonly became the practice in the experimental literature, as Bell (1965) found in his comprehensive survey of 318 f.c.c. single-crystal tests covering a 40-year span. Yet when X-ray measurements of the changing lattice orientation actually were made, there rarely was support for the equal double slip hypothesis. Bell & Green (1967) analyzed 152 experiments from the literature in which axis rotations reaching the symmetry line had been measured; and they conducted tensile tests on 16 aluminum single crystals, each of which reached the symmetry line. From their own experiments and all the published work they examined, they concluded that the predicted double-slip behavior of f.c.c crystals at the symmetry line "has *not* been observed since 1929" (p. 470). Neither had that behavior generally been found by Taylor & Elam (1923, 1925), as discussed previously, nor in many of the experiments by Elam (1926, 1927a, b) on other f.c.c. metal crystals, as briefly reviewed in Havner & Shalaby (1977).

In contrast to the practice in Karnop & Sachs (1927) and many similar experimental papers of making assumptions about the modes of crystal deformation, Taylor (1927a), in work on compression of aluminum discs, applied his general geometric method to the analysis of complete finite distortion measurements. He used the reference frame of Taylor & Farren (1926) previously described (the axis Z normal to the disc faces) and calculated unextended cones for intervals of probable double slip. (Taylor and Farren had established that single slip was the mode of deformation

[8] The general kinematics of proportional double slip will be presented in Chapter 2.

[9] Since a brief mention of Taylor's equations in passing (and then only by implication) in Schmid & Boas (1935, p. 68), their existence appears to have been forgotten. In particular, in the subsequent double-slip analyses of Bowen & Christian (1965), Chin, Thurston & Nesbitt (1966), Shalaby & Havner (1978), Havner (1979), and Sue & Havner (1984) they were overlooked.

up to the symmetry line in compression, reached as the loading axis rotated on a great circle toward the slip-plane normal.) In addition, he derived the exact equation of the unstretched cone in f.c.c. crystals due to equal finite double slip on two systems symmetrically oriented with respect to the compression axis. (A generalization of Taylor's equation is developed in Chapter 2.)

Taylor (1927a) compares these cones in his figure 5 for a specimen in the interval of spacing stretch $\lambda_{(z)}$ from 0.667 to 0.566 (a relative additional thickness decrease of 15 percent after the initial one-third reduction), the lattice orientation at the beginning of the interval having been established by X-ray analysis. The agreement is reasonably good, and Taylor states that in all the cases he examined, "the cones always coincide at one end, namely, the end which contains the intersection of the two slip-planes" (p. 29). The finite distortions were thus unquestionably due to double slip on the assumed planes. However, because of the greater elongation of the measured cone than the theoretical one, which Taylor also found in all cases, he suggests that there may have been "a certain amount of slipping on the same two slip-planes but in one of the other possible directions of slip." One specimen was compressed in five stages (the disc being relubricated and recut circular four times) to a final thickness only 28 percent of the original (an areal stretch in the disc faces of 3.6!). The remarkable uniformity of distortion achieved throughout the test is illustrated by plate 1b in Taylor (1927a).

For the particular additional deformation just discussed Taylor deduced "nearly equal" double slip on the two planes. From measurements of two other specimens in compression, however, he concluded "that the distortion can be accounted for by supposing that double slipping of the type predicted does take place, but that during the earlier stages at any rate, the amount of slip is greater on the original slip-plane than it is on the second slip-plane. This result is similar to that found in the case of tensile tests (Taylor & Elam, 1925)" (p. 24). Thus, unequal double slip beginning at or beyond the symmetry line, preceded by single slip on the crystallographic system with maximum resolved shear stress in the virgin crystal (which system also contributes the greater share in the subsequent double slip), is the central unifying result of G. I. Taylor's carefully done experimental studies and rigorous geometrical analyses of the distortion of f.c.c. crystals in tension and compression, over the period 1923–7. More than 60 years later, the essential experimental position remains unchanged. (The similar but more complex case of b.c.c. crystals is addressed later in this monograph.)

2

THE KINEMATICS OF DOUBLE SLIP

As remarked in Chapter 1, v. Goler & Sachs (1927) derived equations for equal double slip in f.c.c. crystals in tension with the loading axis on a symmetry line, and Taylor (1927a) derived comparable equations in compression (and also gave the equation of the unstretched cone for f.c.c. crystals). Equations applicable to any crystal class for these same symmetry conditions and for both tension and compression were developed by Bowen & Christian (1965), who presented formulas for various specific combinations of slip systems in f.c.c. and b.c.c. crystals. A general equation for the deformation gradient in (proportional) double slip of arbitrary relative amounts was first given in the work of Chin, Thurston, & Nesbitt (1966) mentioned previously. They carried the analysis no further, however, and applied the equation only to cases of equal, symmetric double slip in f.c.c. crystals.

Apparently the first explicit equations for rotation and stretch of a crystal material line in arbitrary (proportional) double slip in f.c.c. and b.c.c. crystals were developed by Shalaby & Havner (1978) (independently of the work of Chin, Thurston, & Nesbitt (1966)). The equations were illustrated for various nonsymmetric axis positions and relative amounts of slip. General equations for material line and areal vectors and both the finite deformation gradient and its inverse in arbitrary (proportional) double slip were derived in Havner (1979). Here we shall follow this last approach to the analysis of double slip in crystals.

2.1 General Solutions for Material Line and Plane

Let **l** denote an embedded material line vector in the gross (macroscopic) crystal and **a** be the areal vector of an embedded material

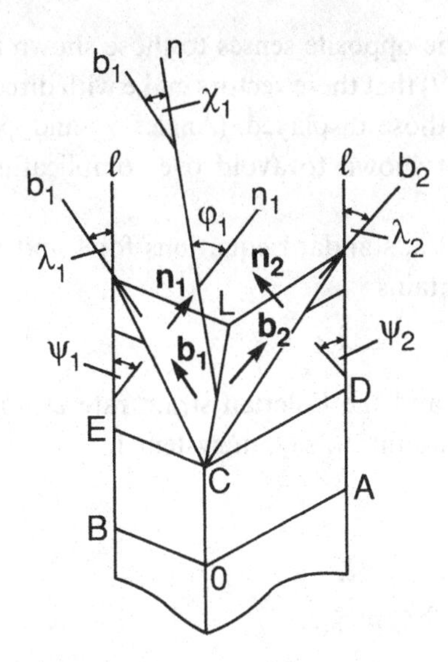

Fig. 2.1. Segment of triangular-prism crystal showing orientations of slip systems.

plane (as distinct from lattice plane) during an interval of double slip. With reference to Fig. 2.1, in which a portion of a triangular-prism crystal region is depicted that has been cut away on lattice planes to reveal the crystallographic slip systems, **l** may be considered to represent line segment *OC* (or *AD* or *BE*). Areal vector **a** then equals the area of material plane *AOB* (or *DCE*) in magnitude and is in the direction of the normal **n** to that plane. The sets of parallel lines drawn on the faces of the crystal region represent (embedded) material lines that both rotate and stretch relative to the underlying lattice. Only the intersection *CL* of the two slip planes and all lines parallel to *CL* remain unstretched and of fixed orientation with respect to the lattice throughout finite double slip. As lattice straining is here disregarded, the unit vectors $\mathbf{b}_k, \mathbf{n}_k$ $(k = 1, 2)$ in the slip and normal directions of the two systems are fixed relative to the lattice and so are of constant relative orientation throughout the deformation.

In the general development that follows, the crystal region depicted in Fig. 2.1 is an arbitrary choice and bears no special relationship to the direction or directions of loading of the overall crystal specimen. (In other words, no prism face is necessarily parallel to an external specimen face or edge.) However, particularly useful applications of the equations are to a direction **l** parallel to the specimen axis in tension and a direction **n** normal to the specimen end faces in compression. In the latter case the slip directions $\mathbf{b}_1$ and $\mathbf{b}_2$ are in the opposite senses to those shown in the figure, and the angles λ_k, χ_k $(k = 1, 2)$ that these vectors make with directions $\mathbf{l}$ and $\mathbf{n}$ are the supplements of those displayed. (Angles χ_2 and φ_2, the latter between $\mathbf{n}_2$ and $\mathbf{n}$, are not shown to avoid overcomplicating the figure.)

From material derivatives of the standard equations for $\mathbf{l}$ and $\mathbf{a}$ (the latter Nanson's equation) one obtains

$$\dot{\mathbf{l}} = \Gamma \mathbf{l}, \quad \dot{\mathbf{a}} = (\operatorname{tr} D)\mathbf{a} - \mathbf{a}\Gamma, \quad \Gamma = \dot{A}A^{-1}, \tag{2.1}$$

where Γ is the velocity gradient and the Eulerian strain rate $D = \operatorname{sym} \Gamma$. In single slip (simple shear) of amount γ_1, say, in system 1,

$$\mathbf{l} = \mathbf{l}_0 + (\mathbf{l}_0 \cdot \mathbf{n}_1)\mathbf{b}_1 \gamma_1,$$

from which

$$A = I + (\mathbf{b} \otimes \mathbf{n})_1 \gamma_1, \quad \Gamma = (\mathbf{b} \otimes \mathbf{n})_1 \dot{\gamma}_1,$$

the second equation requiring that the frame be corotational with the plane and direction of shear.[1] For slipping on two (or more) crystallographic systems the separate contributions may be linearly superimposed in Γ but not in A. Thus, in a lattice-corotational frame

$$\Gamma = \sum (\mathbf{b} \otimes \mathbf{n})_k \dot{\gamma}_k \tag{2.2}$$

and from equations (2.1) (as $\operatorname{tr} D = \operatorname{tr} \Gamma = 0$)

$$\dot{\mathbf{l}} = \sum (l \cos \psi_k) \mathbf{b}_k \dot{\gamma}_k, \quad \dot{\mathbf{a}} = -\sum (a \cos \chi_k) \mathbf{n}_k \dot{\gamma}_k, \tag{2.3}$$

in which $l = |\mathbf{l}|$, $a = |\mathbf{a}|$, and the angles defined in Fig. 2.1 have been adopted (i.e., $l \cos \psi_k = \mathbf{l} \cdot \mathbf{n}_k$, $a \cos \chi_k = \mathbf{a} \cdot \mathbf{b}_k$).

In proportional double slip[2] $\dot{\gamma}_2 = \alpha \dot{\gamma}_1$, the differential equations for $\mathbf{l}$ and $\mathbf{a}$ are most conveniently solved in terms of $\gamma_1 = \gamma$ by forming their scalar products with each of the unit vectors $\mathbf{n}_k, \mathbf{b}_k$. Then equation $(2.3)_1$ provides four coupled equations with constant coefficients for the four variables $l \cos \psi_k$, $l \cos \lambda_k$ $(k = 1, 2)$; and equation $(2.3)_2$ similarly provides four equations for $a \cos \chi_k$, $a \cos \varphi_k$. Consider first the solution for $\mathbf{l}$. With the notation

$$x_k = l \cos \psi_k, \quad y_k = l \cos \lambda_k, \quad c_{kj} = \mathbf{n}_k \cdot \mathbf{b}_j, \quad b = \mathbf{b}_1 \cdot \mathbf{b}_2,$$

[1] The evident relation $\Gamma = \dot{A}$ in simple shear (and in the specified corotational frame) has mistakenly been used in the literature a number of times for the analysis of other finite deformations to which it does not apply.

[2] As remarked in Havner (1979), any double-slip path which is not strictly proportional can be accurately approximated as piecewise proportional.

the equations become

$$x_1' = \alpha c_{12} x_2, \qquad x_2' = c_{21} x_1,$$
$$y_1' = x_1 + \alpha b x_2, \quad y_2' = b x_1 + \alpha x_2,$$

where $(\cdot)'$ signifies differentiation with respect to γ. These equations are readily solved by standard procedures, and the results may be expressed as

$$\begin{aligned}(l/l_0)\cos\psi_1 &= K_1 \exp(c\gamma) + K_2 \exp(-c\gamma),\\ (l/l_0)\cos\psi_2 &= 1/(\alpha\kappa)\{K_1 \exp(c\gamma) - K_2 \exp(-c\gamma)\},\end{aligned} \tag{2.4}$$

with

$$\begin{aligned}&K_1 = \tfrac{1}{2}(\cos\psi_1^0 + \alpha\kappa\cos\psi_2^0), \quad K_2 = \tfrac{1}{2}(\cos\psi_1^0 - \alpha\kappa\cos\psi_2^0),\\ &\kappa = c_{12}/c, \quad c = (\alpha c_{12} c_{21})^{\frac{1}{2}},\end{aligned} \tag{2.5}$$

and

$$\begin{aligned}(l/l_0)\cos\lambda_1 &= \cos\lambda_1^0 + K_3(\exp(c\gamma) - 1) - K_4(\exp(-c\gamma) - 1),\\ (l/l_0)\cos\lambda_2 &= \cos\lambda_2^0 + K_5(\exp(c\gamma) - 1) - K_6(\exp(-c\gamma) - 1),\end{aligned} \tag{2.6}$$

where

$$\begin{aligned}&K_3 = K_1(1/c + b/c_{12}), \quad K_4 = K_1(1/c - b/c_{12}),\\ &K_5 = K_1(b/c + 1/c_{12}), \quad K_6 = K_2(b/c - 1/c_{12}).\end{aligned} \tag{2.7}$$

Upon substituting equations (2.4) into equation $(2.3)_1$ and integrating, one obtains (Havner 1979)

$$\begin{aligned}\mathbf{l} = \mathbf{l}_0 + (l_0/c)\{&[K_1 \exp(c\gamma) - K_2 \exp(-c\gamma) - \alpha\kappa\cos\psi_2^0]\mathbf{b}_1\\ &+ (1/\kappa)[K_1 \exp(c\gamma) + K_2 \exp(-c\gamma) - \cos\psi_1^0]\mathbf{b}_2\}.\end{aligned} \tag{2.8}$$

The stretch $\lambda_l = l/l_0$ of material line "l" may be found by forming the scalar product of $\mathbf{l}$ with itself. The result is

$$\begin{aligned}(l/l_0)^2 = 1 &+ K_7 \exp(2c\gamma) + K_8 \exp(-2c\gamma) + K_9 \exp(c\gamma)\\ &+ K_{10} \exp(-c\gamma) - (K_7 + K_8 + K_9 + K_{10}),\end{aligned} \tag{2.9}$$

where

$$\begin{aligned}K_7 &= (K_1/c)^2(1 + 2b/\kappa + 1/\kappa^2), \quad K_8 = (K_2/c)^2(1 - 2b/\kappa + 1/\kappa^2),\\ K_9 &= 2(K_1/c)\{\cos\lambda_1^0 + (1/\kappa)\cos\lambda_2^0 - (1/c_{12})(b + 1/\kappa)\cos\psi_1^0\\ &\qquad - (\alpha/c)(b + \kappa)\cos\psi_2^0\},\\ K_{10} &= 2(K_2/c)\{-\cos\lambda_1^0 + (1/\kappa)\cos\lambda_2^0 + (1/c_{12})(b - 1/\kappa)\cos\psi_1^0\\ &\qquad - (\alpha/c)(b - \kappa)\cos\psi_2^0\}.\end{aligned} \tag{2.10}$$

For a direction $\mathbf{l}_0$ along the line of intersection CL (Fig. 2.1) of the two slip planes (i.e., $\psi_1^0 = \psi_2^0 = 90°$), it is evident from the definitions (2.5), (2.7), and (2.10) that all constants $K_1, \ldots, K_{10}$ are zero, whence $l = l_0$, $\lambda_1 = \lambda_1^0$, $\lambda_2 = \lambda_2^0$, and line CL neither changes in length nor rotates relative to the lattice, as required.

For areal vector $\mathbf{a}$, the results of a similar procedure are (refer to Havner (1979) for details)

$$\begin{aligned} (a/a_0)\cos\chi_1 &= C_1 \exp(c\gamma) + C_2 \exp(-c\gamma), \\ (a/a_0)\cos\chi_2 &= \kappa\{-C_1 \exp(c\gamma) + C_2 \exp(-c\gamma)\} \end{aligned} \tag{2.11}$$

with

$$C_1 = \tfrac{1}{2}\{\cos\chi_1^0 - (1/\kappa)\cos\chi_2^0\}, \quad C_2 = \tfrac{1}{2}\{\cos\chi_1^0 + (1/\kappa)\cos\chi_2^0\}, \tag{2.12}$$

and

$$\begin{aligned} (a/a_0)\cos\varphi_1 &= \cos\varphi_1^0 - C_3(\exp(c\gamma) - 1) + C_4(\exp(-c\gamma) - 1), \\ (a/a_0)\cos\varphi_2 &= \cos\varphi_2^0 - C_5(\exp(c\gamma) - 1) + C_6(\exp(-c\gamma) - 1), \end{aligned} \tag{2.13}$$

where (with $n = \mathbf{n}_1 \cdot \mathbf{n}_2$)

$$\begin{aligned} C_3 &= C_1(1/c - n/c_{21}), \quad C_4 = C_2(1/c + n/c_{21}), \\ C_5 &= C_1(n/c - 1/c_{21}), \quad C_6 = C_2(n/c + 1/c_{21}). \end{aligned} \tag{2.14}$$

Also, from equations (2.11) and $(2.3)_2$ (after integrating),

$$\begin{aligned} \mathbf{a} = \mathbf{a}_0 - (a_0/c)\{&[C_1 \exp(c\gamma) - C_2 \exp(-c\gamma) + (1/\kappa)\cos\chi_2^0]\mathbf{n}_1 \\ &- \alpha\kappa[C_1 \exp(c\gamma) + C_2 \exp(-c\gamma) - \cos\chi_1^0]\mathbf{n}_2\}, \end{aligned} \tag{2.15}$$

and the areal stretch in direction $\mathbf{n}$ (Fig. 2.1) is

$$\begin{aligned} (a/a_0)^2 = 1 &+ C_7 \exp(2c\gamma) + C_8 \exp(-2c\gamma) + C_9 \exp(c\gamma) \\ &+ C_{10} \exp(-c\gamma) - (C_7 + C_8 + C_9 + C_{10}), \end{aligned} \tag{2.16}$$

in which

$$\begin{aligned} C_7 &= (C_1/c)^2\{1 - 2n\alpha\kappa + (\alpha\kappa)^2\}, \quad C_8 = (C_2/c)^2\{1 + 2n\alpha\kappa + (\alpha\kappa)^2\}, \\ C_9 &= 2(C_1/c)\{-\cos\varphi_1^0 + \alpha\kappa\cos\varphi_2^0 + (\alpha\kappa/c)(n - \alpha\kappa)\cos\chi_1^0 \\ &\qquad + (1/c_{12})(1 - n\alpha\kappa)\cos\chi_2^0\}, \\ C_{10} &= 2(C_2/c)\{\cos\varphi_1^0 + \alpha\kappa\cos\varphi_2^0 - (\alpha\kappa/c)(n + \alpha\kappa)\cos\chi_1^0 \\ &\qquad - (1/c_{12})(1 + n\alpha\kappa)\cos\chi_2^0\}. \end{aligned} \tag{2.17}$$

As is obvious from the definitions of constants c, κ, $K_1, \ldots, K_{10}$, and $C_1, \ldots, C_{10}$, parameters c_{12}, c_{21} must be nonzero in all the above equations. Also, it has been implicitly assumed that they are of the same sign, so that c is real (with $\alpha > 0$). The standard cases of double slip preceded by single slip in f.c.c. crystals in tension and compression satisfy these requirements. For systems $a\bar{2}, b1$ (Fig. 1.6) or any comparable pair in tension, $c_{12} = c_{21} = \frac{\sqrt{6}}{3}$ (and $b = \frac{1}{2}$, $n = -\frac{1}{3}$). For systems $a2, c2$ (stereographic triangles $a\bar{2}, c\bar{2}$ of Fig. 1.6) or any comparable pair in compression, $c_{12} = c_{21} = -\frac{\sqrt{6}}{3}$ (with $b = 0$, $n = \frac{1}{3}$). Common double-slip combinations in b.c.c. crystals also satisfy the criteria. However, the cases (i) $\mathbf{n}_1 \cdot \mathbf{b}_2 = 0$, (ii) $\mathbf{n}_2 \cdot \mathbf{b}_1 = 0$, and (iii) both $\mathbf{n}_1 \cdot \mathbf{b}_2$, $\mathbf{n}_2 \cdot \mathbf{b}_1 = 0$ are physically possible in cubic crystals (the last corresponding to either a common slip plane or a common slip direction), and solutions for $\mathbf{l}$ and $\mathbf{a}$ in each of these special cases may be found respectively in Shalaby & Havner (1978) and Havner (1979). (Case (iii) is almost trivial as the separate contributions of the two systems to the deformation gradient superimpose linearly.)

2.2 Deformation Gradient and Axis Rotation

From equations (2.3), (2.4), and (2.11), as the slips increase without limit,

$$\lim_{\gamma \to \infty} d\mathbf{l} = \tfrac{1}{2} l_0 (\cos \psi_1^0 + \alpha\kappa \cos \psi_2^0)(\mathbf{b}_1 + (1/\kappa)\mathbf{b}_2)\, d\gamma,$$

$$\lim_{\gamma \to \infty} d\mathbf{a} = -\tfrac{1}{2} a_0 (\cos \chi_1^0 - (1/\kappa) \cos \chi_2^0)(\mathbf{n}_1 - \alpha\kappa \mathbf{n}_2)\, d\gamma.$$

Thus, for very large deformations in tension the crystal axis rotates toward $\mathbf{b}_1 + (1/\kappa)\mathbf{b}_2$ (with $\cos \psi_1^0$, $\cos \psi_2^0$, and κ all positive); and in compression the crystal end-plane normal rotates toward $\mathbf{n}_1 - \alpha\kappa \mathbf{n}_2$ (with $\cos \chi_1^0$, $\cos \chi_2^0$, and κ then negative). For double slip on systems $a\bar{2}, b1$ (Fig. 1.6) in f.c.c. crystals in tension, $\kappa = 1/\sqrt{\alpha}$ and the limit direction is $\mathbf{b}_1 + \sqrt{\alpha}\mathbf{b}_2$. For double slip on systems $a2, c2$ in f.c.c. crystals in compression, $\kappa = -1/\sqrt{\alpha}$ and the limit direction is $\mathbf{n}_1 + \sqrt{\alpha}\mathbf{n}_2$. (In the case of equal double slip the corresponding limits are $[\bar{1}12]$ in tension, as mentioned in Chapter 1, and $[011]$ in compression. Refer to Fig. 1.6.)

To determine the deformation gradient, observe from equation (2.8) and the definitions (2.5) that we may equivalently write (after substituting $l_0 \cos \psi_k^0 = \mathbf{n}_k \cdot \mathbf{l}_0$):

$$\mathbf{l} = \mathbf{l}_0 + (1/c)[\{\mathbf{b}_1 \otimes \mathbf{n}_1 + \alpha \mathbf{b}_2 \otimes \mathbf{n}_2\} \sinh(c\gamma) + \{\alpha\kappa \mathbf{b}_1 \otimes \mathbf{n}_2 + (1/\kappa)\mathbf{b}_2 \otimes \mathbf{n}_1\}(\cosh(c\gamma) - 1)]\mathbf{l}_0. \tag{2.18}$$

It then follows immediately that

$$A = I + (1/c)[B\sinh(c\gamma) + (1/c)B^2(\cosh(c\gamma) - 1)], \tag{2.19}$$

where

$$B = (\mathbf{b} \otimes \mathbf{n})_1 + \alpha(\mathbf{b} \otimes \mathbf{n})_2. \tag{2.20}$$

This form for the deformation gradient in proportional double slip was first given in Chin, Thurston & Nesbitt (1966, appendix), where a derivation by E. N. Gilbert, of Bell Telephone Laboratories, using a series approach was presented.

An equivalent expression to equation (2.15) for $\mathbf{a}$ is similarly obtained by substituting equations (2.12) and $a_0 \cos\chi_k = \mathbf{a}_0 \cdot \mathbf{b}_k$:

$$\mathbf{a} = \mathbf{a}_0 - (1/c)\mathbf{a}_0[B\sinh(c\gamma) - (1/c)B^2(\cosh(c\gamma) - 1)], \tag{2.21}$$

with B the second-order tensor defined previously. Thus from Nanson's equation $\mathbf{a} = |A|\mathbf{a}_0 A^{-1}$, since $|A| = 1$, the inverse deformation gradient is (Havner 1979)

$$A^{-1} = I - (1/c)[B\sinh(c\gamma) - (1/c)B^2(\cosh(c\gamma) - 1)]. \tag{2.22}$$

It is straightforward to show from equations (2.19) and (2.22) (making use of $B^4 = c^2 B^2$) that $AA^{-1} = I$ as required.

Let the radius of a stereographic projection in a lattice-corotational frame be set equal to unity, and let ξ, η denote the projection coordinates of a unit vector $\mathbf{v}$ that remains coincident with either a material direction $\mathbf{l}$ or a material plane normal $\mathbf{n}$. Also, let θ, ϕ denote the spherical angles of $\mathbf{v}$ in the chosen frame. (θ is the angle between $\mathbf{v}$ and projection axis ζ ([001], say), and ϕ is the counterclockwise angle between the projection of $\mathbf{v}$ on the stereographic plane and the reference ξ-axis ([100], say).) Thus,

$$\mathbf{v} = (\sin\theta\cos\phi, \sin\theta\sin\phi, \cos\theta), \quad \xi = \tan(\theta/2)\cos\phi, \\ \eta = \tan(\theta/2)\sin\phi. \tag{2.23}$$

For a material line in double slip, the trace of $\mathbf{v}$ on the stereographic projection may be determined from $\mathbf{v}\cdot\mathbf{n}_k = \cos\psi_k$, $\mathbf{v}\cdot\mathbf{b}_k = \cos\lambda_k$, and equations (2.4), (2.6), and (2.9). For a material plane the corresponding equations are $\mathbf{v}\cdot\mathbf{b}_k = \cos\chi_k$, $\mathbf{v}\cdot\mathbf{n}_k = \cos\varphi_k$, and equations (2.11), (2.13), and (2.16).

Consider the case of double slip on systems $a\bar{2}, b1$ (Fig. 1.6) in f.c.c. crystals in tension and substitute equation $(2.23)_1$ and the vectors $\mathbf{n}_k$, $\mathbf{b}_k$ from Table 1 into the preceding expressions for ψ_k, λ_k. The resulting equations for the spherical angles of a material line in the [100][010][001] lattice frame are given by

$$\cos\theta = \frac{\sqrt{3}}{2}(\cos\psi_1 + \cos\psi_2), \quad \tan\phi = -\frac{\sqrt{2}\cos\lambda_2 - \cos\theta}{\sqrt{2}\cos\lambda_1 - \cos\theta}. \tag{2.24}$$

Similarly, for double slip in compression on systems $a2, c2$ in f.c.c. crystals, the spherical angles of a material plane in the (standard) lattice frame are given by

$$\cos\theta = \frac{\sqrt{3}}{2}(\cos\varphi_1 + \cos\varphi_2), \quad \tan\phi = \frac{\sqrt{6}\cos\varphi_1 + 2\cos\chi_2}{\cos\chi_1 - \cos\chi_2}. \tag{2.25}$$

(Formulas for other combinations of slip systems in f.c.c. and b.c.c. crystals in tension and compression may be found in Shalaby & Havner (1978) and Havner (1979), along with a number of illustrations of their application to various initial positions for a range of α from 0.02 to 50.)

2.3 Cones of Unstretched Material Lines

Let $\boldsymbol{\iota}_0$ denote a unit vector in the reference direction of a momentarily unstretched material line after proportional double slip of amounts $\gamma_2 = \alpha\gamma$, $\gamma_1 = \gamma$, with angles ψ_k^0, λ_k^0 determined by direction cosines $\mathbf{n}_k \cdot \boldsymbol{\iota}_0$ and $\mathbf{b}_k \cdot \boldsymbol{\iota}_0$. For double slip in systems $a\bar{2}, b1$ of f.c.c. crystals in tension,

$$\begin{aligned}
\cos\psi_1^0 &= \frac{1}{\sqrt{3}}\{\cos\theta_0 + \sin\theta_0(\sin\phi_0 + \cos\phi_0)\},\\
\cos\lambda_1^0 &= \frac{1}{\sqrt{2}}(\cos\theta_0 - \sin\theta_0\cos\phi_0),\\
\cos\psi_2^0 &= \frac{1}{\sqrt{3}}\{\cos\theta_0 - \sin\theta_0(\sin\phi_0 + \cos\phi_0)\},\\
\cos\lambda_2^0 &= \frac{1}{\sqrt{2}}(\cos\theta_0 + \sin\theta_0\sin\phi_0),
\end{aligned} \tag{2.26}$$

where θ_0, ϕ_0 are the spherical angles of $\boldsymbol{\iota}_0$ in the lattice frame. The equation of the unstretched cone in the reference state may be determined by setting equation (2.9) equal to one and substituting the above relations into equations (2.5) and (2.10) for constants $K_1, K_2, K_7, \ldots, K_{10}$. After a considerable amount of algebra, the following equation is obtained:

$$f_1 \sin^2 \theta_0 + f_2 \sin \theta_0 \cos \theta_0 + f_3 \cos^2 \theta_0 = 0, \tag{2.27}$$

where (with $c = 2\sqrt{(\alpha/6)}$)

$$\begin{aligned} f_1 &= \{(\alpha^2+1)(\cosh(2c\gamma)-1) - \sqrt{\alpha}(\alpha+1)\sinh(2c\gamma)\}(1+\sin 2\phi_0) \\ &\quad + 2\sqrt{\alpha}(\alpha-1)\sinh(c\gamma)\cos 2\phi_0, \\ f_2 &= -2(\alpha-1)\{(\alpha+1)(\cosh(2c\gamma)-1) \\ &\quad + \sqrt{\alpha}\sinh(2c\gamma)\}(\sin\phi_0+\cos\phi_0) \\ &\quad + 4\alpha(\cosh(c\gamma)-1)(\sin\phi_0-\cos\phi_0) \\ &\quad + 4\sqrt{\alpha}\sinh(c\gamma)(\alpha\sin\phi_0-\cos\phi_0), \qquad (2.28) \\ f_3 &= (\alpha^2+1)(\cosh(2c\gamma)-1) + 4\alpha(\cosh(2c\gamma)-\cosh(c\gamma)) \\ &\quad + \sqrt{\alpha}(\alpha+1)(3\sinh(2c\gamma)-2\sinh(c\gamma)). \end{aligned}$$

The deformation-dependent parameters f_1, f_2, f_3 are identically zero in the reference state, as required. For $\theta_0 = 90°$ (the perimeter of the [001] projection; see Fig. 1.6), f_1 must be zero for all α and γ, whence $\phi_0 = -45°$, 135° corresponding to directions $[1\bar{1}0]$ and $[\bar{1}10]$, the opposing senses along the line of intersection of the (111) and $(\bar{1}\bar{1}1)$ slip planes. As already noted, only this material line (*CL* in Fig. 2.1) remains unstretched throughout the deformation, and the two unstretched cones emanating from the origin always pass through the corresponding points $[\bar{1}10]$ and $[1\bar{1}0]$ of the [001] projection (one cone for $\theta_0 \leqslant 90°$, the other for $\theta_0 \geqslant 90°$).

For $\theta_0 \neq 90°$, equation (2.27) obviously is quadratic in $\tan\theta_0$, giving two angles θ_0 for each ϕ_0 and the specified α, γ. Equations (2.27)–(2.28) (developed here for the first time) together with equation (2.9) are the relations that Taylor & Elam (1925) needed for their analysis and assessment of double slip in aluminum crystals but did not have available. Instead, as discussed in Chapter 1, they used as an approximation the cones of zero stretch rates to compare with their experimentally determined unstretched cones in tensile loading.

To compare a theoretical unstretched cone in proportional double slip with an actual cone of unextended directions calculated from experimental measurements, it is necessary to transform either equation (1.6) or equation (2.27). In the former equation the angles of $\mathbf{\iota}_0$ are in the specimen frame *xyz* (Fig. 1.1), whereas in the latter equation θ_0, ϕ_0 are given in the lattice frame. Let ϕ_L, θ_L, ψ_L denote the Euler angles of the lattice axes relative to the specimen frame. (Beginning with coincident axes, the lattice is first rotated by ϕ_L about Z, then by θ_L about the momentary direction $\mathbf{e}_v$ of the [010]-axis, and finally by ψ_L about the new position of the [001]-axis, all rotations being counterclockwise.) Thus

$$\mathbf{e}_1 = (\mathbf{e}_u \cos\theta_L - \mathbf{e}_z \sin\theta_L)\cos\psi_L + \mathbf{e}_v \sin\psi_L,$$
$$\mathbf{e}_2 = -(\mathbf{e}_u \cos\theta_L - \mathbf{e}_z \sin\theta_L)\sin\psi_L + \mathbf{e}_v \cos\psi_L,$$
$$\mathbf{e}_3 = \mathbf{e}_u \sin\theta_L + \mathbf{e}_z \cos\theta_L, \tag{2.29}$$

where

$$\mathbf{e}_u = \mathbf{e}_x \cos\phi_L + \mathbf{e}_y \sin\phi_L, \quad \mathbf{e}_v = -\mathbf{e}_x \sin\phi_L + \mathbf{e}_y \cos\phi_L.$$

We denote the angles of $\mathbf{\iota}_0$ in the specimen frame (from eq. (1.6)) by θ_0^s, ϕ_0^s. The corresponding experimentally determined cone in the lattice frame is given by

$$\cos\theta_0 = \mathbf{\iota}_0 \cdot \mathbf{e}_3, \quad \sin\theta_0 \sin\phi_0 = \mathbf{\iota}_0 \cdot \mathbf{e}_2, \quad \sin\theta_0 \cos\phi_0 = \mathbf{\iota}_0 \cdot \mathbf{e}_1,$$

with

$$\mathbf{\iota}_0 \cdot \mathbf{e}_x = \sin\theta_0^s \cos\phi_0^s, \quad \mathbf{\iota}_0 \cdot \mathbf{e}_y = \sin\theta_0^s \sin\phi_0^s, \quad \mathbf{\iota}_0 \cdot \mathbf{e}_z = \cos\theta_0^s.$$

Conversely, the theoretical double-slip cone from equation (2.27) can be mapped as well into the specimen frame using the transpose of the orthogonal transformation (2.29). In either case, equation (2.9) also must be used, with l/l_0 corresponding to the specimen axis equated to the experimental λ_z of equations (1.3) in order to determine γ for an assumed α. The process can be repeated for various values of α to obtain the best match with an actual unstretched cone.

For double slip in systems *a*2, *c*2 of f.c.c. crystals in compression, with the same notation as in equations (2.26),

$$\cos\psi_1^0 = \frac{1}{\sqrt{3}}\{\cos\theta_0 + \sin\theta_0(\sin\phi_0 + \cos\phi_0)\},$$
$$\cos\lambda_1^0 = -\frac{1}{\sqrt{2}}(\cos\theta_0 - \sin\theta_0\cos\phi_0),$$
$$\cos\psi_2^0 = \frac{1}{\sqrt{3}}\{\cos\theta_0 + \sin\theta_0(\sin\phi_0 - \cos\phi_0)\}, \tag{2.30}$$
$$\cos\lambda_2^0 = -\frac{1}{\sqrt{2}}(\cos\theta_0 + \sin\theta_0\cos\phi_0).$$

Proceeding as before (again after a considerable amount of algebra), one obtains the equation of the unstretched cone

$$g_1 \sin^2\theta_0 + g_2 \sin\theta_0 \cos\theta_0 + g_3 \cos^2\theta_0 = 0, \tag{2.31}$$

where

$$
\begin{aligned}
g_1 = {} & (1+\alpha)(\cosh(2c\gamma)-1)\{1+\alpha+(1-\alpha)\sin 2\phi_0\} \\
& + 2\sqrt{\alpha}(1+\alpha)\sinh(2c\gamma)\cos 2\phi_0 \\
& - 8\alpha(\cosh(c\gamma)-1)\sin^2\phi_0 \\
& + 4\sqrt{\alpha}(1+\alpha)\sinh(c\gamma)\sin^2\phi_0 \\
& + 2\sqrt{\alpha}(1-\alpha)\sinh(c\gamma)\sin 2\phi_0, \qquad (2.32) \\
g_2 = {} & 2(1+\alpha)(\cosh(2c\gamma)-1)\{(1+\alpha)\sin\phi_0 \\
& + (1+\alpha)\cos\phi_0\} - 4\{\sqrt{\alpha}(1+\alpha)\sinh(2c\gamma) \\
& + 2\alpha(\cosh(c\gamma)-1) \\
& - \sqrt{\alpha}(1+\alpha)\sinh(c\gamma)\}\sin\phi_0, \\
g_3 = {} & (1+\alpha)^2(\cosh(2c\gamma)-1) - 2\sqrt{\alpha}(1+\alpha)\sinh(2c\gamma)
\end{aligned}
$$

(again with $c = 2\sqrt{(\alpha/6)}$). The parameters g_1, g_2, g_3 are of course zero at $\gamma = 0$. Moreover, it is readily confirmed that for $\phi_0 = -90°$, $g_1 + g_2 = -g_3$ for all α, γ. Thus $\phi_0 = -90°$, $\theta_0 = 45°$ define an unstretched material line throughout the deformation. This is direction $[0\bar{1}1]$, the intersection of slip planes (111) and $(\bar{1}11)$. ($\phi_0 = 90°$, $\theta_0 = 135°$ also necessarily satisfy eq. (2.31) for all γ corresponding to the opposite direction $[01\bar{1}]$.) As $\theta_0 = 90°$ in general will not satisfy equation (2.31), once again we have a quadratic equation in $\tan\theta_0$ for each ϕ_0 and specified α, γ.

Equations (2.31)–(2.32) are the generalization of Taylor's (1927a) equation of the unstretched cone in symmetric, equal double slip mentioned in Chapter 1. For that case a considerable simplification can be achieved. Not only is $\alpha = 1$, but the original kinematic solution is greatly simplified (Taylor 1927a) by recognizing that, with the present slip systems, the compression axis (i.e., the normal to the specimen end planes) rotates in the (100) plane toward the [011] direction.

For comparison of theoretical and experimental unstretched cones in compression the orthogonal transformation (2.29) again applies. However, equation (1.6) is no longer the equation of the experimental cone if the reference frame of Taylor & Farren (1926) mentioned in Chapter 1 is adopted, because the respective deformation gradients from equations (1.3) (corresponding to Fig. 1.1) and their equations (5)–(7) in the rotated frame have different zero elements. The appropriate expression for the cone of unchanged lengths in compression is their equation (9). Also note that equation (2.16) rather than equation (2.9) must be used, with $\lambda_{(z)} = (a/a_0)^{-1}$

from equation (2.16) set equal to the experimental spacing stretch (ratio of specimen thicknesses) to determine γ for an assumed α.

As illustration of the application of equations (2.9) and (2.27) in tensile loading, we consider an apparent double-slip deformation investigated by Taylor & Elam (1925), namely, their specimen no. 59 in the axial stretch range 1.745 to 1.936. (This crystal deformed in single slip to and somewhat beyond an extension of 56 percent.) From their fig. 3 and chosen labeling of the lattice axes, the primary (initial) and secondary slip systems were $c3$ and $b\bar{1}$ with the lattice located relative to the specimen axes Z and X (Fig. 1.1) by spherical angles of directions $[\bar{1}11]$ and $[100]$. It is evident from Fig. 1.6 that these lattice directions bear the same relation to system $c3$ as directions $[111]$ and $[\bar{1}00]$ bear to system $a\bar{2}$ (i.e., $c3$ can be put in position $a\bar{2}$ by a 180° rotation of the lattice about $[011]$). Thus we may use equations (2.27)–(2.28) based upon active systems $a\bar{2}$, $b1$ simply by interpreting Taylor and Elam's spherical angles (from their table 2) as defining directions $[111]$, $[\bar{1}00]$ rather than $[\bar{1}11]$, $[100]$ in the specimen frame.

The measured angles at axial stretch 1.745 were: $[111]$, $\theta_0^s = 62.8°$, $\phi_0^s = 190.3°$; $[\bar{1}00]$, $\theta_0^s = 65.3°$, $\phi_0^s = 350.6°$. These values give an angle of 124.7° between the two lattice directions whereas the ideal angle is 125.3°, from which one may judge that the X-ray diffraction measurements at this stage of the deformation were quite accurate. To obtain an orthogonal set of lattice axes we shall slightly adjust these values. Accordingly, all calculations henceforth are based upon the following angles, which give the ideal angle of 125.3° between $[111]$ and $[\bar{1}00]$: $[111]$, $\theta_0^s = 63.0°$, $\phi_0^s = 190.0°$; $[\bar{1}00]$, $\theta_0^s = 65.5°$, $\phi_0^s = 350.8°$. After a straightforward application of the standard orthogonality and normality conditions $\mathbf{e}_i \cdot \mathbf{e}_j = \delta_{ij}$ using these angles, the following (right-handed) transformation giving the relative orientation of the lattice and specimen axes is obtained:

$$\begin{aligned}
\mathbf{e}_1 &= -0.8983\mathbf{e}_x + 0.1455\mathbf{e}_y - 0.4147\mathbf{e}_z, \\
\mathbf{e}_2 &= -0.3124\mathbf{e}_x - 0.8751\mathbf{e}_y + 0.3695\mathbf{e}_z, \\
\mathbf{e}_3 &= -0.3091\mathbf{e}_x + 0.4615\mathbf{e}_y + 0.8315\mathbf{e}_z.
\end{aligned}$$

(It is readily confirmed that the (adjusted) angles $\theta_a^s = 63.0°$, $\phi_a^s = 190.0°$ and $\theta_b^s = 65.5°$, $\phi_b^s = 350.8°$ of directions $\mathbf{e}_a = [111]$, $\mathbf{e}_b = [\bar{1}00]$ are given to this accuracy by the above values from

$$\mathbf{e}_0 = (\mathbf{e}_x \cos\phi_0^s + \mathbf{e}_y \sin\phi_0^s)\sin\theta_0^s + \mathbf{e}_z \cos\theta_0^s$$

applied to $\mathbf{e}_a$, $\mathbf{e}_b$ in the specimen frame.) From column three of the transformation matrix, the angles of the specimen axis in the lattice frame are $\theta_z = 33.75°$, $\phi_z = 138.3°$; whence the axis has rotated 3.3° beyond the $a\bar{2}$, $b1$ symmetry line (Fig. 1.6) at stretch 1.745 from its initial position near [011] in triangle $a\bar{2}$ (see fig. 3 of Taylor & Elam (1925)).

From the last two columns of Taylor and Elam's table 1, the values required for evaluation of the relative deformation gradient between stretches 1.745 and 1.936 of their specimen no. 59 are:

$$a/a_0 = 0.9291, \quad b/b_0 = 0.9907, \quad \lambda_z = 1.1095,$$
$$\alpha_0 = 77.93°, \quad \beta_0 = 107.73°, \quad \delta_0 = 100.1°,$$
$$\alpha = 75.48°, \quad \beta = 115.45°, \quad \delta = 102.5°.$$

(To interpret Taylor and Elam's symbols, which are different from these except for the βs, compare Fig. 1.1 here with fig. 1 of Taylor & Elam (1923) (their "Face 1" containing the axis X). Also see Taylor & Elam (1926, fig. 1).) Upon substituting these values into equations (1.3) one obtains

$$x = 0.9291x_0 - 0.05229y_0, \quad y = 0.9824y_0,$$
$$z = -0.08746x_0 + 0.00405y_0 + 1.1095z_0,$$

with x, y, z the coordinates of a crystal point at the final stretch of 1.936. Then, from equation (1.6) and these relations, the equation of the experimental unstretched cone in the reference state is

$$(0.9691\sin^2\phi_0^s - 0.4894\sin 2\phi_0^s - 1.2907)\sin^2\theta_0^s$$
$$+ (0.09\sin\phi_0^s - 1.9407\cos\phi_0^s)\sin\theta_0^s\cos\theta_0^s + 2.3089\cos^2\theta_0^s = 0,$$

which gives the directions of all material lines whose lengths at specimen stretch 1.936 are momentarily equal to their values at stretch 1.745.[3]

To determine a theoretical unstretched cone from equations (2.27)–(2.28), it is first necessary to determine γ for an assumed α using equations (2.9) and (2.26) evaluated for the specimen axis (i.e., with $\lambda_z = 1.936/1.745$, $\theta_z = 33.75°$ and $\phi_z = 138.3°$ substituted for l/l_0, θ_0, and ϕ_0, respectively.) For active systems $a\bar{2}$ and $b1$, the constants in equation (2.9) are most conveniently evaluated for a given initial direction using the following expressions (determined from eqs. (2.5), (2.10), and (2.26), with $b = \frac{1}{2}$ and

[3] From the determinant of the deformation gradient, Taylor and Elam's measurements give a volume increase of 1.27% during this interval. As that seems rather large in an additional extension of only 11%, there perhaps may be small errors in some of their reported values, making the equation of the experimental cone slightly in error as well.

$c_{12}=c_{21}=\sqrt{6}/3$):

$$K_1=\frac{1}{2\sqrt{3}}(c_1-c_2\sqrt{\alpha}),\quad K_2=\frac{1}{2\sqrt{3}}(c_1+c_2\sqrt{\alpha}),$$

$$K_7=\frac{3}{2\alpha}K_1^2(1+\alpha+\sqrt{\alpha}),\quad K_8=\frac{3}{2\alpha}K_2^2(1+\alpha-\sqrt{\alpha}),$$

$$K_9=\frac{\sqrt{3}}{2}K_1c_3(1+1/\sqrt{\alpha}),\quad K_{10}=\frac{\sqrt{3}}{2}K_2c_3(1-1/\sqrt{\alpha}), \tag{2.33}$$

$$c_1=\sin\theta_0(\sin\phi_0+\cos\phi_0)+\cos\theta_0,$$

$$c_2=\sin\theta_0(\sin\phi_0+\cos\phi_0)-\cos\theta_0,$$

$$c_3=\sin\theta_0(\sin\phi_0-\cos\phi_0)-\cos\theta_0.$$

Choosing Taylor and Elam's estimated value of $\frac{1}{3}$ for $\alpha=\gamma_2/\gamma_1$ and solving equation (2.9), we obtain $\gamma_1=\gamma=0.1929$. The functions f_1, f_2, f_3 in equation (2.28) then may be expressed solely in terms of angles θ_0 and ϕ_0, from which the equation of the theoretical unstretched cone in the lattice frame is determined. The result is

$$-\{0.1224(1+\sin 2\phi_0)+0.07011\cos 2\phi_0\}\sin^2\theta_0$$
$$+(0.2459\sin\phi_0-0.04556\cos\phi_0)\sin\theta_0\cos\theta_0+0.3172\cos^2\theta_0=0.$$

Finally, from the matrix of direction cosines at axial stretch 1.745, the angles in the specimen frame are related to the angles in the lattice frame by

$$\sin\theta_0^s\cos\phi_0^s=-(0.8983\cos\phi_0+0.3124\sin\phi_0)\sin\theta_0$$
$$-0.3091\cos\theta_0,$$

$$\sin\theta_0^s\sin\phi_0^s=(0.1455\cos\phi_0-0.8751\sin\phi_0)\sin\theta_0$$
$$+0.4615\cos\theta_0,$$

$$\cos\theta_0^s=(-0.4147\cos\phi_0+0.3695\sin\phi_0)\sin\theta_0$$
$$+0.8315\cos\theta_0.$$

(For example, the intersection of the slip planes, direction $[\bar{1}10]$ ($\theta_0=90°$, $\phi_0=135°$), which identically satisfies the theoretical equation, has angles $\theta_0^s=56.32°$, $\phi_0^s=-60.14°$ in the specimen frame.)

In Fig. 2.2 are shown the Z-axis stereographic projections of the experimental and theoretical unstretched cones together with the locations of intersection $[\bar{1}10]$ of the a and b slip planes and three other lattice directions defined in figure 7 of Taylor & Elam (1925). These latter directions (for the present choice of active systems $a\bar{2}$ and $b1$) are $[0\bar{1}1]$,

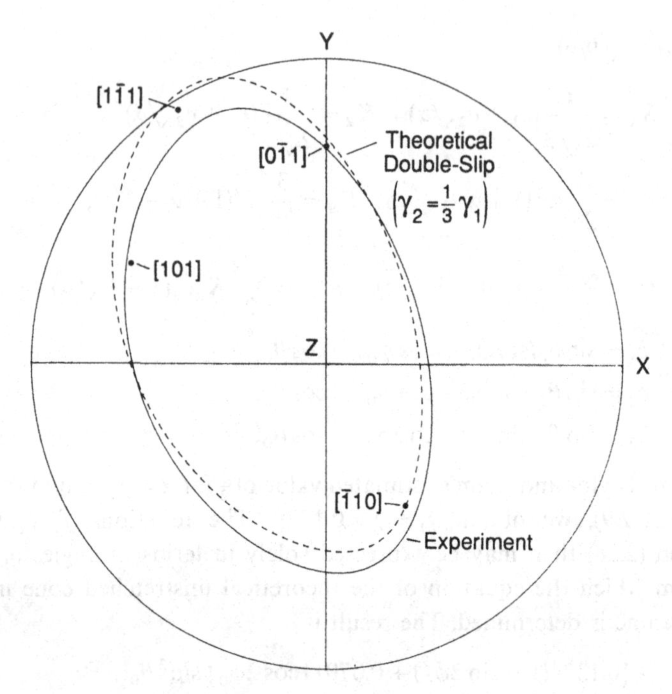

Fig. 2.2. Z-axis stereographic projections, in an aluminum crystal extended 74.5%, of the positions of theoretical and experimental unstretched cones for extension from 74.5 to 93.6%.

[101], and [1$\bar{1}$1], respectively labeled *B*, *C*, and *D* in Taylor and Elam's figure 7. All cones of double slip in planes (111) and ($\bar{1}\bar{1}$1) must contain axis [$\bar{1}$10], whatever the deformation and relative slips; and the calculated locus of the experimental cone is seen to pass very close to that point of the projection. The other three points were established by Taylor and Elam as the additional directions through which the cone of *differential* double slip (that is, the cone of zero stretch rates) must pass at stretch 1.745. However, the exact theoretical locus for additional relative stretch $\lambda_z = 1.936/1.745$ and $\alpha = \frac{1}{3}$ only lies near those points because the slips obviously are finite in the interval (0.1929 and 0.0643, respectively). Nonetheless, Taylor and Elam's estimate of the relative amounts of slip, based on their comparisons of solutions of the incremental strain problem and the finite-deformation experimental cone, is a very good approximation. The corresponding finite theoretical and experimental loci are seen to be close, differing principally by a small relative displacement in the direction of their long axes. A slightly improved theoretical locus corresponds to $\alpha = 0.29$, but the deviation from the double-slip curve in

Fig. 2.2 is minor. (From other calculations, proportionately larger double slip in the second system than the first produces theoretical loci that are finitely rotated counterclockwise from that of the figure.)

Multiple-slip kinematic solutions in the *channel die compression test*, both with and without lattice rotation, are presented in Chapter 5. In the next chapter we turn to the development of a comprehensive theoretical framework for the analysis of finite plastic deformation in crystals.

3

A GENERAL THEORY OF ELASTOPLASTIC CRYSTALS

In setting down a general continuum mechanics description of finite deformation processes in metal crystals that takes into account both lattice straining and gross crystallographic slip, it is useful to begin with an assessment of the minimum physical scale at which such a description has meaning. The following discussion, based upon similar discussions in Havner (1973a,b; 1982a), is pertinent to the determination of that minimum scale.

As seen by an observer resolving distances to 10^{-3} mm, the deformation of a crystal grain (of typical dimensions 10^{-3}–10^{-2} cm within a finitely strained polycrystalline metal) may be considered relatively smooth. At this level of observation, which for convenience we shall call microscopic, one can just distinguish between slip lines on crystal faces after extensive distortion of a specimen. In contrast, a *sub*microscopic observer resolving distances to 10^{-5} mm (the order of 100 atomic spacings) is aware of highly discontinuous displacements within crystals. The microscopic observer's slip lines appear to the submicroscopic observer as slip *bands* of order 10^{-4} mm thickness, containing numerous glide lamellae between which amounts of slip as great as 10^3 lattice spacings have occurred, as first reported by Heidenreich (1949); hence a continuum perspective at this second level would seem untenable. Accordingly, we adopt a continuum model in which a material "point" has physical dimensions of order 10^{-3} mm. This is greater than 10^3 lattice spacings yet at least an order of magnitude smaller than typical grain sizes in polycrystalline metals.[1]

With this scale in mind, the dominant phenomenological aspect of the plastic deformation of metal crystals, as evident from the preceding

[1] Single-crystal grain sizes are of course vastly larger. Taylor & Elam's (1923) original aluminum specimen, for example, was 1.0 cm × 1.0 cm × 20.0 cm.

chapters, is the movement of the gross material (as defined, say, by lines scribed on crystal surfaces) relative to the underlying crystalline lattice. Material lines and planes translate and rotate with respect to the (averaged) lattice orientation over volumes of typical dimensions of order 10^{-2} mm. The inelastic behavior of metal crystals is thereby distinguished by kinematics alone from that of polymers and geomaterials.

The interaction and movement of dislocations (whose analysis is outside the scope of this monograph) takes place at any instant in and through the current deformed lattice, but the averaged lattice strain at a material point is in general only slightly affected by the dislocations present; and their net effect at the continuum scale is simply crystallographic slip and the relative motion of material and lattice. Thus, excluding the mechanism of twinning from consideration, the appropriate kinematic representation of gross incremental motion from the current deformed state of a crystalline solid is a linear superposition of additional lattice straining and rotation with incremental slips on well-defined crystallographic slip systems. Each of the latter deformation mechanisms translates material lines (i.e., glide lamellae) relative to one another, resulting in gross incremental shearing, but leaves the averaged lattice structure unchanged. We analyze this dual kinematics of material and lattice in the following section.

3.1 Fundamental Kinematics of Crystal and Lattice

A general kinematics of the finite deformation of crystals that includes lattice distortion without approximation apparently was first given by Rice (1971) and may be found in equivalent forms in Hill & Rice (1972), Rice (1975), Havner (1974, 1982a, 1986, 1987a), and Hill & Havner (1982). Both the account here of kinematics and the theoretical developments in following sections are largely based upon presentations in the last three of those works.

In Fig. 3.1 are depicted the geometric decomposition and basic transformations, or mappings, of crystal and lattice deformations that are our starting point. Here an asterisk refers to quantities related to the underlying lattice, but on a continuum scale such that "1*" represents, say, 10^2 atomic spaces. In contrast, "1" corresponds to the differential neighborhood of a crystal material point, considered homogeneously deformed, and is greater than 10^{-3} mm (of order 10^4 atomic spacings). The deformation gradient $\tilde{A}$, due to the accumulation of incremental slips, maps the reference-crystal neighborhood into an imagined intermediate

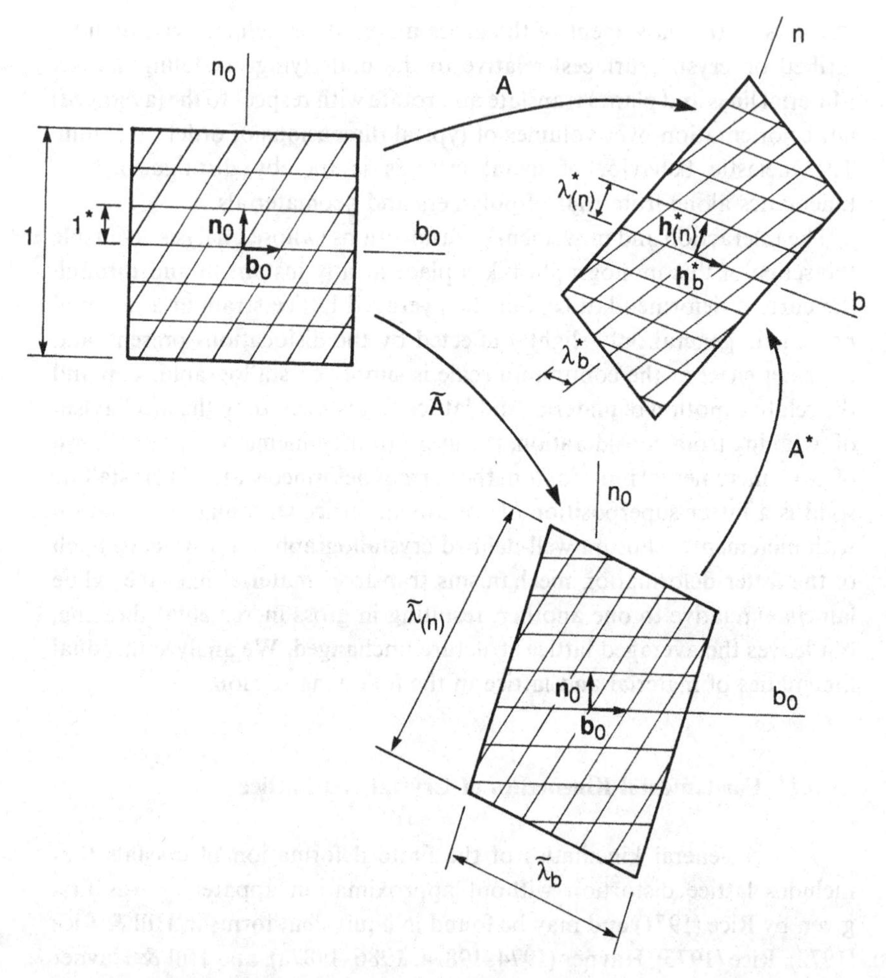

Fig. 3.1. Geometric decomposition and mappings of crystal and lattice deformations.

configuration, as shown, with no change in volume (area in the figure) and no distortion or rotation of the lattice in the reference frame. This configuration, which contains the reference state B^* of the lattice, is mapped by the deformation gradient A^* into the current configuration, with crystal and lattice deforming and rotating as one in that transformation (constructed to scale in Fig. 3.1).

The vectors $\mathbf{b}_0$, $\mathbf{n}_0$ are unit vectors in the referential slip and normal directions of a representative slip system; $\mathbf{h}_b^*$ is an embedded basis vector for the lattice which equals $\mathbf{b}_0$ in the reference state; and $\mathbf{h}_{(n)}^*$ is a reciprocal basis vector, identically the lattice inverse spacing–stretch vector, which initially equals $\mathbf{n}_0$(i.e., $\mathbf{h}_b^*, \mathbf{h}_{(n)}^*$ are respectively covariant and contravariant vectors). Thus

$$\mathbf{h}_b^* = A^*\mathbf{b}_0 = \lambda_b^*\mathbf{b}, \quad \mathbf{h}_{(n)}^* = \mathbf{n}_0 A^{*-1} = \lambda_{(n)}^{*-1}\mathbf{n}, \tag{3.1}$$

the latter from Nanson's equation using $a_{(n)}^*\lambda_{(n)}^* = v^*$, where $\lambda_{(n)}^*$, $a_{(n)}^*$ are the spacing and areal stretches in the family of lattice planes defined by normal $\mathbf{n}$, and $v^* = \det A^*$ is the relative specific volume of the lattice in its current configuration B^*. (As shown in Fig. 3.1, $\tilde{\lambda}_b, \tilde{\lambda}_{(n)}$ are the line and spacing stretches in the deformation $\tilde{A}$.)

From the polar decomposition theorem $A^* = R^*\Lambda^*$, where R^* is the rotation of the triad of principal directions of lattice strain and Λ^* is the symmetric, right-stretch tensor of the lattice. In cubic metals, typically $\Lambda^* = \lambda^*(p)I + \xi^*$, with the elements of symmetric tensor ξ^* of $O(10^{-3})$ or less and $\lambda^*(p)$ significantly different from unity only at extremely high pressure p. Consequently, $A^* \approx R^*$ (with R^* now signifying the rotation of the lattice axes) is a simplification that commonly may be used in the analysis of experiments and was implicit in the kinematics of the preceding chapters. We also shall make use of this simplification in later chapters, but for the present development of a general theory shall retain A^* without approximation (noting at various points, however, the interpretation of equations that would result from setting $A^* \approx R^*$).

By analogy with equations $(2.1)_3$ and (2.2), the evolution of the deformation gradient $\tilde{A}$ is given by

$$\mathrm{d}\tilde{A} = \sum(\mathbf{b}_0 \otimes \mathbf{n}_0 \, \mathrm{d}\tilde{\gamma})_j \tilde{A}, \tag{3.2}$$

in which $\mathrm{d}\tilde{\gamma}$ is the invariant measure of slip introduced by Rice (1971). In a given system $\mathrm{d}\tilde{\gamma}$ is the gradient with respect to the slip-plane normal direction, in units of spacing stretch $\lambda_{(n)}^*$, of the incremental slip displacement in units of lattice stretch λ_b^*; hence its integrated value is independent of lattice strain. In terms, therefore, of the spatial gradient $\mathrm{d}\gamma$ of incremental slip displacement (in system $\mathbf{b}$, $\mathbf{n}$) in ordinary units of measurement,

$$\mathrm{d}\tilde{\gamma} = \lambda_{(n)}^* \, \mathrm{d}\gamma / \lambda_b^*. \tag{3.3}$$

That this is the necessary meaning of the $\mathrm{d}\tilde{\gamma}$s in order for equation (3.2) to be the precise equation of $\tilde{A}$ may be established as follows.

Let $\mathrm{d}\mathbf{x}_0$, $\mathrm{d}\mathbf{x}$, and $\mathrm{d}\tilde{\mathbf{x}}$ denote relative position vectors of material points (infinitesimally close on the continuum scale of B_0) in the reference, current, and intermediate configurations, respectively, where $\mathrm{d}\tilde{\mathbf{x}}$ results from the transformation of $\mathbf{dx}_0$ by the accumulated invariant slips alone. Then, by definition of A, A^*, and $\tilde{A}$,

$$\mathbf{dx} = A\,\mathbf{dx}_0 = A^*\,\mathbf{d\tilde{x}}, \quad \mathbf{d\tilde{x}} = \tilde{A}\,\mathbf{dx}_0, \quad A = A^*\tilde{A} \tag{3.4}$$

(with $A \approx R^*\tilde{A}$ at ordinary pressures). This "multiplicative decomposition" (as it occasionally is called) of the overall deformation gradient A was introduced into macroscopic plasticity theory by Lee & Liu (1967) and Lee (1969) and into crystal plasticity by Rice (1971) and Kratochvil (1971). Only in crystal mechanics, however, is equation $(3.4)_3$ unambiguously a pure kinematic relation.

Alternately, from the discussion earlier in this chapter of a physically appropriate incremental kinematics of crystal and lattice in the current state, we may write

$$\Gamma = \Gamma^* + \sum(\mathbf{b}\otimes\mathbf{n}\dot{\gamma})_j, \tag{3.5}$$

where

$$\Gamma^* = \varepsilon + \omega \tag{3.6}$$

is the contribution of the additional lattice straining and rotation to the velocity gradient, with ε and ω, respectively, the lattice Eulerian strain-rate and spin tensors. Upon differentiating equation $(3.4)_3$ and substituting that result together with equations (3.2), (3.5) and the inverse of equation $(3.4)_3$ into the basic kinematic relation $\Gamma = \dot{A}A^{-1}$, one obtains

$$\Gamma^*\,d\theta + \sum(\mathbf{b}\otimes\mathbf{n}\,d\gamma)_j = dA^*A^{*-1} + A^*\sum(\mathbf{b}_0\otimes\mathbf{n}_0\,d\tilde{\gamma})_j A^{*-1}, \tag{3.7}$$

with θ representing time or a timelike variable. Finally, after substitution of equation (3.1), since we must have $\Gamma^* = \dot{A}^*A^{*-1}$ the connection (3.3) in each slip system is proved. (In cubic metals, whose response to pure pressure is isotropic, the ratio $\lambda^*_{(n)}/\lambda^*_b$ will differ from unity by a term of $O(10^{-3})$ or less. Thus, for practical purposes $d\tilde{\gamma}$ and $d\gamma$ are indistinguishable in any slip system. Nevertheless, for exactness in the general theory we shall continue to use $d\tilde{\gamma}$.)

3.2 Basic Stress and Strain Analysis

The differential kinematics of material and lattice together with a representation of nominal traction vectors and their changes, resolved on embedded basis vectors, are shown in Fig. 3.2 (based upon Havner (1987a)). This figure encompasses five configurations (those for material and lattice being separately defined) and three differential mappings that

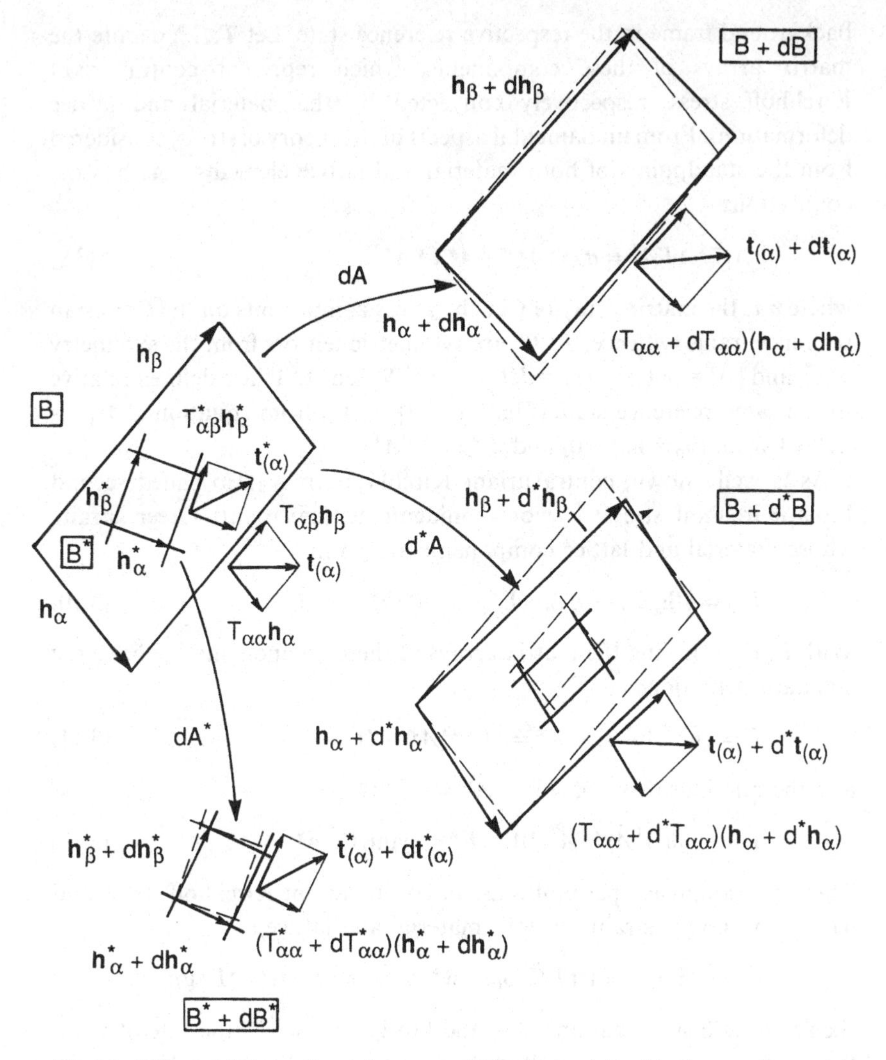

Fig. 3.2. Differential kinematics and stress analysis of material and lattice.

are explained subsequently. The vectors $\mathbf{t}_{(\alpha)}, \mathbf{t}^*_{(\alpha)}$ are forces per unit reference area (i.e., "nominal" tractions) acting on the respective material and lattice planes (defined by referential normal directions), and $T_{\alpha\beta}, T^*_{\alpha\beta}$ are the components of these forces on the embedded bases $\mathbf{h}_\alpha, \mathbf{h}^*_\alpha$ ($\alpha, \beta = 1, 2, 3$ or x, y, z). Thus

$$\mathbf{t}_\alpha = T_{\alpha\beta}\mathbf{h}_\beta, \quad \mathbf{t}^*_\alpha = T^*_{\alpha\beta}\mathbf{h}^*_\beta, \tag{3.8}$$

with $\mathbf{h}^{(0)}_\alpha$, $\mathbf{h}^{*(0)}_\alpha$ each forming an orthonormal basis coincident with the background frame in the respective reference state. Let T, T^* denote the matrix arrays of these components, which represent contravariant Kirchhoff stress, respectively convected by the material and lattice deformations. From fundamental aspects of the theory of stress, considered from the standpoints of both material and lattice elements, one has the connections

$$|A|^{-1}ATA^{\mathrm{T}} = \sigma = |A^*|^{-1}A^*T^*A^{*\mathrm{T}}, \tag{3.9}$$

where σ is the matrix array of Cauchy stress components on the Cartesian reference frame (whence, T, T^* are symmetric tensors from the symmetry of σ), and $|A| \equiv \det A$, $|A^*| \equiv \det A^* = v^*$. When A, A^* are defined relative to the same reference state (Fig. 3.1), $|A| = |A^*|$ from equation $(3.4)_3$ (as $|\tilde{A}| = 1$ from $|\mathbf{b}_0 \otimes \mathbf{n}_0| = 0$) and $T^* = ATA^{\mathrm{T}}$.

As is well known, contravariant Kirchhoff stress (also called second Piola–Kirchhoff stress) is work-conjugate to (covariant) Green strain, whose material and lattice components are given by

$$E_{\alpha\beta} = \tfrac{1}{2}(\mathbf{h}_\alpha \cdot \mathbf{h}_\beta - \delta_{\alpha\beta}), \quad E^*_{\alpha\beta} = \tfrac{1}{2}(\mathbf{h}^*_\alpha \cdot \mathbf{h}^*_\beta - \delta_{\alpha\beta}). \tag{3.10}$$

With E, E^* denoting the matrix arrays of these components, we have the standard definitions

$$E = \tfrac{1}{2}(A^{\mathrm{T}}A - I), \quad E^* = \tfrac{1}{2}(A^{*\mathrm{T}}A^* - I), \tag{3.11}$$

and the relations (from $\dot{A} = \Gamma A$, $\dot{A}^* = \Gamma^* A^*$)

$$\dot{E} = \mathrm{sym}(A^{\mathrm{T}}\dot{A}) = A^{\mathrm{T}}DA, \quad \dot{E}^* = \mathrm{sym}(A^{*\mathrm{T}}\dot{A}^*) = A^{*\mathrm{T}}\varepsilon A^*. \tag{3.12}$$

The work-conjugacy per unit mass of contravariant Kirchhoff stress and (covariant) Green strain for both material and lattice,

$$\dot{w} \equiv \mathrm{tr}(\sigma D)/\rho = \mathrm{tr}(T\dot{E})/\rho_0, \quad \dot{w}^* \equiv \mathrm{tr}(\sigma\varepsilon)/\rho = \mathrm{tr}(T^*\dot{E}^*)/\rho_0^*, \tag{3.13}$$

then follows from equations (3.9) and (3.12), with $\dot{w}$, $\dot{w}^*$ the rate of work (or "stress working") per unit mass in crystal and lattice deformations, respectively, and ρ the mass density in the current state. (Distinguishing between ρ_0 and ρ_0^* permits one to choose different reference states for material and lattice, in which case $vT^* = \tilde{A}T\tilde{A}^{\mathrm{T}}$, with $v = |A|/|A^*| = \rho_0/\rho_0^*$.)

With regard to the differential mappings in Fig. 3.2, the following explanations taken from Havner (1987a) (see also Hill & Havner 1982, p. 12) of the distinctions among ordinary differentials of unstarred variables, differentials of starred (*) variables, and starred differentials of unstarred variables should be carefully noted: dA is the ordinary material differential of deformation gradient A (referred to initial reference state B_0) from current configuration B of the crystal element; dA^* is the differential of the deformation gradient of the underlying lattice from its current state B^* (with A^* referred to reference state B_0^* of the lattice); d^*A is the differential change of a lattice element chosen to have the same configuration (shown dashed in state $B + d^*B$ of the figure) as the crystal element in current state B (as if the lattice element had deformed with the material from the reference state). Thus, dA takes material state B into $B + dB$; dA^* takes lattice state B^* into $B^* + dB^*$; and d^*A takes a lattice element having the configuration of material state B into configuration $B + d^*B$, so both dB^* and d^*B are purely lattice changes, but from different momentary reference configurations. Equivalently, $B + d^*B$ would be the new state if material and lattice deformed incrementally as one from the current state (or the differential change with all $\tilde{\gamma}$s, hence $\tilde{A}$, held fixed). Both the stress changes dT, dT^*, d^*T indicated in Fig. 3.2 and the strain changes dE, dE^*, d^*E may be similarly understood. Therefore the starred differentials of A, E, and T are connected to the differentials dA^*, dE^*, and dT^* through the slip transformation $\tilde{A}$ from equations $(3.4)_3$, (3.11), and (3.9) (for the same initial reference state) by

$$d^*A = dA^*\tilde{A}, \quad d^*E = \tilde{A}^{\mathrm{T}} dE^* \tilde{A}, \quad dT^* = \tilde{A}\, d^*T \tilde{A}^{\mathrm{T}}. \tag{3.14}$$

(For simplicity, dA^* and d^*A are constructed to scale as pure rotations in Fig. 3.2.)

3.3 General Relationships for Arbitrary Measures

Although one could continue with the exclusive use of Green strain and contravariant Kirchhoff stress in all equations, these variables are not always the most convenient ones in which to express constitutive relations for crystals. Moreover, a valuable perspective is gained by adopting arbitrary strain measures and their work-conjugate stresses, as in the pioneering work of Hill (1968, 1970, 1972, 1978).[2]

Let e denote a symmetric tensor measure of strain in Hill's general family

$$e = \sum e_\alpha \mathbf{u}_\alpha \otimes \mathbf{u}_\alpha, \quad e_\alpha = f(\lambda_\alpha), \tag{3.15}$$

where $f(\lambda)$ is a smooth, monotone function satisfying $f(1) = 0,\ f'(1) = 1$

[2] Hill's general theory of conjugate stress and strain measures and work invariants, one of the major contributions to basic solid mechanics during the latter half of this century, is reviewed in the Appendix for readers not well familiar with that part of his work.

but otherwise arbitrary, and the $\mathbf{u}_\alpha$ are referential (Lagrangian) directions of principal stretches λ_α ($\alpha = 1, 2, 3$). Following Havner (1986) (also see Hill (1984)) we define a fourth-order, kinematic transformation tensor $\mathscr{K}$ for measure e by

$$\mathrm{d}e = \mathscr{K}\,\mathrm{d}A, \quad \mathscr{K} = \frac{\partial e}{\partial A^{\mathrm{T}}}, \tag{3.16}$$

with symmetry $\mathscr{K}_{ijkl} = \mathscr{K}_{jikl}$ (only) and eigenvectors $\mathrm{d}A = WA\,\mathrm{d}\theta$ corresponding to pure spin W. The latter result becomes transparent by expressing $\mathscr{K}$ in terms of the gradient of e with Green strain. From equation $(3.12)_1$ and the chain rule of partial differentiation one has

$$\mathscr{K}_{ijkl} \equiv \frac{\partial e_{ij}}{\partial A_{lk}} = \frac{\partial e_{ij}}{\partial E_{km}} A_{lm},$$

whence

$$\mathscr{K}(WA) = \frac{\partial e}{\partial E}(A^{\mathrm{T}}WA) = 0 \quad \text{(inner products)}$$

from the symmetry of E and skew-symmetry of W. Note that for $f(\lambda) = \frac{1}{2}(\lambda^2 - 1)$, $e = E$ and $\mathscr{K}$ equals tensor $\mathscr{G}$ of Hill (1984), given by $\mathscr{G}_{ijkl} = \frac{1}{2}(A_{li}\delta_{jk} + A_{lj}\delta_{ik})$.

In changes from current state B to states $B + \mathrm{d}B$ and $B + \mathrm{d}^*B$ of material and lattice (fig. 3.2), we may write for any measure

$$\mathrm{d}e - \mathrm{d}^*e = \mathscr{K}(\mathrm{d}A - \mathrm{d}^*A) \tag{3.17}$$

where from equations (3.2), $(3.4)_3$, and $(3.14)_1$

$$\mathrm{d}A - \mathrm{d}^*A = A\sum(B\,\mathrm{d}\tilde{\gamma})_j, \quad B_j = \tilde{A}^{-1}(\mathbf{b}_0 \otimes \mathbf{n}_0)_j \tilde{A}, \tag{3.18}$$

the latter tensor introduced in Hill & Havner (1982). Thus

$$\mathrm{d}e - \mathrm{d}^*e = \sum(v\,\mathrm{d}\tilde{\gamma})_j, \quad v_j = \mathscr{K}(AB_j). \tag{3.19}$$

For the Green measure, v_j may of course be determined by substituting $\mathscr{K} = \mathscr{G}$ into the Cartesian-component form of equation $(3.19)_2$. Alternately, as

$$\mathrm{d}E = \mathrm{sym}(A^{\mathrm{T}}\mathrm{d}A), \quad \mathrm{d}^*E = \mathrm{sym}(A^{\mathrm{T}}\mathrm{d}^*A) \tag{3.20}$$

(the latter from eqs. $(3.14)_{1,2}$ with eq. $(3.4)_3$), one obtains directly (using eq. (3.18))

$$v_j = \mathrm{sym}(A^{\mathrm{T}}AB_j) = \tilde{A}^{\mathrm{T}}\,\mathrm{sym}\{A^{*\mathrm{T}}A^*(\mathbf{b}_0 \otimes \mathbf{n}_0)_j\}\tilde{A} \tag{3.21}$$

first given as equations $(5.7)_2$ and (5.5) of Hill & Havner (1982). For $A^* \approx R^*$, we evidently have $v_j \approx \tilde{A}^{\mathrm{T}} \operatorname{sym}(\mathbf{b}_0 \otimes \mathbf{n}_0)_j \tilde{A}$ for the Green measure.

Let t denote the symmetric stress tensor conjugate to arbitrary measure e such that the scalar product $t\,de/\rho_0$ is differential work per unit mass, an invariant under change in strain measure and reference state:

$$dw = \operatorname{tr}(T\,dE)/\rho_0 = t\,de/\rho_0. \tag{3.22}$$

We define $\tilde{\tau}_j$ as the scalar variable conjugate to the invariant slip measure $\tilde{\gamma}_j$ in the sense that

$$\sum(\tilde{\tau}\,d\tilde{\gamma})_j \equiv \rho_0(dw - dw^*) = \rho_0\,d\tilde{w}, \tag{3.23}$$

the differential stress work in the incremental slips per unit reference volume. From equations (3.9) and $(3.13)_2$, with a common reference state for material and lattice,

$$dw^* = \operatorname{tr}(T\,d^*E)/\rho_0 = t\,d^*e/\rho_0 \tag{3.24}$$

(as dE, de in eq. (3.22) equal d^*E, d^*e at fixed slips). Thus (Hill & Havner 1982, eq. (4.1))

$$\rho_0\,d\tilde{w} = t(de - d^*e) = \sum(\tilde{\tau}\,d\tilde{\gamma})_j, \tag{3.25}$$

with

$$\tilde{\tau}_j = t v_j \quad \text{(scalar product)} \tag{3.26}$$

from equation $(3.19)_1$.

As $\rho_0\,d\tilde{w}$ is invariant under change in strain measure, so too are the $\tilde{\tau}$s. Consequently, equation (3.26) may be evaluated using t and v_j of the Green measure. It follows from equations (3.9), $(3.18)_2$, and (3.21) (with $|A| = |A^*|$) that

$$\tilde{\tau}_j = |A^*| \operatorname{tr}\{A^{*-1}\sigma A^*(\mathbf{b}_0 \otimes \mathbf{n}_0)_j\}.$$

Finally, from equations (3.1) we have

$$\tilde{\tau}_j = |A^*|(\mathbf{h}^*_{(n)}\sigma\mathbf{h}^*_b)_j, \tag{3.27}$$

which for $A^* \approx R^*$ may be approximated by the resolved Cauchy shear stress $(\sigma_{nb})_j = \mathbf{n}_j\sigma\mathbf{b}_j$. Henceforth we shall call $\tilde{\tau}_j$ the *generalized Taylor–Schmid stress* of the jth slip system.

We further define symmetric tensors α_j by[3]

$$dt - d^*t = \sum(\alpha\,d\tilde{\gamma})_j, \tag{3.28}$$

[3] Tensors which are $(\lambda^*_{(n)}/\lambda^*_b)$ multiples of α, v in a given system were introduced by Hill & Rice (1972).

for we evidently must have $\mathrm{d}t = \mathrm{d}^*t$ at fixed slips (i.e., when the material and lattice deform incrementally as one). Hill & Havner (1982) established the following key connection among t, e, v_j, and α_j for any measure:

$$\alpha_j = -t\frac{\partial v_j}{\partial e}, \tag{3.29}$$

where the gradient with e is at constant $\tilde{A}$. The proof here closely follows that in Hill & Havner (1982, section 3). Let the δ-operator signify a variation (whether real or virtual) at fixed slips $\tilde{\gamma}_j + \mathrm{d}\tilde{\gamma}_j$ of states $B + \mathrm{d}B$, $B + \mathrm{d}^*B$ of material and lattice after deformations $\mathrm{d}A, \mathrm{d}^*A$ of a common reference element in configuration B (see fig. 3.2). By definition of conjugate measures,

$$(t + \mathrm{d}t)\delta(e + \mathrm{d}e) = (t + \mathrm{d}^*t)\delta(e + \mathrm{d}^*e),$$

as each side is equivalent to the work per unit volume of Cauchy stress in the δ-variations of the simultaneous states $B + \mathrm{d}B$, $B + \mathrm{d}^*B$. Thus, to consistent second-order in products,

$$t\delta(\mathrm{d}e) + \mathrm{d}t\delta e = t\delta(\mathrm{d}^*e) + \mathrm{d}^*t\delta e$$

or, with equations $(3.19)_1$ and (3.28),

$$t\sum(\delta v\,\mathrm{d}\tilde{\gamma})_j + \sum(\alpha\,\mathrm{d}\tilde{\gamma})_j\delta e = 0. \tag{3.30}$$

As $\delta v_j = (\partial v_j/\partial e)\delta e$ by definition of the δ-variation, and as equation (3.30) must hold for arbitrary δe whatever the incremental slips $\mathrm{d}\tilde{\gamma}_j$, equation (3.29) immediately follows.

Adopting the fourth-order tensor (at fixed slips)

$$\mathcal{N}_j^{\mathrm{T}} = \frac{\partial v_j}{\partial e} \tag{3.31}$$

introduced in Hill & Havner (1982), we may express equations (3.29) and (3.28) by

$$\alpha_j = -\mathcal{N}_j t \tag{3.32}$$

and

$$\mathrm{d}t - \mathrm{d}^*t = -\sum(\mathcal{N}\,\mathrm{d}\tilde{\gamma})_j t. \tag{3.33}$$

For the Green measure, as

$$\delta v_j = 2\,\mathrm{sym}(\delta E B_j) = \mathcal{N}_j^{\mathrm{T}}\delta E$$

from equations $(3.11)_1$ and (3.21), we find (recognizing that $\mathcal{N}_{ijkl}$ in an arbitrary system must be symmetric with respect to separate i, j and k, l interchanges)

$$\mathcal{N}_{ijkl} = \tfrac{1}{2}(B_{ik}\delta_{jl} + B_{jk}\delta_{il} + B_{il}\delta_{jk} + B_{jl}\delta_{ik}) \tag{3.34}$$

and (from eq. (3.34))

$$\alpha_j = -2\,\mathrm{sym}(B_j T). \tag{3.35}$$

The preceding equations as well as algebraic rules for transformation of tensors ν, α, and $\mathcal{N}$ under change in strain measure and reference state were first given by Hill & Havner (1982).[4] Specific equations for $\mathcal{K}$, $\boldsymbol{\nu}$, $\boldsymbol{\alpha}$, and $\mathcal{N}$ corresponding to the Almansi strain measure $E_A = \frac{1}{2}(I - A^{-1}A^{-T})$ (i.e., $f(\lambda) = \frac{1}{2}(1 - \lambda^{-2})$) may be found in Havner (1986).

3.4 Lattice Elasticity and Existence of Plastic Potentials

We now make the basic assumption that the averaged lattice deformation is *Green-elastic*:

> In whatever way the elements of any material system may act upon each other, if all the internal forces exerted be multiplied by the elements of their respective directions, the total sum for any assigned portion of the mass will always be the exact differential of some function. (Green 1839, p. 1).

A further prescient statement of George Green's with respect to the science of thermodynamics then yet to be developed may be found in that same paper: "Indeed, if $\delta\phi$ were not an exact differential, a perpetual motion would be possible, and we have every reason to think, that the forces of nature are so disposed as to render this a natural impossibility" (p. 4).

Let $\phi^*(e^*)$ (which also may be expressed $\phi^*(\Lambda^*)$) denote the Green-elastic potential function of the crystal lattice, where e^* is any symmetric measure of lattice strain, with t^* its work-conjugate stress. We then may write

$$t^*\,\mathrm{d}e^* = \rho_0^*\,\mathrm{d}\phi^* \tag{3.36}$$

[4] In that work the derivation of general relationships (such as eqs. $(3.19)_1$, (3.26), and (3.28) here) proceeds from a somewhat different basis. An elastic domain in e-space is presupposed, and changes $\mathrm{d}t$, $\mathrm{d}e$ take place via a finite elastic path (of lattice straining) from a point e within that domain to a boundary point. There an infinitesimal elastoplastic deformation takes place, incrementally changing the domain. This is followed by another finite elastic path to a point $e + \mathrm{d}e$ in the new domain.

(with $d\phi^*$ here playing the role of Green's $\delta\phi$). Consequently it follows that

$$t^* = \rho_0^* \frac{\partial \phi^*}{\partial e^*}, \quad dt^* = \mathscr{L}^* de^*, \quad \mathscr{L}^* = \rho_0^* \frac{\partial^2 \phi^*}{\partial e^* \partial e^*}, \tag{3.37}$$

where $\mathscr{L}^*$ is the fourth-order tensor of instantaneous elastic moduli of the lattice for measure e^*, with (Green-elastic) symmetries $\mathscr{L}^*_{ijkl} = \mathscr{L}^*_{jikl} = \mathscr{L}^*_{klij}$. Moreover, from the lattice work-equivalence (recall eq. (3.24))

$$dw^* = t\, d^*e/\rho_0 = t^*\, de^*/\rho_0^*, \tag{3.38}$$

we have (Hill & Havner 1982; Havner 1986)

$$t = \rho_0 \frac{\partial \phi}{\partial e}, \quad d^*t = \mathscr{L} d^*e, \quad \mathscr{L} = \rho_0 \frac{\partial^2 \phi}{\partial e \partial e}, \quad \phi(e, \tilde{A}) = \phi^*(e^*), \tag{3.39}$$

with all gradients of the scalar potential ϕ with respect to e taken at constant $\tilde{A}$ (i.e., at fixed slips $\tilde{\gamma}_j$). It is evident that $\mathscr{L}$, the fourth-order tensor of instantaneous elastic moduli relative to the material for measure e, has the same symmetries as $\mathscr{L}^*$.

For arbitrary strain measures e, e^* the connection between $\mathscr{L}$ and $\mathscr{L}^*$ may be established as follows. We define $\mathscr{K}^*$ for the lattice (analogously to $\mathscr{K}$ for the material) by

$$de^* = \mathscr{K}^*\, dA^*, \quad \mathscr{K}^* = \frac{\partial e^*}{\partial A^{*\mathrm{T}}} \tag{3.40}$$

(i.e., $\mathscr{K}^*_{ijkl} = \partial e^*_{ij}/\partial A^*_{lk}$). Then, using equation (3.38) (with $\rho_0 = \rho_0^*$) and equaton $(3.14)_1$, we have

$$t^* \mathscr{K}^*\, dA^* = t\mathscr{K}(dA^* \tilde{A}) \quad \text{(scalar product)}$$

from which

$$t^*_{ij} \mathscr{K}^*_{ijkl} = t_{ij} \mathscr{K}_{ijml} \tilde{A}_{km}. \tag{3.41}$$

For the Green measure this reduces to $T^* = \tilde{A} T \tilde{A}^{\mathrm{T}}$, as previously obtained from equations $(3.4)_3$ and (3.9) (with $|A| = |A^*|$). Taking an elastic variation of equation (3.41) (that is, a differential change at fixed $\tilde{A}$) and eliminating dA^* from both sides of the resulting equation with the aid of equations $(3.14)_1$, (3.16), and (3.40), one finds

$$\left(\mathscr{K}^{*\mathrm{T}} \mathscr{L}^* \mathscr{K}^* + t^* \frac{\partial \mathscr{K}^*}{\partial A^{*\mathrm{T}}}\right)_{ijkl} = \tilde{A}_{im} \left(\mathscr{K}^{\mathrm{T}} \mathscr{L} \mathscr{K} + t \frac{\partial \mathscr{K}}{\partial A^{\mathrm{T}}}\right)_{mjnl} \tilde{A}_{kn} \tag{3.42}$$

which is the desired connection for arbitrary measures, where

$$\left(\mathcal{K}^{*\mathrm{T}}\mathcal{L}^{*}\mathcal{K}^{*}+t^{*}\frac{\partial\mathcal{K}^{*}}{\partial A^{*\mathrm{T}}}\right)_{ijkl}=\mathcal{K}^{*}_{mnij}\mathcal{L}^{*}_{mnpq}\mathcal{K}^{*}_{pqkl}+t^{*}_{mn}\frac{\partial\mathcal{K}^{*}_{mnij}}{\partial A^{*}_{lk}},\tag{3.43}$$

with an equivalent meaning for the tensor sum within parentheses on the right-hand side of equation (3.42).

For each of Green strain and Almansi strain, the second terms on the two sides of equation (3.42) may be shown to be equal, and the equation remaining upon subtracting them is equivalent to $\mathrm{d}t^{*}\,\mathrm{d}e^{*}=\mathrm{d}^{*}t\,\mathrm{d}^{*}e$ (scalar product) for those measures (Havner 1986). For Green strain this relation follows immediately from equations $(3.14)_{2,3}$. For the Almansi measure the relevant equations are (with the same reference state for material and lattice)

$$\begin{aligned}&\mathrm{d}E_{\mathrm{A}}=-\tfrac{1}{2}\mathrm{sym}\{A^{-1}\,\mathrm{d}(A^{-\mathrm{T}})\},\\&\mathrm{d}E^{*}_{\mathrm{A}}=A\,\mathrm{d}^{*}E_{\mathrm{A}}A^{\mathrm{T}},\quad \mathrm{d}^{*}T_{\mathrm{c}}=A^{\mathrm{T}}\,\mathrm{d}T^{*}_{\mathrm{c}}A,\end{aligned}\tag{3.44}$$

with T_{c}, T^{*}_{c} here representing the symmetric arrays of covariant Kirchhoff stress respectively convected by the material and lattice deformations:

$$T_{\mathrm{c}}=|A|A^{\mathrm{T}}\sigma A,\quad T^{*}_{\mathrm{c}}=|A^{*}|A^{*\mathrm{T}}\sigma A^{*}.\tag{3.45}$$

The relation $\mathrm{d}t^{*}\,\mathrm{d}e^{*}=\mathrm{d}^{*}t\,\mathrm{d}^{*}e$ is again seen to follow immediately from equation (3.44). The specific connections between $\mathcal{L}$ and $\mathcal{L}^{*}$ for the Green and Almansi measures may be determined directly from this scalar equality by substituting both equations $(3.37)_2$ and $(3.39)_2$ and respectively equations $(3.14)_{2,3}$ and $(3.44)_{2,3}$. The results are

$$\begin{aligned}&\mathcal{L}^{*}_{ijkl}=\tilde{A}_{im}\tilde{A}_{jn}\tilde{A}_{kp}\tilde{A}_{lq}\mathcal{L}_{mnpq}\quad\text{(Green)},\\&\mathcal{L}_{ijkl}=\tilde{A}_{mi}\tilde{A}_{nj}\tilde{A}_{pk}\tilde{A}_{ql}\mathcal{L}^{*}_{mnpq}\quad\text{(Almansi)}.\end{aligned}\tag{3.46}$$

As is evident from equations (3.37) and (3.39), it has been assumed here, following Hill & Havner (1982), that the Green-elastic potential function of the lattice is unaffected by incremental slips. A more general theory which does not initially make this assumption is discussed in Rice (1971), Hill & Rice (1972), and Havner (1973b, 1974). However, the effects of plastic deformation and dislocation multiplication on the elastic properties of metal crystals apparently are quite small (see, for example, McLean (1962, chapter 1)), and the above authors subsequently also neglect those effects as a physically appropriate simplification and adopt equations equivalent to equations (3.37) or (3.39). Further discussion of the matter may be found in Havner (1982a, pp. 283–4).

The existence of plastic potential equations for local crystal behavior, and precisely what they are, now can be established. We define the *plastic decrement in stress* and *plastic increment in strain* for arbitrary conjugate measures t, e and incremental (virtual) strain and stress cycles by (Hill 1972; Hill & Rice 1972; Hill & Havner 1982)

$$\mathrm{d}^{\mathrm{p}}t = \mathscr{L}\,\mathrm{d}e - \mathrm{d}t, \quad \mathrm{d}^{\mathrm{p}}e = \mathrm{d}e - \mathscr{M}\,\mathrm{d}t, \quad \mathscr{M} = \mathscr{L}^{-1} \tag{3.47}$$

($\mathscr{M}$ being the fourth-order tensor of elastic compliances relative to the material for measure e). Thus, $t - \mathrm{d}^{\mathrm{p}}t$ would be the stress state after an imagined differential cycle of strain in which incremental slips $\mathrm{d}\tilde{\gamma}_j$ accumulated only during the addition of $\mathrm{d}e$. Similarly, $e + \mathrm{d}^{\mathrm{p}}e$ would be the strain state after an imagined differential cycle of stress with incremental slipping only during the addition of $\mathrm{d}t$. (Also note that $\mathrm{d}^{\mathrm{p}}t = \mathscr{L}\,\mathrm{d}^{\mathrm{p}}e$.) Upon substituting equations $(3.19)_1$, $(3.39)_2$, and (3.28) into each of equations $(3.47)_{1,2}$, we obtain

$$\mathrm{d}^{\mathrm{p}}t = \sum(\Lambda\,\mathrm{d}\tilde{\gamma})_j, \quad \mathrm{d}^{\mathrm{p}}e = \sum(M\,\mathrm{d}\tilde{\gamma})_j \tag{3.48}$$

with

$$\Lambda_j = \mathscr{L}v_j - \alpha_j, \quad M_j = v_j - \mathscr{M}\alpha_j. \tag{3.49}$$

From the definition (3.26) of the invariant generalized Taylor–Schmid stress $\tilde{\tau}_j$ it follows that

$$\frac{\partial \tilde{\tau}_j}{\partial e} = \mathscr{L}v_j + t\frac{\partial v_j}{\partial e}, \quad \frac{\partial \tilde{\tau}_j}{\partial t} = v_j + t\frac{\partial v_j}{\partial e}\mathscr{M},$$

in which use has been made of the relations

$$\mathscr{L} = \mathscr{L}^{\mathrm{T}} = \frac{\partial t}{\partial e}, \quad \mathscr{M} = \mathscr{M}^{\mathrm{T}} = \frac{\partial e}{\partial t} \tag{3.50}$$

at fixed slips (from eq. (3.39) and $\mathscr{M} = \mathscr{L}^{-1}$). Then, after substitution of the key connection (3.29) among t, e, v_j, and α_j for arbitrary measures into the gradients, we have (from the definitions (3.49) and the symmetry of $\mathscr{M}$)[5]

$$\frac{\partial \tilde{\tau}_j}{\partial e} = \Lambda_j, \quad \frac{\partial \tilde{\tau}_j}{\partial t} = M_j, \tag{3.51}$$

[5] We note in passing that without a Green-elastic potential for the lattice the right-hand side of eq. $(3.51)_2$ would be $v_j - \mathscr{M}^{\mathrm{T}}\alpha_j$, with $\mathscr{M}^{\mathrm{T}} \neq \mathscr{M}$.

whence

$$\mathrm{d}^{\mathrm{p}}t = \frac{\partial}{\partial e}\sum(\tilde{\tau}\,\mathrm{d}\tilde{\gamma})_j, \quad \mathrm{d}^{\mathrm{p}}e = \frac{\partial}{\partial t}\sum(\tilde{\tau}\,\mathrm{d}\tilde{\gamma})_j, \tag{3.52}$$

which are the *crystal plastic potential equations* in the form derived by Hill & Havner (1982), dependent only upon the slip geometry and the Green-elasticity of the lattice. Thus, the generalized Taylor–Schmid stress $\tilde{\tau}_j$ serves as a plastic potential in strain space for the contribution of incremental slip in the jth system to the plastic decrement in stress (for any measure), and likewise in stress space for the $\mathrm{d}\tilde{\gamma}_j$ contribution to the plastic increment in strain.

It is important to recognize, as first remarked by Hill & Rice (1972, p. 404) and again by Hill & Havner (1982, pp. 10–11), that the increment $\mathrm{d}^{\mathrm{p}}e$ of strain for which a plastic potential exists contains a net contribution from the lattice caused by its slip-induced incremental rotation relative to the material during the addition of $\mathrm{d}t$. Consequently, even with the current state as reference, so long as lattice strain is taken into account $\mathrm{d}^{\mathrm{p}}e/\mathrm{d}\theta$ does not equal the plastic strain rate (recall eq. (3.5))

$$D^{\mathrm{p}} = \sum N_j\dot{\gamma}_j, \quad N_j = \mathrm{sym}\,(\mathbf{b}\otimes\mathbf{n})_j, \tag{3.53}$$

for any measure (although then $\dot{e} = D$ momentarily for all measures).

3.5 Critical Strengths and Hardening Laws: Normality

As remarked early in Chapter 1, the first explicit introduction of the concept that there is a deformation-dependent physical property in a crystallographic slip system which the resolved shear stress must equal for continuation of slip in that system is found in Taylor & Elam (1925) (although implied in Taylor's 1923 Bakerian Lecture to the Royal Society). Named by them the "resistance to slipping" per unit area of the slip plane, this material parameter we rather shall call the critical strength of a given system. As additionally discussed in Chapter 1, Taylor (1925) and Schmid (1926) independently introduced an experimentally founded hardening law in single slip wherein the critical strength is expressible as a function of the slip. In all the experimental work of the 1920s, as well as (apparently) in most single-crystal experiments performed since then, lattice strains were not separately determined nor taken into account.

To encompass and extend those basic ideas into a general theory of elastoplastic crystals, as was first done in a precise manner by Hill & Rice (1972), we introduce N parameters $\tilde{\tau}_k^c$ which are the *critical strengths* of the crystallographic slip systems. (Opposite senses of slip in the same glide plane are represented by distinct ks so that $d\gamma_k$ is always either positive or zero; thus $N = 24$ in f.c.c. crystals.) Furthermore, we shall take these critical strengths to depend only upon the evolution of the invariant measures of slip $\tilde{\gamma}_j$ (which would have been indistinguishable from the accumulations of the $d\gamma$s in essentially all the finite-deformation experiments that have been reported). Thus, we express a general hardening law for all N systems by (Hill & Havner 1982)

$$d\tilde{\tau}_k^c = \sum_j H_{kj}\, d\tilde{\gamma}_j, \quad j = 1, \ldots, n, \quad k = 1, \ldots, N, \tag{3.54}$$

with $n < N/2$ the number of potentially active systems in the current state and the H_{kj} the *physical slip-systems hardening moduli*. These are in general functions of the slip history of all systems, and may be functions of the current stress state relative to the lattice as well.

On the basis of Taylor & Elam's (1923) second experimental law, which established the resolved Cauchy shear stress σ_{nb} as the determiner of the active slip system in a virgin crystal, a *critical* (i.e., potentially active) system seemingly should be defined as one in which σ_{nb} equals the current critical strength $\tilde{\tau}^c$ of that system. However, the $\tilde{\tau}^c$s as expressed previously are independent of lattice strain, whereas it is known from experiment (Haasen & Lawson 1958) that a critical strength in cubic metals defined in terms of resolved Cauchy stress is affected by very large pressure, hence is dependent upon lattice volume change. We already have noted from equation (3.27) that the generalized Taylor–Schmid stress $\tilde{\tau}_j$, unequivocally established as a plastic potential, is very nearly $(\sigma_{nb})_j$ at ordinary pressures and lattice strains. Moreover, in the experiments in the 1920s of Taylor, Elam, and Farren in England and Schmid and his colleagues in Germany, and in most other experiments through the years on finite deformation of metal crystals, $\tilde{\tau}_j$ and $(\sigma_{nb})_j$ could not have been readily differentiated. Accordingly, for clarity and precision in the theory that generally would not be distinguishable from experiment, we instead choose

$$\tilde{\tau}_j(\Lambda^*) = \tilde{\tau}_j^c, \quad j = 1, \ldots, n, \tag{3.55}$$

to define each of the n critical slip systems (with $\tilde{\tau}_j < \tilde{\tau}_j^c$ in a noncritical system). A much more important and difficult issue, taken up in Chapter 4, is the specification of the changing moduli H_{kj}.

Adoption of equation (3.55), which, as shortly will be shown, guarantees a "normality structure" relative to elastic domains that is invariant under change in measure (Hill & Rice 1972; Hill & Havner 1982), has the following consequence. In cubic metals, from equation (3.27), a critical strength expressed in terms of the resolved Cauchy shear stress is inversely proportional to $|A^*|$ and so increases with increasing pressure. The question remains whether this increase is in reasonable agreement with experiment at large pressures. The only precise analysis of the matter of which I am aware is found in Havner (1973b), which is reviewed in the next section. Here, limiting consideration to ordinary pressures, we turn to slip-system inequalities based upon equations (3.54) and (3.55).

Define the "net strength" $\tilde{f}_k = \tilde{\tau}_k^c - \tilde{\tau}_k$, with $\tilde{f}_k \geqslant 0$ in every system. In a critical system $\tilde{f}_k = 0$, $\mathrm{d}\tilde{f}_k \geqslant 0$ and we have the following *critical slip-system inequalities* from equations (3.26), (3.54), and (3.55), independent of strain measure:

$$\mathrm{d}(t\nu_k) \leqslant \sum_j H_{kj}\,\mathrm{d}\tilde{\gamma}_j, \quad j, k = 1, \ldots, n. \tag{3.56}$$

(For the typical case of infinitesimal lattice strain, $A^* \approx R^*$, $\mathrm{d}\tilde{\gamma}_j \approx \mathrm{d}\gamma_j$, and $\mathrm{d}\tilde{\tau}_k \approx (\mathbf{n}\mathscr{D}^*\sigma\mathbf{b})_k$, with $\mathscr{D}^*$ signifying differential change in a lattice-corotational frame, from which we may write

$$(\mathbf{n}\mathscr{D}^*\sigma\mathbf{b})_k \leqslant \sum_j H_{kj}\,\mathrm{d}\gamma_j,$$

the general inequalities introduced by Hill (1966).) As $\tilde{\tau}_k = t\nu_k$ is a scalar invariant, we may express its differential change in the frame of a lattice observer, whence

$$\mathrm{d}\tilde{\tau}_k = \Lambda_k\,\mathrm{d}^*e = M_k\,\mathrm{d}^*t \tag{3.57}$$

from equations (3.51). Then, upon substituting each of the equalities (3.57) in turn (with eqs. (3.19) and (3.28), respectively) into equation (3.56), we obtain the slip-system inequalities expressed in the equivalent forms (Hill & Havner 1982; see also Hill & Rice 1972)

$$\Lambda_k\,\mathrm{d}e \leqslant \sum_j g_{kj}\,\mathrm{d}\tilde{\gamma}_j, \quad g_{kj} = H_{kj} + \Lambda_k \nu_j, \tag{3.58}$$

$$M_k\,\mathrm{d}t \leqslant \sum_j h_{kj}\,\mathrm{d}\tilde{\gamma}_j, \quad h_{kj} = H_{kj} + M_k \alpha_j, \tag{3.59}$$

$j, k = 1, \ldots, n$. Inequalities (3.58) and (3.59) imply normality rules for $\mathrm{d}^{\mathrm{p}}t$ and $\mathrm{d}^{\mathrm{p}}e$ in the respective strain and stress spaces, as follows.

Let $\delta t, \delta e$ denote increments in conjugate stress and strain measures producing purely elastic response from the current state (i.e., $\delta t = \mathscr{L}\delta e$), requiring an elastic domain in each of t- and e-space. As all $\mathrm{d}\tilde{\gamma}_j$ would be zero in this elastic unloading, we have

$$\frac{\partial \tilde{\tau}_k}{\partial e} \delta e \leqslant 0, \quad \frac{\partial \tilde{\tau}_k}{\partial t} \delta t \leqslant 0, \quad k = 1, \ldots, n, \tag{3.60}$$

in every critical system (from eqs. (3.51) and eqs. (3.58)–(3.59)). Upon multiplying each of inequalities (3.60) by the nonnegative $\mathrm{d}\tilde{\gamma}_k$ corresponding to interconnected increments $\mathrm{d}t$, $\mathrm{d}e$ that produce a general elastoplastic response, summing over all critical systems, and substituting the plastic potential equations (3.52), one obtains

$$\mathrm{d}^{\mathrm{p}}t\,\delta e \leqslant 0, \quad \mathrm{d}^{\mathrm{p}}e\,\delta t \leqslant 0 \text{ (scalar products)}, \tag{3.61}$$

which are equivalent relations from $\mathrm{d}^{\mathrm{p}}t = \mathscr{L}\,\mathrm{d}^{\mathrm{p}}e$. These are the *crystal normality laws* in e- and t-space, respectively, which are invariant under change in strain measure as established by Hill (1972).[6] It is evident from the foregoing development that they are a consequence of the Green-elasticity of the lattice, the assumption that the critical strengths depend only upon the history of the invariant slips, and the definition (3.55) of critical slip systems. Their standard geometric interpretation is expressed in the following.

In strain space the range of all instantaneous rates of change $\delta e/\delta\theta$ corresponding to purely elastic response from the current state (with n critical systems) defines a pyramid whose n hyperplanes are tangent to the respective level surfaces (from eqs. (3.55), (3.26), (3.39)$_1$, (3.19)$_2$, and (3.16)$_2$)

$$\rho_0 \frac{\partial \phi}{\partial A^{\mathrm{T}}}(AB_j) = \tilde{\tau}_j^{\mathrm{c}}, \quad j = 1, \ldots, n, \tag{3.62}$$

(where we have used $\partial\phi/\partial A^{\mathrm{T}} = (\partial\phi/\partial e)(\partial e/\partial A^{\mathrm{T}})$). These functions are the plastic potentials for $\mathrm{d}^{\mathrm{p}}t$ and define the *yield surface* in the vicinity of point e. The nonpositive scalar product $\mathrm{d}^{\mathrm{p}}t\,\delta e$ of the normality rule (3.61)$_1$ is equivalently a statement that the plastic decrement in stress $\mathrm{d}^{\mathrm{p}}t$ after a differential strain cycle must lie within an *orthogonal* pyramid that is the pyramid of outward normals to the yield surface at e. A parallel statement for $\delta t/\delta\theta$ and $\mathrm{d}^{\mathrm{p}}e$ applies in conjugate stress space. (In the latter the normals to the level surfaces $\tilde{\tau}_j = \tilde{\tau}_j^{\mathrm{c}}$ are the tensors M_j from eq. (3.51)$_2$. In strain space they are the tensors $\Lambda_j = \mathscr{L}M_j$.)

[6] The invariance of inequalities (3.61) with change in measure is an example of Hill's invariant bilinear form (Hill 1968, 1972) as follows: From eqs. (3.47), $\delta t = \mathscr{L}\delta e$, and the symmetry of $\mathscr{L}$ (and $\mathscr{M}$) we have, from either of the inequalities, $\delta t\,\mathrm{d}e - \mathrm{d}t\,\delta e \leqslant 0$, which is invariant in value in any virtual cycle of stress and strain (see Appendix, Sect. A.3).

3.5.1 *Pressure Dependence of Cauchy Flow Stress*

From equations (3.1) and (3.27) and $a^*_{(n)}\lambda^*_{(n)} = |A^*|$, the generalized Taylor–Schmid stress $\tilde{\tau}_j$ may be expressed

$$\tilde{\tau}_j = (a^*_{(n)}\sigma_{nb}\lambda^*_b)_j. \tag{3.63}$$

Henceforth denoting the resolved Cauchy shear stress $(\sigma_{nb})_j$ by τ_j, we have

$$\left(\frac{\mathrm{d}\tau}{\tau}\right)_j = \left(\frac{\mathrm{d}\tilde{\tau}}{\tilde{\tau}}\right)_j - \left(\frac{\mathrm{d}a^*_{(n)}}{a^*_{(n)}} + \frac{\mathrm{d}\lambda^*_b}{\lambda^*_b}\right)$$

(cf. eq. (5.11) of Havner (1973b), evaluated with the current state as reference). Upon substituting the lattice kinematic relations

$$\dot{a}^*_{(n)}/a^*_{(n)} = \operatorname{tr}\varepsilon - \mathbf{n}\varepsilon\mathbf{n}, \quad \dot{\lambda}^*_b/\lambda^*_b = \mathbf{b}\varepsilon\mathbf{b}, \tag{3.64}$$

which follow from equations (3.1) and (3.6) and $\dot{A}^* = \Gamma^* A^*$ (as $\operatorname{tr}\omega$, $\mathbf{n}\omega\mathbf{n}$, and $\mathbf{b}\omega\mathbf{b}$ are zero), the above equation of incremental change in conventional resolved shear stress τ_j may be alternately written

$$\left(\frac{\dot{\tau}}{\tau}\right)_j = \left(\frac{\dot{\tilde{\tau}}}{\tilde{\tau}}\right)_j - (\operatorname{tr}\varepsilon + \varepsilon_{bb} - \varepsilon_{nn})_j \tag{3.65}$$

(with $\varepsilon_{bb} = \mathbf{b}\varepsilon\mathbf{b}$, $\varepsilon_{nn} = \mathbf{n}\varepsilon\mathbf{n}$).

Consider an interval of elastic loading or unloading during which the jth system remains critical according to the normality rule. The generalized Taylor–Schmid stress $\tilde{\tau}_j$ then remains constant from equations (3.54) and (3.55) (as all dγs are zero throughout the interval), and it follows that

$$\dot{\tau}^c_j = -(\dot{a}^*_{(n)}/a^*_{(n)} + \dot{\lambda}^*_b/\lambda^*_b)\tau^c_j$$

or (3.66)

$$\dot{\tau}^c_j = -(\operatorname{tr}\varepsilon + \varepsilon_{bb} - \varepsilon_{nn})\tau^c_j,$$

where τ^c_j is the critical value of the resolved Cauchy shear stress, or "flow stress" as it is commonly named in the metallurgical literature. For cubic crystals, changes in pressure producing only volume dilatation can be imposed separately from (small) changes in additional, infinitesimal lattice strain due to a general superimposed stress state. As such strains typically are of $O(10^{-3})$ or less, the only significant variation in flow stress according to equation (3.66) will be that corresponding to large pressure change.

Let $\lambda(p)$ denote the joint material and lattice isotropic stretch ratio under purely hydrostatic stress. After neglecting second-order terms in the additional (infinitesimal) lattice strain, the rate of change of τ^c_j according to the normality law (i.e., eq. (3.66)) reduces to (Havner 1973b)

$$\dot{\tau}^c_j = -3(\dot{\lambda}/\lambda)\tau^c_j,$$

whence

$$(\tau^c_0/\tau^c)_j = \lambda^3, \tag{3.67}$$

where $(\tau^c_0)_j$ is the value of the critical Cauchy shear strength, or flow stress, at zero pressure. Thus, as remarked in Section 3.5, the theoretical flow stress in a cubic metal is inversely proportional to the volume dilatation. To evaluate equation (3.67) for comparison with experiment, an elastic equation for λ in terms of pressure p is required.

For (isothermal) elastic deformation of a single crystal, equations $(3.37)_1$ and $(3.39)_1$ are identical and may be expanded in a Taylor series about the unstressed reference state. Thus, for the Green measure,

$$T = \mathscr{E}^{(2)}_0 \cdot E + \frac{1}{2!}(\mathscr{E}^{(3)}_0 \cdot E) \cdot E + \frac{1}{3!}(\mathscr{E}^{(4)}_0 \cdot E) \cdot (E \otimes E) + \cdots, \tag{3.68}$$

where

$$\mathscr{E}^{(m)}_0 = \rho_0 \left(\frac{\partial^m \phi}{\partial E \partial E \cdots \partial E} \right)_0 \tag{3.69}$$

is the tensor (of rank $2m$) of mth-order elastic moduli as traditionally defined. Let $\eta = \frac{1}{2}(\lambda^2 - 1)$ denote the isotropic Green strain in cubic crystals under pure pressure, with $t^c = -\lambda p$ the corresponding scalar value of (contravariant Kirchhoff) stress T. There follows the equation

$$t^c = (\mathscr{E}_{11kk})_0 \eta + \tfrac{1}{2}(\mathscr{E}_{11kkmm})_0 \eta^2 + \cdots$$

or

$$-\lambda p = c_1 \eta + \tfrac{1}{2} C_1 \eta^2 + \cdots, \tag{3.70}$$

where from the cubic symmetry

$$c_1 = (c_{11} + 2c_{12})_0, \quad C_1 = (C_{111} + 6C_{112} + 2C_{123})_0, \tag{3.71}$$

with the c_{ij} and C_{ijk} the second- and third-order (isothermal) elastic moduli in the reference state, use having been made of the generalized Voigt notation introduced by Brugger (1964). (Subscript pairs in the respective fouth- and sixth-order tensors $\mathscr{E}^{(2)}_0$ and $\mathscr{E}^{(3)}_0$ are replaced by single indices according to the standard scheme $(11) \to 1$, $(22) \to 2$, $(33) \to 3$, $(23, 32) \to 4$, $(31, 13) \to 5$, and $(12, 21) \to 6$). Inverting equation (3.70) (after substituting $\lambda^{-1} = 1 - \eta + \frac{3}{2}\eta^2 + \cdots$), one obtains (Havner 1973b, eq. (2.18))

$$\eta = -(p/c_1) + \tfrac{1}{2}(2 - C_1/c_1)(p/c_1)^2 + \cdots. \tag{3.72}$$

Finally, substituting this into $\lambda^3 = 1 + 3\eta + \frac{3}{2}\eta^2 + \cdots$, we have

$$\lambda^3 = 1 - 3(p/c_1) + \tfrac{3}{2}(3 - C_1/c_1)(p/c_1)^2 + \cdots. \tag{3.73}$$

Consequently, from equation (3.67) and the above, the normality criterion for the pressure dependence of Cauchy flow stress in cubic crystals becomes (with $\Delta\tau^c = \tau^c - \tau_0^c$ in any system)

$$\Delta\tau^c/\tau_0^c = 3(p/c_1) + \tfrac{3}{2}(3 + C_1/c_1)(p/c_1)^2 + \cdots \tag{3.74}$$

(equivalent to eq. (6.8) in Havner (1973b)).

Haasen & Lawson (1958) conducted full-range tensile tests on several metal crystals at 5×10^3 atm as well as at zero pressure. They concluded that average increases in the flow stress for copper and aluminum at 300 K were approximately 2 and 6 percent, respectively. For these metals the isothermal values of c_1 and C_1 may be computed from the second- and third-order adiabatic and isothermal elastic moduli tabulated in Barsch & Chang (1967). They are, in units of 10^5 MPa (Havner 1973b): Cu, $c_1 = 3.99$ and $C_1 = -67.7$; Al, $c_1 = 2.195$ and $C_1 = -36.6$. Therefore the predicted increases from equation (3.74) in going from 0 to 5×10^3 atm are 0.38 percent for copper and 0.68 percent for aluminum. Thus, as pointed out in Havner (1982a, p. 286), the normality criterion (3.74) is qualitatively consistent with experiment in that it predicts an increase in Cauchy flow stress with increasing pressure and gives the relative changes for the two metals investigated in correct order and roughly correct proportion at 5×10^3 atm (2 to 1 as compared with 3 to 1 experimentally). Quantitatively, however, the agreement obviously is less satisfactory. Although a single point on the pressure axis for each metal is scant datum as a basis for assessment, nevertheless it may be reasonable to expect larger increases in flow stress with high pressure than would be predicted by the normality criterion. (The predicted increases also are less than changes in theoretical Peierls–Nabarro stress from dislocation theory, which changes were calculated approximately by Haasen & Lawson (1958) as the percentage increases in shear modulus with increasing pressure.)

3.6 Constitutive Inequalities: Uniqueness

Consider again the normality rules (3.58)–(3.59), which as we have seen are the consequence of the general hardening law (3.54) and the definition (3.55) ($\tilde{\tau}_j = \tilde{\tau}_j^c$) of a critical system. (Henceforth we shall assume there are no large changes in pressure, so that any deviations from normality should be insignificant and essentially unobservable in single crystals finitely deforming by crystalline slip.) From

$$\tilde{f}_k = \tilde{\tau}^c_k - \tilde{\tau}_k \geqslant 0, \quad k = 1, \ldots, N, \tag{3.75}$$

(the definition of net strength) we have in each of the n critical systems $\tilde{f}_k = 0$ and

$$\begin{aligned} \mathrm{d}\tilde{f}_k &= 0 \quad \text{if } \mathrm{d}\tilde{\gamma}_k > 0, \\ \mathrm{d}\tilde{f}_k &\geqslant 0 \quad \text{if } \mathrm{d}\tilde{\gamma}_k = 0. \end{aligned} \tag{3.76}$$

Thus $\mathrm{d}\tilde{f}_k \, \mathrm{d}\tilde{\gamma}_k = 0$ in every slip system, and upon multiplying each of equations $(3.58)_1$ and $(3.59)_1$ by the corresponding nonnegative $\mathrm{d}\tilde{\gamma}_k$ and summing over all (critical) systems, one obtains (with eqs. (3.48))

$$\mathrm{d}^{\mathrm{p}}t \, \mathrm{d}e = \sum\sum g_{kj} \, \mathrm{d}\tilde{\gamma}_k \, \mathrm{d}\tilde{\gamma}_j, \quad \mathrm{d}t \, \mathrm{d}^{\mathrm{p}}e = \sum\sum h_{kj} \, \mathrm{d}\tilde{\gamma}_k \, \mathrm{d}\tilde{\gamma}_j. \tag{3.77}$$

Because each Λ_k in inequality $(3.58)_1$ is the gradient with strain of a scalar potential (the level surface of $\tilde{\tau}_k$ from eq. $(3.51)_1$, the Λs transform as stress t under change in strain measure e (for which see the Appendix). Consequently, from equation $(3.48)_1$, as the $\mathrm{d}\tilde{\gamma}_j$ are invariants, the plastic decrement in stress $\mathrm{d}^{\mathrm{p}}t$ also transforms as t. This is equally evident from the initial definition (3.47) since $t - \mathrm{d}^{\mathrm{p}}t$ is the stress after a differential cycle of strain (with elastic unloading during the removal of $\mathrm{d}e$). Therefore $\mathrm{d}^{\mathrm{p}}t \, \mathrm{d}e$ is invariant under change in measure. The physical reason for this invariance for any work-conjugate pair t, e is that $\frac{1}{2}\mathrm{d}^{\mathrm{p}}t \, \mathrm{d}e$ is the second-order work (the first-order work being identically zero) in a differential strain cycle with elastic unloading (see Hill 1972). In contrast, in a differential cycle of stress, the scalar product $\mathrm{d}t \, \mathrm{d}^{\mathrm{p}}e$ is not invariant under change in measure, which may be understood from the formal equivalence

$$\mathrm{d}t \, \mathrm{d}^{\mathrm{p}}e = \mathrm{d}^{\mathrm{p}}t \, \mathscr{M} \, \mathrm{d}t$$

(from $\mathrm{d}^{\mathrm{p}}e = \mathscr{M} \, \mathrm{d}^{\mathrm{p}}t$ and $\mathscr{M} = \mathscr{M}^{\mathrm{T}}$). Although $\mathrm{d}^{\mathrm{p}}t$ transforms as t, $\mathscr{M} \, \mathrm{d}t$ does not transform as $\mathrm{d}e$ except under purely elastic response (when of course it equals $\mathrm{d}e$, and $\mathrm{d}^{\mathrm{p}}t$ is identically zero).

From equation $(3.77)_1$ and the above, it follows that the parameters

$$g_{kj} = M_k \mathscr{L} v_j + H_{kj} \tag{3.78}$$

(from eq. $(3.58)_2$, $\Lambda_k = \mathscr{L} M_k$, and $\mathscr{L} = \mathscr{L}^{\mathrm{T}}$) are invariant under change in measure, as also are the physical hardening moduli H_{kj} from equation

(3.54). However, again from the previous analysis, the parameters

$$h_{kj} = H_{kj} - M_k \mathcal{N}_j t \tag{3.79}$$

(from eqs. $(3.59)_2$ and (3.32)) are measure dependent, as initially pointed out by Hill & Rice (1972). Therefore a postulated inequality of the form $dt\, d^p e > 0$, commonly incorporated into infinitesimal plasticity theory, is meaningless in the context of finite strain without specifying the measure.[7] Rather, we postulate (Havner 1986; see also Hill 1978, p. 44)

$$d^p t\, de > 0, \tag{3.80}$$

which henceforth is taken to be a fundamental constitutive inequality in crystal plasticity, being a specialization of Ilyushin's postulate $\oint t\, de > 0$ (Ilyushin 1961) to a differential strain cycle. (For an analysis of Ilyushin's postulate in a more general context of inelastic material behavior, see Hill (1968, part II).) Inequality (3.80) encompasses but is not required by the normality conditions (3.61),[8] which in turn encompass but are not required by the plastic potential equations (3.52) that follow from basic crystal kinematics and the assumed Green-elasticity of the lattice. The respective invariance and noninvariance of the two scalar products $d^p t\, de$ and $dt\, d^p e$ in equation (3.77), together with the invariance and equivalence of the normality rules (3.61), are illustrated in Fig. 3.3, which depicts differential cycles of (uniaxial) stress and strain for two different strain measures, one having a falling stress–strain curve at finite deformation. The work interpretation of $d^p t\, de > 0$ (for either measure) also is evident from the figure since this product obviously is twice the net triangular area in the differential strain cycle. (For a falling curve, a differential cycle of stress must begin rather than end with an incremental elastic change, as shown.)

The inequality (3.80), equivalently a requirement of positive second-order stress work in every differential cycle of strain, would of course be satisfied from equation $(3.77)_1$ if all the parameters g_{kj} were positive (since no $d\tilde{\gamma}_j$ is negative). However, as slip-systems hardening moduli H_{kj} typically are three orders of magnitude smaller than incremental elastic moduli, the first term dominates in equation (3.78). Consequently, for some metal crystals and loading configurations (aluminum in a [111] tensile

[7] With the current state as reference, it well may happen at finite strain that the sign of $dt\, d^p e$ changes from positive to negative in going from the logarithmic to the Green measure, although all $\dot{e}$ then equal D and all t equal Cauchy stress σ.

[8] As $-de$ is taken to be an elastic unloading, such that the previous yield point e is within the subsequent elastic domain (or on its surface) at strain $e + de$, we have $-de = \delta e$, whence $d^p t \delta e < 0$ from ineq. (3.80).

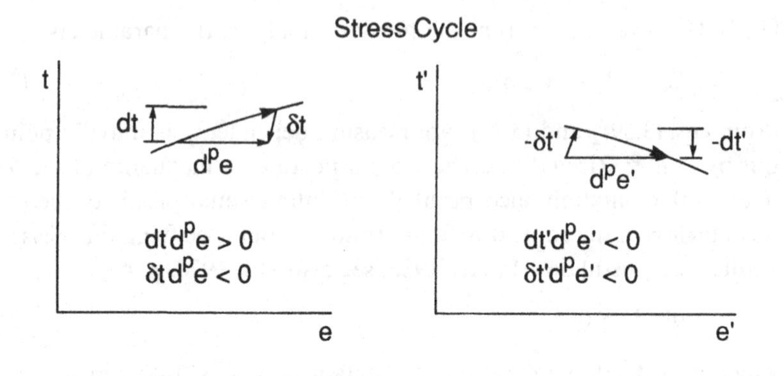

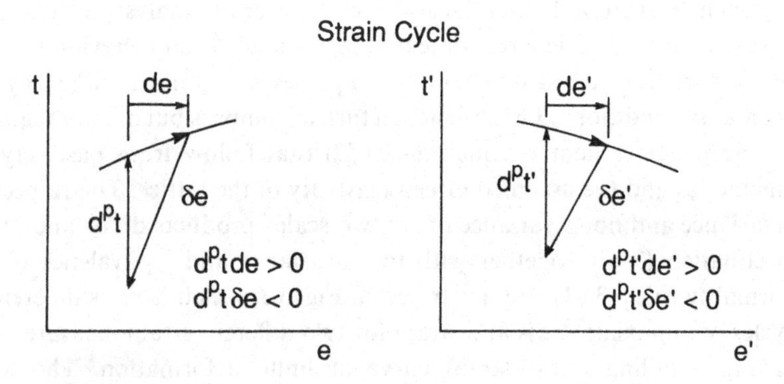

Fig. 3.3. Differential cycles in uniaxial stress and strain.

test, for example) certain of the g_{kj} are negative (see Havner (1985) for specific equations). Instead, to insure the physically motivated postulate (3.80) of positive second-order work, we take the matrix g_{kj} over the n critical systems to be *positive definite*, as first proposed by Hill & Rice (1972). This also guarantees a unique set of incremental slips (and thereby increments in lattice strain and stress) corresponding to a prescribed incremental deformation dA, hence incremental strain de from equation (3.16). The uniqueness is immediately evident for positive-definite g_{kj} from the inequality (Hill & Rice 1972; Sewell 1972)

$$\Delta(\mathrm{d}^p t)\Delta(\mathrm{d}e) \geqslant \sum\sum g_{kj}\Delta(\mathrm{d}\tilde{\gamma}_k)\Delta(\mathrm{d}\tilde{\gamma}_j) \qquad (3.81)$$

(with Δ signifying the difference between the first and second incremental values of a quantity), which may be established as follows.

In a critical system "activated" (i.e., $\mathrm{d}\tilde{\gamma}_k > 0$) by either of two strain increments de, $\Delta(\mathrm{d}\tilde{f}_k) = 0$ from equations (3.76); in another (critical) system activated by the first but not the second $\mathrm{d}e$, $\Delta(\mathrm{d}\tilde{f}_k) \leqslant 0$ while $\Delta(\mathrm{d}\tilde{\gamma}_k) > 0$; in a critical system activated by the second but not the first increment, $\Delta(\mathrm{d}\tilde{f}_k) \geqslant 0$ and $\Delta(\mathrm{d}\tilde{\gamma}_k) < 0$; finally, in a system activated by neither, $\Delta(\mathrm{d}\tilde{\gamma}_k) = 0$. Thus in every case

$$\Delta(\mathrm{d}\tilde{f}_k)\Delta(\mathrm{d}\tilde{\gamma}_k) \leqslant 0, \quad k = 1, \ldots, n. \tag{3.82}$$

But

$$\mathrm{d}\tilde{f}_k = \sum_j g_{kj}\,\mathrm{d}\tilde{\gamma}_j - \Lambda_k\,\mathrm{d}e \tag{3.83}$$

from equations (3.75), (3.54), (3.57), (3.19), and (3.58). Therefore, upon substituting equation (3.83) and summing equation (3.82) over all critical systems (using eq. $(3.48)_1$), the inequality (3.81) is proved. (Note: This final result requires that parameters g_{kj}, hence hardening moduli H_{kj} from eq. (3.78), be independent of strain rate.)

The basic inequality (3.82) leads to several other inequalities relevant to uniqueness of solution for local slip rates. First, from equations (3.75), (3.54), and (3.57), the incremental net strength satisfies

$$\mathrm{d}\tilde{f}_k = \sum_j H_{kj}\,\mathrm{d}\tilde{\gamma}_j - M_k\,\mathrm{d}^*t, \tag{3.84}$$

whence (with eq. $(3.48)_2$)

$$\Delta(\mathrm{d}^*t)\Delta(\mathrm{d}^{\mathrm{p}}e) \geqslant \sum\sum H_{kj}\Delta(\mathrm{d}\tilde{\gamma}_k)\Delta(\mathrm{d}\tilde{\gamma}_j). \tag{3.85}$$

Thus, if the physical hardening moduli H_{kj} are positive definite, the response to a prescribed increment in stress as calculated relative to the lattice will be unique (Hill 1966). Second, from equations (3.28), (3.59) (or eqs. (3.33), (3.79)), and (3.84),

$$\mathrm{d}\tilde{f}_k = \sum_j h_{kj}\,\mathrm{d}\tilde{\gamma}_j - M_k\,\mathrm{d}t. \tag{3.86}$$

Therefore (again with eq. $(3.48)_2$)

$$\Delta(\mathrm{d}t)\Delta(\mathrm{d}^{\mathrm{p}}e) \geqslant \sum\sum h_{kj}\Delta(\mathrm{d}\tilde{\gamma}_k)\Delta(\mathrm{d}\tilde{\gamma}_j) \tag{3.87}$$

(Havner 1977), and the response to a prescribed increment in stress as calculated relative to the material will be unique for a given measure if the parameters h_{kj}, henceforth called the *effective slip-systems hardening moduli*, are positive definite for that measure. (Recall that $\mathrm{d}^*t\,\mathrm{d}^{\mathrm{p}}e$ is a scalar invariant but $\mathrm{d}t\,\mathrm{d}^{\mathrm{p}}e$ is measure dependent.) Additional inequalities that are related to the foregoing ones may be found in Havner (1977, 1986) and Hill (1978).

3.7 Saddle Potential Functions and Boundary Value Problems

As the parameters g_{kj} are taken to be positive definite over the n critical systems at a crystal material point (to ensure positive second-order work in a differential cycle of strain), their $n \times n$ matrix in the set of identities (3.83) can be inverted to obtain

$$d\tilde{\gamma}_k = \sum_j g_{kj}^{-1}(\Lambda_j de + d\tilde{f}_j), \quad j, k = 1, \ldots, n. \tag{3.88}$$

Thus, from equations (3.47) and (3.48)$_1$,

$$dt = \mathscr{L}_p de - \sum\sum \Lambda_k g_{kj}^{-1} d\tilde{f}_j, \tag{3.89}$$

with

$$\mathscr{L}_p = \mathscr{L} - \sum\sum g_{kj}^{-1} \Lambda_k \otimes \Lambda_j \tag{3.90}$$

the tensor modulus of *fully plastic response* for measure e (requiring $d\tilde{f}_j = 0$ in each critical system). Therefore twice the second-order work in a strain increment (which may be negative for some measures) is expressible as

$$dt\, de = de\, \mathscr{L}_p de + \sum\sum g_{kj}^{-1} d\tilde{f}_k d\tilde{f}_j - \sum\sum\sum g_{ki} g_{kj}^{-1} d\tilde{f}_j d\tilde{\gamma}_i$$

(eq. (3.83) again having been used). It is obvious that if the g_{kj} are *symmetric* the last term vanishes because $d\tilde{f}_j d\tilde{\gamma}_j = 0$ in every system. Furthermore, considering two different incremental responses, for symmetric g_{kj} we have (temporarily dropping the tildes for simplicity)

$$\Delta\dot{t}\Delta\dot{e} = \Delta\dot{e}\mathscr{L}_p\Delta\dot{e} + \sum\sum g_{kj}^{-1}\Delta\dot{f}_k\Delta\dot{f}_j - \sum\Delta\dot{f}_j\Delta\dot{\gamma}_j. \tag{3.91}$$

Since the second term on the right-hand side is nonnegative from the positive definiteness of the g_{kj} and the last term (including sign) is nonnegative from inequality (3.82), there follows Sewell's (1972) inequality

$$\Delta\dot{t}\Delta\dot{e} \geqslant \Delta\dot{e}\mathscr{L}_p\Delta\dot{e} \tag{3.92}$$

as well as

$$(dt - \mathscr{L}_p de)\, de \geqslant 0 \tag{3.93}$$

independently of measure. Thus the incremental response for any measure e is at least as "stiff" as would be that of fully plastic response in that measure, which is certainly a reasonable physical expectation. Put differently, symmetric positive-definite parameters g_{kj} in critical systems guarantee that the second-order work in a given strain increment is not less than it would be with all critical systems active.

For the balance of this chapter we shall take the parameters g_{kj} to be symmetric as well as positive definite. However, the following consequence of symmetric g_{kj} that was pointed out in Havner & Shalaby (1977) should be noted. From equations (3.78) and (3.79) and the basic equations (3.32) and (3.49), the moduli H_{kj}, h_{kj} may be expressed in terms of these parameters by

$$H_{kj} = (g_{kj} - v_k \mathscr{L} v_j) - v_j \mathscr{N}_k t, \tag{3.94}$$

$$h_{kj} = g_{kj} - M_k \mathscr{L} M_j. \tag{3.95}$$

Thus, for symmetric g_{kj} the h_{kj} also are symmetric from $\mathscr{L} = \mathscr{L}^{\mathrm{T}}$; but the physical hardening moduli H_{kj} are not, differing according to

$$H_{kj} - H_{jk} = v_k \mathscr{N}_j t - v_j \mathscr{N}_k t.$$

Consequently, Taylor's classical hardening rule discussed in Chapter 1, in which all moduli are hypothesized to be equal (Taylor & Elam 1923, 1925), is not admissible within a theoretical structure adopting symmetric g_{kj}. However, the more physically based specification of only positive-definite g_{kj} is not, in general, in conflict with Taylor hardening; and predictions from Taylor's rule are investigated in comparison with both experiment and other theories in later chapters.

As emphasized in Havner & Shalaby (1977) and utilized by Havner (1977, 1978), adoption of a "symmetry postulate" for the matrix of effective hardening moduli h_{kj}, and hence also for the g_{kj} from equation (3.95), enables use to be made of the theoretical framework of saddle-shaped generating functions of Sewell (1974) (see Sewell (1987) for a comprehensive treatment) and leads to minimum principles that are a consequence of uniqueness theorems. Again suppressing the tildes for simplicity of notation, we introduce the saddle potential function (Havner 1986; see also Havner (1977) and Sewell (1974) for equivalent forms)

$$2\Psi(\dot{e}, \dot{f}) = \dot{e} \mathscr{L}_p \dot{e} - \sum\sum (2\Lambda_k \dot{e} + \dot{f}_k) g_{kj}^{-1} \dot{f}_j. \tag{3.96}$$

It is then straightforward to show that Ψ is a potential function for both objective stress rate $\dot{t}$ in measure e and invariant slip rates $\dot{\gamma}_k$, $k = 1, \ldots, n$. We have (compare eqs. (3.88)–(3.90), using $\mathscr{L}_p = \mathscr{L}_p^{\mathrm{T}}$ from $g_{kj}^{-1} = g_{jk}^{-1}$ and the Green-elasticity of the lattice)

$$\dot{t} = \frac{\partial \Psi}{\partial \dot{e}}, \quad \dot{\gamma}_k = -\frac{\partial \Psi}{\partial \dot{f}_k}, \quad \mathscr{L}_p = \frac{\partial^2 \Psi}{\partial \dot{e} \partial \dot{e}}. \tag{3.97}$$

Moreover, for related differences $\Delta\dot{e}, \Delta\dot{f}_k$ between different possible responses from the current state,

$$\Delta \dot{t} = \frac{\partial \Psi}{\partial \Delta \dot{e}}, \quad \Delta \dot{\gamma}_k = -\frac{\partial \Psi}{\partial \Delta \dot{f}_k}, \tag{3.98}$$

from which[9]

$$2\Psi(\Delta \dot{e}, \Delta \dot{f}) = \frac{\partial \Psi}{\partial \Delta \dot{e}} \Delta \dot{e} + \sum \left(\frac{\partial \Psi}{\partial \Delta \dot{f}} \Delta \dot{f} \right)_k. \tag{3.99}$$

Additionally, from equations (3.82) and (3.98) it is evident that

$$\sum \left(\frac{\partial \Psi}{\partial \Delta \dot{f}} \Delta \dot{f} \right)_k \geqslant 0, \quad 2\Psi(\Delta \dot{e}, \Delta \dot{f}) \geqslant \Delta \dot{t} \Delta \dot{e}. \tag{3.100}$$

We now turn to the issue of uniqueness in boundary value problems (for which also see Sewell (1972), Havner (1977, 1978), and Hill (1978)), the analysis here closely following that in Havner (1986). Consider the class of quasistatic, "rate type" (i.e., incremental) problems defined in Hill (1958), and let S denote the unsymmetric nominal stress tensor referred to reference configuration B_0 of a crystalline solid, which may be a single crystal or a polycrystalline body of arbitrary shape. We denote by ∂B_F that part of the bounding surface on which forces (nominal traction vectors) are prescribed and by ∂B_D that part on which displacements are prescribed. (In practice there of course may be component-by-component specification of force or displacement.) Let $\Delta \dot{S}$ and $\Delta \dot{A} = \partial \Delta \dot{\mathbf{x}} / \partial \mathbf{x}_0$ represent differences between statically admissible and kinematically admissible fields, respectively, corresponding to the same body forces and boundary data. Thus $\mathrm{Div}(\Delta \dot{S}) = \mathbf{0}$ in B_0 (except at interfaces of crystal grains across which nominal traction rates $\mathbf{n}_0 \dot{S}$ must be continuous); $\mathbf{da}_0 \Delta \dot{S} = \mathbf{0}$ on ∂B_F (with $\mathbf{da}_0$ the referential areal vector of a surface element); and $\Delta \dot{\mathbf{x}} = \mathbf{0}$ on ∂B_D. There follows the well-known result

$$\int_{B_0} \mathrm{tr}(\Delta \dot{S} \Delta \dot{A}) \, \mathrm{d}V_0 = 0 \tag{3.101}$$

connecting the respective "self-stressing" and "self-straining" fields of rates of change of these conjugate variables, sometimes called the equation of virtual work rates (a consequence of the divergence theorem, or Green–Gauss transformation, and the separate definitions of static and kinematic admissibility).

[9] Because Ψ obviously is a homogeneous function of second degree in its arguments, this result also follows from Euler's theorem on homogeneous functions.

From $\mathrm{d}e = \mathscr{K}\,\mathrm{d}A$ and $\mathrm{tr}(S\,\mathrm{d}A) = t\,\mathrm{d}e$ (with the same reference configuration for all sets of conjugate variables), we have

$$S = \mathscr{K}^{\mathrm{T}}t, \tag{3.102}$$

whence

$$\mathrm{d}S = \mathscr{K}^{\mathrm{T}}\,\mathrm{d}t + \mathscr{T}\,\mathrm{d}A, \quad \mathscr{T} = t\frac{\partial^2 e}{\partial A^{\mathrm{T}}\partial A^{\mathrm{T}}} = \mathscr{T}^{\mathrm{T}}. \tag{3.103}$$

That is, $\mathscr{T}_{ijkl} = t_{mn}(\partial^2 e_{mn}/\partial A_{ji}\partial A_{lk})$, which reduces to $T_{ik}\delta_{jl}$ for the Green measure $e = E$. (For that case, upon substituting $\mathscr{K} = \mathscr{G}$ and simplifying, we recover the result $\mathrm{d}S = \mathrm{d}\,TA^{\mathrm{T}} + T\mathrm{d}A^{\mathrm{T}}$, which of course follows directly from the standard equation $S = TA^{\mathrm{T}}$.) The constitutive equations in the nominal variables S, A may be obtained from equations $(3.47)_1$, $(3.48)_1$, $(3.52)_1$, (3.16), and (3.103). They are

$$\mathrm{d}S = \mathscr{C}\,\mathrm{d}A - \mathrm{d}^{\mathrm{p}}S, \quad \mathrm{d}^{\mathrm{p}}S = \sum(X\,\mathrm{d}\tilde{\gamma})_j = \frac{\partial}{\partial A^{\mathrm{T}}}\sum(\tilde{\tau}\,\mathrm{d}\tilde{\gamma})_j, \tag{3.104}$$

with

$$\mathscr{C} = \mathscr{K}^{\mathrm{T}}\mathscr{L}\mathscr{K} + \mathscr{T} = \mathscr{C}^{\mathrm{T}}, \quad X_j = \mathscr{K}^{\mathrm{T}}\Lambda_j \tag{3.105}$$

(where we have used

$$\mathscr{K}^{\mathrm{T}}\frac{\partial}{\partial e}(\cdot) = \frac{\partial}{\partial e}(\cdot)\frac{\partial e}{\partial A^{\mathrm{T}}} = \frac{\partial}{\partial A^{\mathrm{T}}}(\cdot) \tag{3.106}$$

from the chain rule, as all gradients are at fixed slips). Thus, we see that $\tilde{\tau}_j$ also acts as a plastic potential (in nine-dimensional, deformation-gradient space) for the contribution of $\mathrm{d}\tilde{\gamma}_j$ to the plastic decrement in unsymmetric nominal stress. (Choosing $\mathscr{L}$ in eq. (3.105) to be the instantaneous elastic moduli for Green strain, we may write $\mathscr{C}_{ijkl} = A_{jm}\mathscr{L}^{\mathrm{G}}_{imnk}A_{ln} + T_{ik}\delta_{jl}$.)

Following Havner (1986), we introduce a second saddle potential function defined over the n critical slip systems at a crystal point:

$$2U(\dot{A}, \dot{f}) = \dot{A}\mathscr{C}_{\mathrm{p}}\dot{A} - \sum\sum(2\,\mathrm{tr}(X_k\dot{A}) + \dot{f}_k)g^{-1}_{kj}\dot{f}_j, \tag{3.107}$$

with

$$\mathscr{C}_{\mathrm{p}} = \mathscr{C} - \sum\sum g^{-1}_{kj}X_k \otimes X_j = \mathscr{C}^{\mathrm{T}}_{\mathrm{p}}. \tag{3.108}$$

The diagonally symmetric fourth-order tensors $\mathscr{C}$ and $\mathscr{C}_{\mathrm{p}}$ are the "nominal moduli" of elastic and fully plastic response, respectively. Analogously to equations (3.89) and (3.97) we find (again for symmetric g^{-1}_{kj})

$$\mathrm{d}S = \mathscr{C}_{\mathrm{p}}\mathrm{d}A - \sum\sum g^{-1}_{kj}X_k\mathrm{d}\tilde{f}_j \tag{3.109}$$

and

$$\dot{S} = \frac{\partial U}{\partial \dot{A}^{\mathrm{T}}}, \quad \dot{\gamma}_k = -\frac{\partial U}{\partial \dot{f}_k}, \quad \mathscr{C}_{\mathrm{p}} = \frac{\partial^2 U}{\partial A^{\mathrm{T}} \partial A^{\mathrm{T}}}. \tag{3.110}$$

Furthermore, for the differences between assumed distinct responses (compare eqs. (3.98)–(3.99)),

$$\Delta\dot{S} = \frac{\partial U}{\partial \Delta\dot{A}^{\mathrm{T}}}, \quad \Delta\dot{\gamma}_k = -\frac{\partial U}{\partial \Delta\dot{f}_k}, \\ 2U = \mathrm{tr}(\Delta\dot{S}\Delta\dot{A}) - \sum(\Delta\dot{f}\Delta\dot{\gamma})_k. \tag{3.111}$$

Thus, from equations (3.88), (3.109), and (3.111),

$$\mathrm{tr}(\Delta\dot{S}\Delta\dot{A}) = 2U_{\mathrm{L}}(\Delta\dot{A}) + \sum\sum g_{kj}^{-1}\Delta\dot{f}_k\Delta\dot{f}_j - \sum(\Delta\dot{f}\Delta\dot{\gamma})_k, \tag{3.112}$$

with

$$2U_{\mathrm{L}}(\Delta\dot{A}) = \Delta\dot{A}\mathscr{C}_{\mathrm{p}}\Delta\dot{A} = \Delta\dot{e}\mathscr{L}_{\mathrm{p}}\Delta\dot{e} + \Delta\dot{A}\mathscr{T}\Delta\dot{A} \tag{3.113}$$

the potential function of a fully plastic "linear comparison solid" (a concept introduced by Hill (1958)). There follows from inequality (3.82) and the assumed positive definiteness of the g_{kj}

$$\int_{B_0} \mathrm{tr}(\Delta\dot{S}\Delta\dot{A})\mathrm{d}V_0 \geqslant 2\int_{B_0} U_{\mathrm{L}}(\Delta\dot{A})\mathrm{d}V_0. \tag{3.114}$$

For $\Delta\dot{S}$ a statically admissible self-stressing rate field, the left-hand side of this inequality is identically zero from equation (3.101). Consequently, if $\int U_{\mathrm{L}}(\Delta\dot{A})\mathrm{d}V_0 > 0$ for all kinematically admissible $\Delta\dot{A}$ (that is, for the gradient with $\mathbf{x}_0$ of every continuous velocity field $\Delta\mathbf{v}$ that vanishes on ∂B_{D}), then there cannot be different statically admissible $\dot{S}(\mathbf{x})$ that satisfy the constitutive equations (3.110) and (3.76), and the solution to the boundary value problem is unique.

Symmetric, positive-definite moduli g_{kj} also ensure that if the *uniqueness condition*

$$\int_{B_0} U_{\mathrm{L}}(\Delta\dot{A})\mathrm{d}V_0 > 0 \tag{3.115}$$

is satisfied, the (unique) solution minimizes the scalar functional

$$2I(\mathbf{v}) = \int_{B_0} \{\mathrm{tr}(\dot{S}(\mathbf{v})\dot{A}(\mathbf{v})) - 2\dot{\mathbf{f}}_0\cdot\mathbf{v}\}\mathrm{d}V_0 - 2\int_{\partial B_{\mathrm{F}}} \dot{\mathbf{t}}_{\mathrm{N}}\cdot\mathbf{v}\,\mathrm{d}a_0 \tag{3.116}$$

on the class of continuous velocity fields $\mathbf{v}(\mathbf{x})$ taking the prescribed values on ∂B_{D}. Here $\mathbf{f}_0$ is the body force per unit reference volume, $\mathbf{t}_{\mathrm{N}}$ is the prescribed nominal traction vector (force per unit reference area) on ∂B_F, and $\dot{S}(\mathbf{v})$ is determined from $\mathbf{v}(\mathbf{x})$ according to the constitutive equations (hence $\dot{S}(\mathbf{x})$ is not necessarily a statically admissible field). This result was established by Sewell (1972) (although not specifically with regard to elastoplastic crystals) and is the generalization of a theorem first given by Hill (1958) (see also Hill 1978). A proof follows.

Let velocity field $\mathbf{u}(\mathbf{x})$ be a solution to the boundary value problem, whether or not unique. Then $\dot{S}(\mathbf{u})$ is statically admissible by definition, and we have from the principle of virtual work rates (equivalently, the Green–Gauss transformation

$$\int_{B_0} \mathrm{tr}(\dot{S}(\mathbf{u})\dot{A}(\mathbf{v}))\mathrm{d}V_0 = \int_{B_0} \dot{\mathbf{f}}_0\cdot\mathbf{v}\,\mathrm{d}V_0 + \int_{\partial B_0} \dot{\mathbf{t}}_N(\mathbf{u})\cdot\mathbf{v}\,\mathrm{d}a_0, \tag{3.117}$$

with $\dot{\mathbf{t}}_N(\mathbf{u}) = \mathbf{n}_0\dot{S}(\mathbf{u}) = \dot{\mathbf{t}}_N$ (prescribed) on ∂B_F and $\mathbf{v} = \mathbf{u}$ (prescribed) on ∂B_D. Thus, at solution $\mathbf{u}(\mathbf{x})$,

$$2I(\mathbf{u}) = -\int_{B_0} \mathrm{tr}(\dot{S}(\mathbf{u})\dot{A}(\mathbf{u}))\mathrm{d}V_0 + 2\int_{\partial B_D} \dot{\mathbf{t}}_N(\mathbf{u})\cdot\mathbf{u}\,\mathrm{d}a_0, \tag{3.118}$$

and upon subtracting equation (3.118) from equation (3.116) we obtain

$$2\Delta I = \int_{B_0} \mathrm{tr}(\Delta\dot{S}\Delta\dot{A})\,\mathrm{d}V_0 + \int_{B_0} \mathrm{tr}\{\dot{S}(\mathbf{v})\dot{A}(\mathbf{u}) - \dot{S}(\mathbf{u})\dot{A}(\mathbf{v})\}\,\mathrm{d}V_0, \tag{3.119}$$

with $\Delta I = I(\mathbf{v}) - I(\mathbf{u})$, and so forth. (This equation applies to the entire class of quasistatic boundary value problems defined at the outset and so is a general result in solid mechanics for the functional $I(\mathbf{v})$.) Considering now the present constitutive equations, we may readily establish from equations $(3.104)_1$ and (3.83) and the diagonal symmetry of $\mathscr{C}$ that

$$\begin{aligned}\mathrm{tr}\{\dot{S}(\mathbf{v})\dot{A}(\mathbf{u}) - \dot{S}(\mathbf{u})\dot{A}(\mathbf{v})\} &= \textstyle\sum\{\dot{f}_k(\mathbf{u})\dot{\gamma}_k(\mathbf{v}) - \dot{f}_k(\mathbf{v})\dot{\gamma}_k(\mathbf{u})\} \\ &\quad + \textstyle\sum\sum(g_{kj} - g_{jk})\dot{\gamma}_k(\mathbf{v})\dot{\gamma}_j(\mathbf{u}).\end{aligned} \tag{3.120}$$

Thus for symmetric g_{kj}, after substituting equation (3.112) and the reduced equation (3.120) into equation (3.119) and using $\Delta\dot{f}_k\Delta\dot{\gamma}_k = -\dot{f}_k(\mathbf{u})\dot{\gamma}_k(\mathbf{v}) - \dot{f}_k(\mathbf{v})\dot{\gamma}_k(\mathbf{u})$, we obtain

$$\begin{aligned}2\Delta I = 2\int_{B_0} U_L(\Delta\dot{A})\,\mathrm{d}V_0 &+ \int_{V_0} \textstyle\sum\sum g_{kj}^{-1}\Delta\dot{f}_k\Delta\dot{f}_j\,\mathrm{d}V_0 \\ &+ 2\int_{V_0} \textstyle\sum\dot{f}_k(\mathbf{u})\dot{\gamma}_k(\mathbf{v})\,\mathrm{d}V_0.\end{aligned} \tag{3.121}$$

The second integral is of course nonnegative from the positive definiteness of the g_{kj}^{-1}, and the last is nonnegative from inequalities (3.76) (i.e., because $\dot{f}_k$ and $\dot{\gamma}_k$ are individually positive or zero for all fields). Thus, we have proved that for symmetric, positive-definite g_{kj}

$$\Delta I = I(\Delta \mathbf{v}) \geqslant \int_{B_0} U_{\mathrm{L}}(\Delta \dot{A})\, \mathrm{d}V_0 \tag{3.122}$$

(with $\Delta\mathbf{v}$ the difference between any kinematically admissible field and a solution). Consequently, if inequality (3.115) is satisfied for all $\Delta\dot{A}$, solution $\mathbf{u}(\mathbf{x})$ is unique and minimizes the functional $I(\mathbf{v})$ over kinematically admissible fields. (For proofs of comparable minimum principles requiring somewhat less strict uniqueness criteria, see Havner (1977).)

4

AXIAL-LOAD EXPERIMENTS AND LATENT HARDENING IN SINGLE CRYSTALS

The specification of symmetric parameters g_{kj} (and h_{kj}) in Section 3.7 delimits the general family of hardening rules given by equation (3.54) (excluding, for example, Taylor hardening as noted). However, the requirement that the matrix g_{kj} only be positive definite on critical systems (to ensure positive second-order work in a differential strain cycle) does not significantly restrict crystal response. Positive-definite g_{kj} can encompass strain-softening as well as strain-hardening behavior in any measure; and this requirement merely serves to exclude so-called locking materials. Consequently, physically meaningful progress beyond the theoretical framework of Chapter 3 requires the consideration and interpretation of experimental information from single-crystal tests. The most comprehensive source of such information is the body of work on axial loading of f.c.c. and b.c.c. crystals to which we now turn.

4.1 Axial Loading of F.C.C. Crystals

As discussed in Chapter 1, overshooting by the tension axis across the symmetry line between an initially active (primary) slip system ($a\bar{2}$, say) and the secondary system ($b1$) toward which the axis rotates[1] in single slip was first observed and measured in the classic experiments of Taylor & Elam (1923, 1925) on aluminum. Overshooting also was evident in experiments on other f.c.c. metals by Elam (1926, 1927a,b) and in b.c.c. crystals by Fahrenhorst & Schmid (1932). Nonetheless, Taylor's original hypothesis of equal double slip at the symmetry line apparently became

[1] The latter slip system is commonly called the "conjugate" system in the metallurgical literature.

the common wisdom among many experimentalists. Load–extension data often was resolved by assuming its validity and (for f.c.c. crystals) using the corresponding kinematic and shear stress equations of v. Göler & Sachs (1927), based upon rotation of the axis along that line. (See Bell (1965) for an extensive survey of 40 years of experimental research encompassing more than 300 f.c.c. crystal tests.)

One of the earliest studies to strongly reemphasize the overshooting phenomenon was that of Piercy, Cahn, & Cottrell (1955) who concluded "that, contrary to the usual view, overshooting should be regarded as a normal characteristic of the plastic deformation of crystals, and that the intriguing feature is that it ever should be absent" (p. 337). They found some conjugate slip bands beyond the symmetry line in their specimens of α-brass (and in one specimen 7 degrees before the axis reached the symmetry position). Nonetheless, the axis extensions which they calculated from the changes of orientation on the assumption of single slip agreed very closely with the directly measured values to deformations of 80 percent (see their fig. 10). More generally, as Bell & Green (1967, p. 475) remarked in a study of 168 experiments (including 16 which they performed) wherein measured axis rotations reached the symmetry line and the crystal was then further extended, although surface traces of multiple slip often are "present on both sides of the symmetry line... the main features of the resolved deformation are given by the initial primary plane behaviour." The axis positions determined by X-ray diffraction in Bell and Green's deadweight tensile tests of aluminum crystals of 99.47 percent purity (which they chose for historical reasons) are shown in Fig. 4.1 (reproduced from their fig. 1). Overshooting and axis rotation consistent with (dominant) single slip clearly are evident in the 14 tests which reached the symmetry line.

Closely related to the phenomenon of overshooting, and undoubtedly its cause, is the deformation-induced anisotropy of latent slip-system hardening in crystals initially oriented for single slip. This anisotropy has been investigated by a number of experimentalists, among them Kocks (1964), Krisement, de Vries, & Bell (1964), Ramaswami, Kocks, & Chalmers (1965), Kocks & Brown (1966), Jackson & Basinski (1967), and Franciosi, Berveiller, & Zaoui (1980) in f.c.c. crystals and Nakada & Keh (1966) in b.c.c. crystals. A standard procedure for determining the hardening of latent systems is the following (see Jackson & Basinski (1967) or Franciosi et al. (1980), for example). An annealed parent crystal (assumed statistically isotropic with respect to the distribution of dislocations) is axially loaded in an initial single-slip orientation and deformed to some predetermined

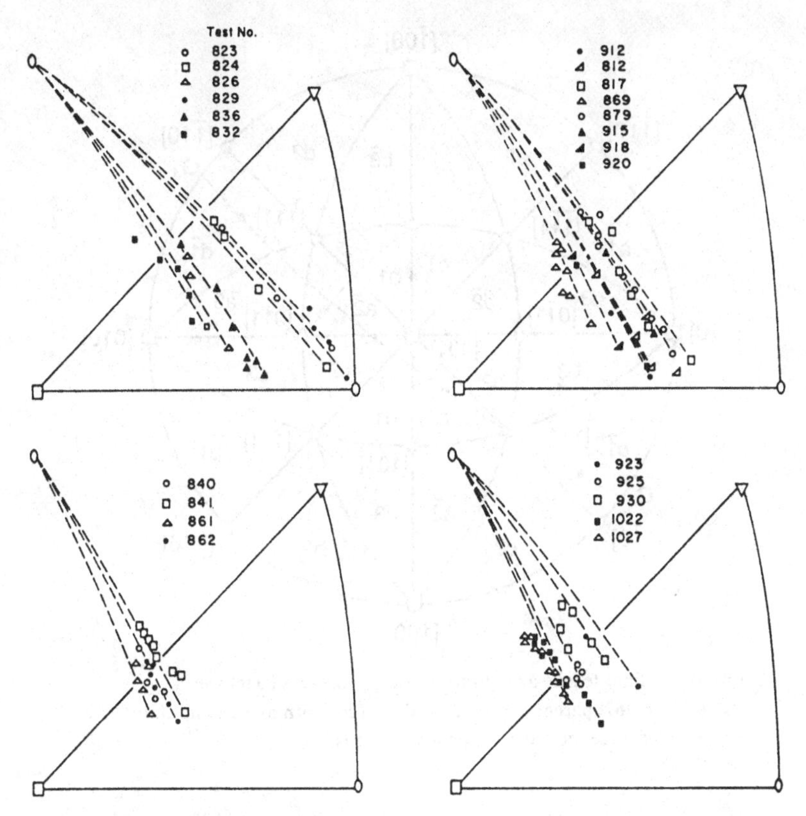

Fig. 4.1. Tensile axis rotations in aluminum crystals (from Bell & Green (1967, fig. 1), as presented in Bell (1973, fig. 4.65)).

level of strain, at which the resolved shear stress/critical strength is calculated for the active system from the measured axial load, extension (or compression), and lattice rotation, care being taken to ensure single slip. After unloading, variously oriented specimens are removed from the prestrained parent crystal by "spark-cutting" techniques. The orientations of these secondary specimens are so chosen that each deforms in single slip on a previously latent system during subsequent loading (at least initially). The unknown critical strengths developed in these latent systems, as a consequence of slip in the original system during deformation of the parent crystal, are taken to equal their respective resolved shear stresses at the onset of measurable slip in the reloaded specimens.

Several general conclusions about latent hardening were drawn by Jackson & Basinski (1967) based upon their extensive program of experiments on copper crystals in tension. Those conclusions of a

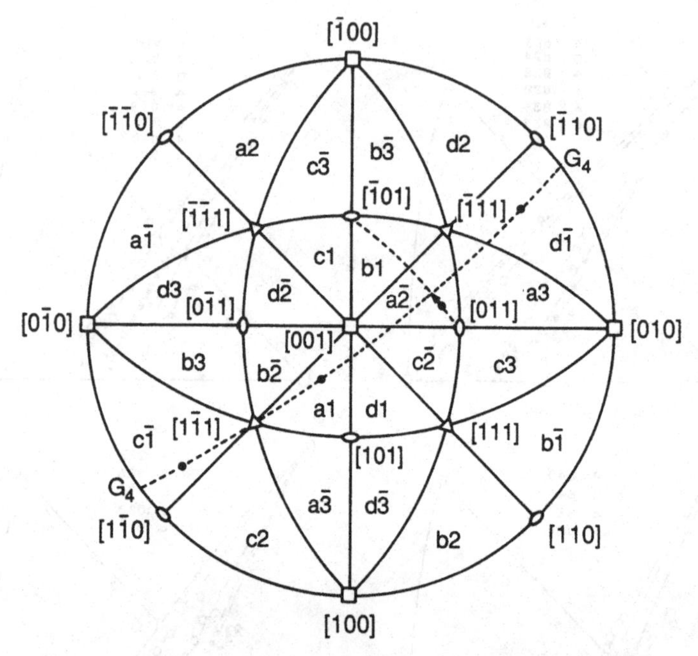

Fig. 4.2. Initial and final tensile axis positions (to $\gamma = 0.245$) of Jackson & Basinski's (1967) parent copper crystal G_4 (dot with arrow) and initial axis positions of three secondary crystal specimens.

comparative nature may be expressed as follows: (i) Hardening of latent slip systems having a different glide plane than the primary system (called intersecting systems) is substantially greater than that of the primary system; (ii) latent systems having only a different slip direction than the primary system harden at very nearly the same rate as that system; and (iii) hardening of the so-called unrelated slip systems, whose slip directions are orthogonal to that of the primary system, is less than the hardening of the other intersecting systems. These conclusions are generally consistent with results from other latent hardening experiments on copper (Krisement et al. 1964; Basinski & Jackson 1965a,b; Franciosi et al. 1980) and from similar experiments on aluminum (Kocks 1964; Kocks & Brown 1966; Franciosi et al. 1980), silver (Kocks 1964; Ramaswami et al. 1965), and (b.c.c.) iron (Nakada & Keh 1966).

Jackson & Basinski's (1967) virgin crystal specimens were thin rectangular plates from 6 to 9 cm long and 2.0 × 0.2 cm in cross section, with the long (tensile) axis oriented 10° from [011] on the great circle connecting [011] and primary ($a\bar{2}$) slip direction [$\bar{1}$01]. (As noted in

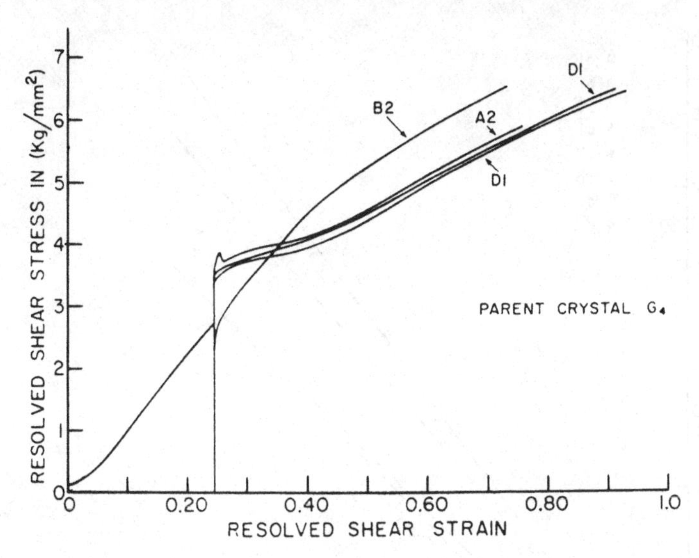

Fig. 4.3. Subsequent resolved shear stress–strain curves in secondary crystal specimens of parent crystal G_4 for slip systems $c\bar{1}$ and $d\bar{1}$ ($A2$ and $D1$) and $a1$ ($B2$) (from Jackson & Basinski 1967, fig. 11).

Chapter 1, this was approximately the position of the tensile axis in the seminal experiment of Taylor & Elam (1923) on an aluminum crystal bar.) In Fig. 4.2 are shown the initial and final axis positions on an [001] stereographic projection of Jackson and Basinski's most highly strained parent crystal "G_4" ($\gamma_{max} = 0.245$), and the tensile axis positions in triangles $a1$, $c\bar{1}$ and $d\bar{1}$ of the secondary specimens spark-cut from crystal G_4. Subsequent resolved shear stress–strain curves[2] for intersecting slip systems $c\bar{1}$ and $d\bar{1}$ (two specimens) and coplanar system $a1$ (which Jackson and Basinski labeled $A2$, $D1$, and $B2$, respectively) are given in their figure 11, included here as Fig. 4.3. The substantially greater latent hardening in the noncoplanar systems due to primary slip in system $a\bar{2}$ is obvious, as is the longer "stage I" range of self-hardening in these intersecting systems during subsequent deformation; whereas the coplanar slip system continues to harden in "stage II" in essentially the same manner as the original system. (Similar results for these same systems at a much lower level of prestrain may be seen in Jackson and Basinski's fig. 10.)

[2] These and other curves in Jackson & Basinski (1967) apparently were calculated on the assumption of single slip in the respective secondary systems. That the anticipated latent system indeed became the active one during loading of a secondary specimen was judged by traces of slip lines on crystal faces, and no shape-change measurements were made.

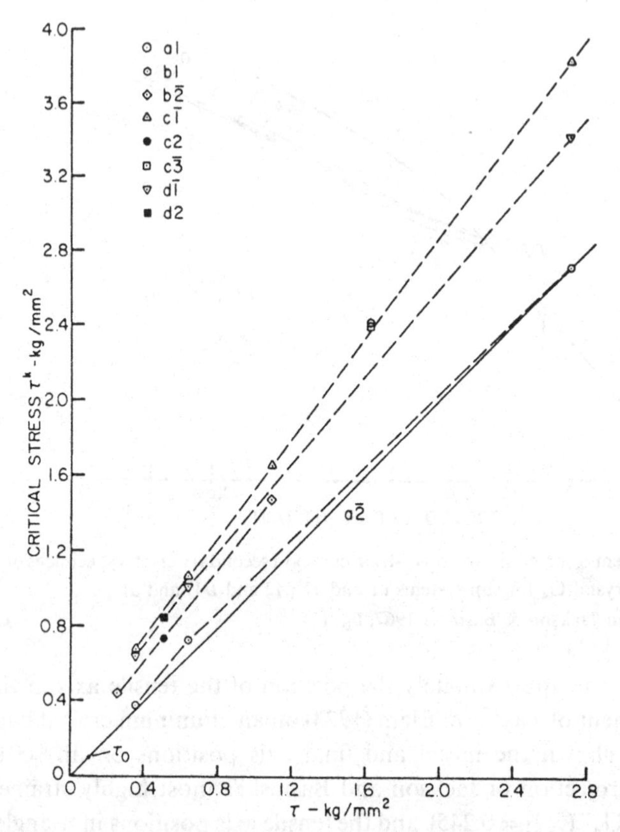

Fig. 4.4. Experimental results of Jackson & Basinski (1967) for latent critical strengths of copper crystals deformed in tension (as presented in Havner, Baker & Vause 1979, fig. 3). Reprinted with permission from *J. Mech. Phys. Solids* 27, K.S. Havner, G.S. Baker & R. F. Vause, Theoretical latent hardening in crystals I, Copyright 1979, Pergamon Press PLC.

Results for all crystals prestrained to shears γ between 5 and 25 percent (stage II hardening) by Jackson and Basinski are presented as latent versus active-system critical strengths in figure 3 of Havner, Baker & Vause (1979), reproduced here as Fig. 4.4. The values of γ at the end of prestrain were provided by Professor P. J. Jackson (private communication, 1978), for which see table 2 of Havner et al. (1979). From Fig. 4.4 it is evident that one may rank the nine systems investigated in this range into four groups in descending order of hardening: $b1$, $c\bar{3}$, $c\bar{1}$; $d\bar{1}$, $d2$; $c2$, $b\bar{2}$; and $a1$, $a\bar{2}$ (primary system). Although there is but a single data point for each of systems $b1$ and $c\bar{3}$ (at 12.8 percent prestrain), with hardening of negligible difference as seen, Franciosi et al. (1980) also found these to be the two most hardened latent systems among eight in aluminum and four in copper crystals that they investigated. (Their crystal prestrains were all in stage I, of order 1 percent in shear.) It also is evident from Fig. 4.4 that the various latent-system critical strengths are approximately linear functions of the active-system strength (at least beyond prestrains of 1 or 2 percent).

The aforementioned experiments on copper and aluminum crystals in tension were not carried through to the point that the loading axis reached the symmetry line with the conjugate system. Thus the issue of overshooting did not arise and was not considered. However, the fact that measured increases in critical strength in conjugate system $b1$ were substantially greater than in active system $a\bar{2}$, at widely varying levels of prestrain, is consistent with the occurrence of overshooting in tests to larger deformations (or from initial orientations nearer the conjugate-system symmetry line) in which it ordinarily is observed.

Compression tests on high-purity aluminum and silver single-crystal cubes (including some latent hardening experiments) were conducted by Kocks (1964) using Teflon film as lubricant. The loading axis (end-face normal) was initially at 45° with each of the slip-plane normal [111] and direction [$\bar{1}$01] (in opposite sense to the slip direction of system $a2$) so that in single slip the axis would rotate toward [111] on the great circle connecting these two lattice directions. With system $a2$ active in compression, the secondary system was the unrelated system $c2$ (represented by triangles $a\bar{2}$, $c\bar{2}$, respectively, in the tensile-loading stereographic projections of Figs 1.6 and 4.2). An example of the highly uniform deformations achieved in these tests, even to large strains, may be seen in figure 2 of Kocks (1964, p. 1162), which shows two views of a finitely sheared aluminum crystal. Kocks stated, "The shape of the crystal after deformation could thus be analyzed rather accurately, and it was always consistent with the single slip expected... even after 70 pct primary glide strain...." He further wrote, "It is evident from fig. 2 and the X-ray experiments... that overshoot in compression of aluminum is large and apparently unlimited. Therefore, the latent hardening of the unrelated system must be higher than the work hardening of the primary system." The latent hardening of system $c2$ was not directly determined by Kocks. However, experimental points for the hardening of latent systems $b3$ and $d3$ in both aluminum and silver were given in his figure 5 (also see figures 13–16 of Vause & Havner (1979), where the data are separately displayed by slip system and metal). According to the experiments, hardening of these latent systems ranged between 10 and 40 percent above that of the primary system *a*2. Kocks, Nakada & Ramaswami (1964) and Ramaswami et al. (1965) also reported overshooting in compression (into the triangle of the unrelated system) for gold and silver crystals, although the overshoot was limited to about 6° in silver.

In marked contrast to the behavior of crystals initially oriented for single slip (whether in tension or compression) is the deformation response of axially loaded specimens in the high-symmetry positions of (potential) sixfold and eightfold multiple slip (i.e., $\langle 111 \rangle$ and $\langle 100 \rangle$ axis orientations). There, as has been found for (b.c.c.) iron crystals by Keh (1965), aluminum by Kocks (1964), gold by Nakada, Kocks & Chalmers (1964), silver by Ramaswami et al. (1965), and copper by Vorbrugg, Goetting & Schwink (1971), Ambrosi, Göttler & Schwink (1974), and Ambrosi & Schwink (1978), the axis position in general is stable (relative to the lattice) and the crystal deforms axisymmetrically to large strains. Franciosi & Zaoui (1982a, b) reported other deformation modes in f.c.c. crystals in these high-symmetry positions, particularly for $\langle 100 \rangle$ loading. However, it may be that the actual axis orientations in the experiments they cited (including their tests on copper crystals at 200 K) deviated too much from the ideal symmetry positions to achieve axis stability. For example, they stated (Franciosi & Zaoui 1982b) that their actual orientations were only within 2° of the theoretical ones; whereas the misorientation of the copper crystals in nominal $\langle 100 \rangle$ and $\langle 111 \rangle$ orientations reported by Ambrosi & Schwink (1978) was less than 1°.

In the various tests by Schwink and his colleagues (at both 77 K and 295 K) of high purity copper crystals in the form of long cylindrical rods, the tensile axis was stable and the deformations consistently axisymmetric to nominal extensional strains as great as 36 percent. In addition, for both $\langle 100 \rangle$ and $\langle 111 \rangle$ orientations, Vorbrugg et al. (1971, p. 260) found from electron microscopy (EM) studies of slip lines and "step heights" that on the average all equivalent slip systems "are active to the same extent in accord with the observed stability of the crystal axis" in each test (see also Ambrosi et al. 1974).

The hardening of all slip systems is, most probably, very nearly isotropic in axisymmetric sixfold and eightfold multiple slip. Wonsiewicz & Chin (1970b, p. 2721) have suggested that the greater number of slip interactions in multiple slip would "provide a more uniform dislocation distribution and hence the dislocations of each system, active or not, would face similar hurdles in their movements." This postulated uniformity of dislocation structure, which would be consistent with isotropic hardening, appears to be borne out by transmission electron microscopy (TEM) studies by

Göttler (1973) of thin foils taken from finitely deformed copper crystals in ⟨100⟩ orientations. He found a cell structure "of nearly dislocation-free regions... surrounded by walls of dislocation networks without preferred orientations" (p. 1057). Moreover, he determined that "dislocations of all 6 Burgers vectors [that is, slip directions [110], [$1\bar{1}0$], [101], [$\bar{1}01$], [011], and [$0\bar{1}1$]] – including those which do not operate under the external stress [namely, [110] and [$1\bar{1}0$] for [001]-loading] – occur with equal frequency." The same statement applies to the "different characters of dislocations" (Göttler 1973, p. 1057). Similar results were obtained by Ambrosi et al. (1974) for ⟨111⟩ loading of copper crystals.

Isotropic (Taylor) hardening in high-symmetry orientations is further supported by unpublished experiments on aluminum crystals by Nakada (1970). His research is briefly described by Wonsiewicz & Chin (1970b, p. 2721) as follows: "He observed that crystals pulled along [111] continued to deform at the same resolved shear stress when subsequently pulled along [100]. Six slip systems were active in the [111] deformation, and eight systems, with only two previously active and two in the reverse sense, became active during the subsequent [100] deformation. Similarly, there was no change of stress when the crystal was first deformed along [100] and then in the [111] direction." (The implications are that each initial and subsequent deformation was axisymmetric, but Wonsiewicz and Chin do not explicitly state this.)

From the investigations discussed in this section and other experimental studies in the literature, the qualitative features of axial-loading behavior of f.c.c. (and also b.c.c.) crystals may be aptly summarized as follows:

1. In virgin crystals oriented for single slip in either tension or compression: (a) the evolving dislocation structures are anisotropic; (b) the intersecting latent systems in general harden more than the active system; (c) significant overshooting is the norm in tests carried to that stage of axis rotation; and (d) typical deformation beyond the symmetry line is consistent with dominant single slip on the original system.
2. In virgin crystals carefully oriented in high-symmetry (six- or eightfold) multiple-slip positions: (a) dislocation networks remain essentially isotropic; (b) the loading axis is stable relative to the underlying lattice; (c) the deformation ordinarily is axisymmetric; and (d) the hardening of all systems (both active and latent) is very nearly equal. (There is limited data regarding this last point, however.)

This rather diverse hardening behavior of cubic crystals subsequently will be related to (and embedded in) the mathematical framework of the preceding chapter. First, however, we review the more complex state of affairs regarding identification of slip systems in finite deformation of b.c.c. crystals, and further review aspects of latent hardening in such metals.

4.2 Slip Systems in B.C.C. Crystals

As remarked in Havner & Baker (1979), the study of finite deformation and latent hardening in b.c.c. crystals is complicated by long-standing ambiguities and contradictions regarding identification of the planes of macroscopic (gross) slip. This is illustrated by the following survey of various noteworthy experimental investigations (from Havner & Baker (1979, sect. 3)).

In their classic paper on distortion of iron crystals, Taylor & Elam (1926) introduced the concept of "pencil glide" to explain their observation that slip lines on polished crystal faces which made appreciable angles with the slip direction were jagged, whereas slip lines on faces parallel to that direction were continuous and straight. They found that slip always took place in a $\langle 111 \rangle$ crystallographic direction, but that the mean plane of slip was the plane of maximum shear stress containing that direction.[3] This slip plane was determined from external measurements of macroscopic deformation (in the manner described in Chapter 1) and was consistent with the general direction of the jagged slip lines in both tension and compression.

Taylor and Elam's investigations were criticized by Fahrenhorst & Schmid (1932), who considered $\{123\}$ planes the most likely slip planes in iron. However, as pointed out by Elam (1936, pp. 273–4), Fahrenhorst and Schmid's work "was confined to the measurement of the movement of the test-piece axis relative to the crystal axes during a tensile test, which they compared with the calculated movement, assuming that the slip-plane was $\{110\}$, $\{112\}$, $\{123\}$, or a random plane,[4] all in a $[111]$ direction." In other words, no external measurements of crystal deformation were made. Moreover, from examination of the axis rotations displayed on a stereographic projection in Fahrenhorst and Schmid's figure 1 (which

[3] These conclusions perhaps may be called *the experimental laws of Taylor and Elam for b.c.c. crystals* as their work also was the pioneering study of finite deformation of this crystal class.

[4] Here is meant the plane of maximum shear stress containing a $\langle 111 \rangle$ direction.

includes several examples of overshooting), it would appear they are accounted for equally well by the assumption that the maximum shear-stress plane is the plane of slip.

Barrett, Ansel & Mehl (1937) conducted torsion and bending tests on alpha-iron and Si–Fe crystals over a wide range of temperatures, from −195 to 700°C, and analyzed a large number of slip-plane traces. They concluded that "every trace tabulated – 194 in all – could be explained by one or more planes of these three types" (p. 711), namely, $\{110\}$, $\{112\}$, and $\{123\}$. The strains in their specimens were of course nonuniform, and they identified a crystal plane as the plane of slip merely on the basis that its predicted trace was within 7° of the mean of the observed traces on the specimen faces.

The (initial) critical shear stresses on $\{112\}$ and $\{123\}$ planes of Si–Fe crystals were determined from tensile experiments by Opinsky & Smoluchowski (1951b) to be nearly equal and from 7 to 16 percent higher than the critical strength of the $\{110\}$ planes. However, they (as others) did not make measurements of external shape changes sufficient to confirm that gross single slip on the crystallographic planes indicated by surface traces was actually the mode of deformation.

In experiments on tantalum crystals at 300 K Ferriss, Rose & Wulff (1962, p. 980) found that the slip planes "were clearly not associated with any particular crystallographic plane or planes." They concluded "that composite slip is taking place and that the traces observed on the surface are merely the average or net result of composite slip." They further suggested that their data support the view that two "cooperating" $\{110\}$ planes are the glide planes contributing to this composite microslip. Chen & Maddin (1954, p. 50) already had argued "from theoretical considerations of the energy of dislocations that the process of alternate $\{110\}$ slip is energetically favored," so apparent gross slip on $\{112\}$ and $\{123\}$ planes (and presumably on noncrystallographic planes as well) would actually consist of alternating segments of microslip on two nonparallel $\{110\}$ planes.

The work of Sestak & Libovicky (1963) also should be noted. They performed pure bending tests of Fe–3-percent-Si single crystals at 78 K and observed "noncrystallographic slip approximately along the maximum resolved shear stress planes" (p. 1191) in both the tension and compression zones of each specimen. (They also found slip along portions of $\{110\}$ planes in the tension zone.)

In attempting to draw conclusions regarding finite deformation of b.c.c. crystals from the preceding survey of experimental investigations (which

I believe to be representative, based upon additional papers examined but not cited), the following point should be recognized. There appears to be no work on b.c.c. crystals since Taylor and Elam's classic paper of 1926, with the exception of their separate studies on β-brass (Taylor 1928; Elam 1936), in which complete external deformation measurements were made. Consequently, whether or not slip traces can be attributed to localized slip on $\{110\}$, $\{112\}$, or $\{123\}$ planes, or to composite microslip on $\{110\}$ planes alone, insofar as iron crystals are concerned Taylor and Elam's conclusion that the mean plane of gross single slip is the plane of maximum resolved shear stress in a $\langle 111 \rangle$ direction has, to my knowledge, never been refuted.

The final remarks by Taylor (1956, p. 10) on this matter are significant. He pointed out that the maximum possible variation between the greatest resolved shear stress on one of the three families of planes $\{110\}$, $\{112\}$, $\{123\}$ and that on an arbitrary plane containing the corresponding $\langle 111 \rangle$ slip direction is only 1.4 percent. He concluded that his experimental law regarding the slip plane, based upon external measurements of the total macroscopic crystal strain, "is so slightly at variance with the hypothesis that the shear on an arbitrary plane is due to an appropriate combination of two slips on crystal planes that no existing technique would be capable of distinguishing between them."

In Fig. 4.5 is shown the subdivision of the standard stereographic triangle $a\bar{2}$ (on an [001] projection) into six slip regions in tension corresponding to the assumed occurrence of slip in a virgin crystal on the most highly stressed of the planes $\{110\}$, $\{112\}$, or $\{123\}$ in a $\langle 111 \rangle$ direction. This subdivision was first presented in Opinsky & Smoluchowski (1951a) (save for the mislabeling of slip region $(\bar{2}\bar{1}3)$ [111] in their fig. 5). The equations of all common boundaries and coordinates of all triple points of the six slip systems are given in Havner & Baker (1979, appendix). For an initial tensile axis lying within either of the slip regions nearest the [001] corner, the axis will rotate toward slip direction $[\bar{1}11]$ in single slip on the respective plane $(1\bar{1}2)$ or $(2\bar{1}3)$. (Note: Line [001]–$[\bar{1}11]$ is not a bounary of slip system $(1\bar{1}2)$ $[\bar{1}11]$ but rather bisects its slip region, the other half being in triangle $b1$. Symmetry line [001]–[011], however, is a boundary of all six slip systems.) For a loading-axis position initially within any one of the other four regions, the axis will rotate toward [111] in tension. If the initial critical strength associated with $\{110\}$ planes is less for a given metal than that for gross slip on $\{112\}$ and $\{123\}$ planes (but the percentage differences are no greater than those indicated by the

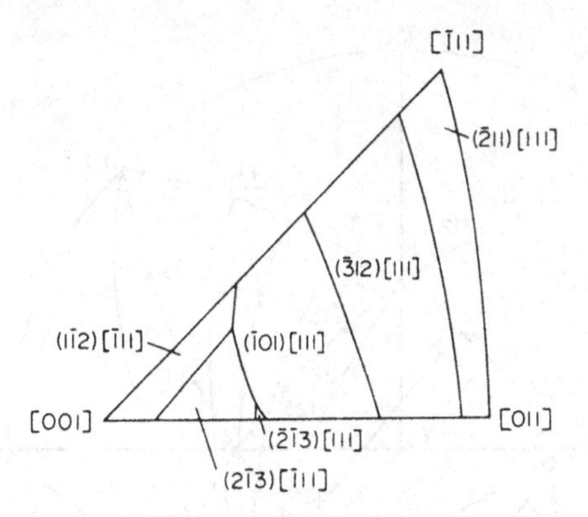

Fig. 4.5. Slip-system regions in stereographic triangle $a\bar{2}$ corresponding to equal $\langle 111 \rangle$ critical strengths on $\{110\}$, $\{112\}$, and $\{123\}$ planes in b.c.c. crystals in tension (from Havner & Baker 1979, fig. 2). Reprinted with permission from *J. Mech. Phys. Solids* 27, K. S. Havner & G. S. Baker, Theoretical latent hardening in crystals - II, Copyright 1979, Pergamon Press, PLC.

experiments of Opinsky & Smoluchowski (1951b) on Si–Fe crystals), the $(\bar{1}01)$ $[111]$ region will simply expand in breadth within triangle $a\bar{2}$, eliminating only the very small $(\bar{2}\bar{1}3)$ $[111]$ region.

If one considers, as in Fig. 4.5, that the $\{110\}$, $\{112\}$, and $\{123\}$ families are equally likely as potential slip planes, there are 96 different slip systems (distinguishing between opposite senses of slip as before): 12 planes each of types $\{110\}$ and $\{112\}$, and 24 planes of type $\{123\}$. These 96 systems may be conveniently identified from the $[001]$ projection in Fig. 4.6 (taken from fig. 3 of Havner & Baker (1979)), which shows the positive poles of the 48 slip planes and the four $\langle 111 \rangle$ slip directions. (The negative poles, of course, lie immediately beneath the positive ones on the opposite or $[00\bar{1}]$ projection.) For clarity only the great circles through the 12 positive slip-plane normals corresponding to slip direction $[111]$ are displayed, together with their continuations for 5 of the 12 negative poles (viz. $(2\bar{1}\bar{1})$, $(10\bar{1})$, $(01\bar{1})$, $(\bar{1}2\bar{1})$, and $(\bar{1}10)$, as seen). The negative slip-plane normals lie on the arc of an $[00\bar{1}]$ projection that is the continuation of arc $(1\bar{1}0)$–$(\bar{1}\bar{1}2)$–$(\bar{1}10)$ on the opposite side of the sphere (i.e., arc $(\bar{1}10)$–$(11\bar{2})$–$(1\bar{1}0)$). The sense of rotation during deformation of a specimen axis lying on one of these great circles corresponds, as marked, to the tensile test and must be reversed in compression (when, of course, the axis rotates toward the slip-plane normal). The various slip-plane poles for the other three slip directions in b.c.c. crystals, all lying on the dashed arcs (and

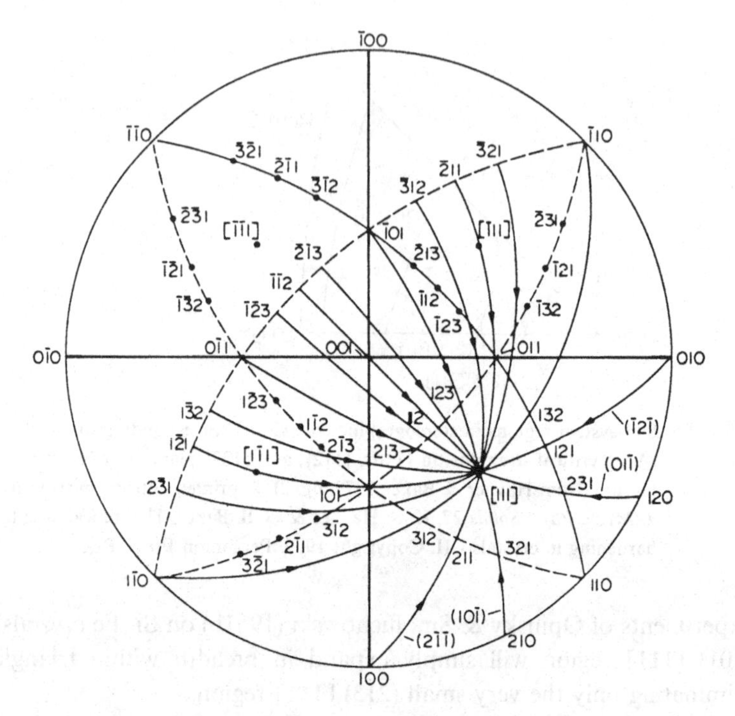

Fig. 4.6. Identification of 96 slip systems in b.c.c. crystals corresponding to slip on {110}, {112}, and {123} planes (from Havner & Baker 1979, fig. 3). Reprinted with permission from *J. Mech. Phys. Solids* 27, K. S. Havner & G. S. Baker, Theoretical latent hardening in crystals - II, Copyright 1979, Pergamon Press, PLC.

their opposites), are similarly identified from the fourfold symmetry about [001].

To consider the alternate assumption that gross single slip takes place on the plane of greatest resolved shear stress containing a ⟨111⟩ direction, we first develop the equations defining the boundaries of the planes of equal maximum stress for all ⟨111⟩ pairs. Let **b**, **b′** denote any two slip directions whose common boundary we wish to determine. Because a slip-plane normal must lie in the plane of **b** and load axis **ι** (unit vector) to achieve the maximum shear stress in a given direction, the boundary is defined by

$$(\mathbf{b}\cdot\boldsymbol{\iota})\{1-(\mathbf{b}\cdot\boldsymbol{\iota})^2\}^{1/2}=(\mathbf{b}'\cdot\boldsymbol{\iota})\{1-(\mathbf{b}'\cdot\boldsymbol{\iota})\}^{1/2}.$$

Let [111] be the primary slip direction, with $\mathbf{b}'=[\bar{1}11]$, $[1\bar{1}1]$, and $[11\bar{1}]$ in turn (the positive poles in a [111] projection). As scalar products $\mathbf{b}\cdot\boldsymbol{\iota}$, $\mathbf{b}'\cdot\boldsymbol{\iota}$ are positive in tension, the common boundaries of the maximally

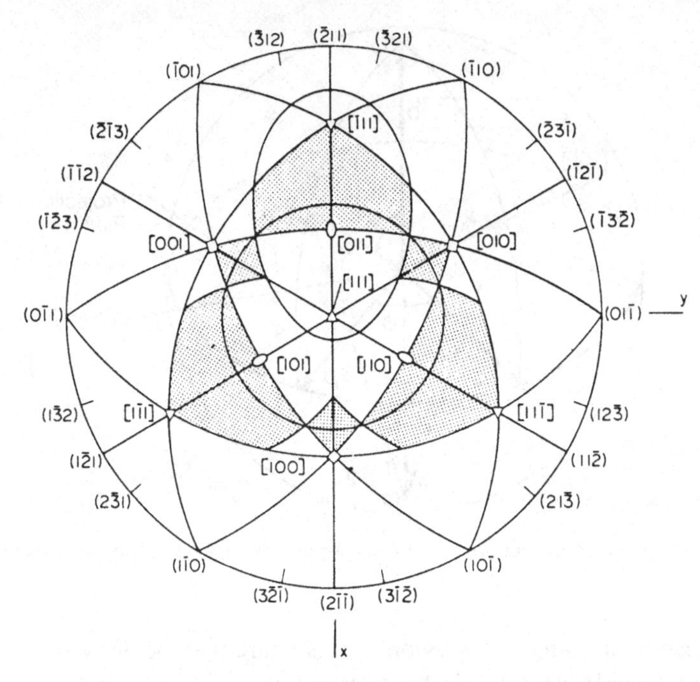

Fig. 4.7. [111] stereographic projection showing initial region (shaded) of [111] single slip in b.c.c. crystals in tension corresponding to Taylor & Elam's (1926) experimental law (from Havner & Baker 1979, fig. 11). Reprinted with permission from *J. Mech. Phys. Solids* 27, K. S. Havner & G. S. Baker, Theoretical latent hardening in crystals - II, Copyright 1979, Pergamon Press, PLC.

stressed systems are found to be (Havner & Baker 1979):

$$\begin{aligned}
&\text{between } [111] \text{ and } [\bar{1}11],\\
&\qquad \iota_2 + \iota_3 > 0, \quad \iota_1 = 0, \quad \text{and} \quad 4\iota_2\iota_3 = 1;\\
&\text{between } [111] \text{ and } [1\bar{1}1],\\
&\qquad \iota_1 + \iota_3 > 0, \quad \iota_2 = 0, \quad \text{and} \quad 4\iota_1\iota_3 = 1;\\
&\text{between } [111] \text{ and } [11\bar{1}],\\
&\qquad \iota_1 + \iota_2 > 0, \quad \iota_3 = 0, \quad \text{and} \quad 4\iota_1\iota_2 = 1.
\end{aligned} \tag{4.1}$$

The nonlinear loci $4\iota_i\iota_j = 1$ and the region within which [111] is the preferred slip direction are most elegantly displayed on a [111] stereographic projection, as was first done by von Mises (1928, fig. 9). A similar display was given independently (even though a half-century later!) by Havner & Baker (1979) in their figure 11, presented here as Fig. 4.7. The loci and

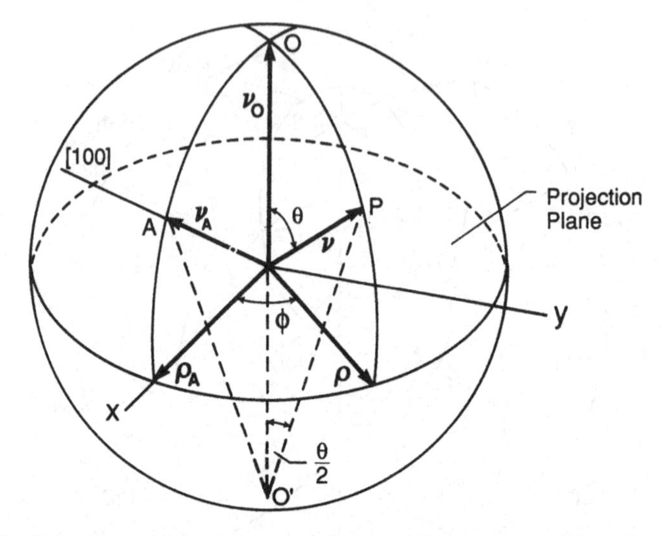

Fig. 4.8. Unit sphere and projection plane for an arbitrary stereographic projection.

corresponding single-slip region of this figure may be derived as follows (taken from Havner and Baker's section 5).

Let $\mathbf{v}_0$ denote the position vector of the point on the unit sphere that is to be mapped into the origin of a stereographic projection (in our case [111]). Define point A as the intersection of lattice direction [100] with this sphere and choose the x-axis of the projection plane to pass through the [100] projected point. Also let $\boldsymbol{\rho}$ denote the unit vector orthogonal to $\mathbf{v}_0$ and lying in the plane of $\mathbf{v}_0$ and a direction $\mathbf{v}$ whose projection coordinates we wish to determine (subject only to $\mathbf{v}\cdot\mathbf{v}_0 \geqslant 0$). With ϕ the counterclockwise angle of $\boldsymbol{\rho}$ with the x-axis and θ the angle between $\mathbf{v}$ and pole axis $\mathbf{v}_0$, we have from these definitions and the simple geometry of projection onto the midplane of the sphere (see Fig. 4.8):

$$\begin{aligned}&\boldsymbol{\rho}\cdot\mathbf{v}_0 = 0, \quad \boldsymbol{\rho}\cdot(\mathbf{v}\wedge\mathbf{v}_0) = 0,\\ &\cos\theta = \mathbf{v}\cdot\mathbf{v}_0, \quad \cos\phi = \boldsymbol{\rho}\cdot\boldsymbol{\rho}_A, \\ &x = \tan(\theta/2)\cos\phi, \quad y = \tan(\theta/2)\sin\phi.\end{aligned} \tag{4.2}$$

These equations are completely general and apply to any stereographic projection, with $\boldsymbol{\rho}_A$ defining the intersection of the x-axis with the projection perimeter. For $\mathbf{v}_0 = [111]$ and $\mathbf{v}_A = [100]$ one has $\boldsymbol{\rho}_A = [2\bar{1}\bar{1}]$, and after straightforward but somewhat lengthy algebra the final equations of the [111] stereographic projection are obtained. These are (Havner &

Baker 1979)

$$x = (2\nu_1 - \nu_2 - \nu_3)d/e, \quad y = \sqrt{3}(\nu_2 - \nu_3)d/e, \tag{4.3}$$

in which

$$\begin{aligned} d &= \{(\sqrt{3} - \textstyle\sum \nu_i)/(\sqrt{3} + \sum \nu_i)\}^{1/2}, \\ e &= 2(1 - \nu_1\nu_2 - \nu_1\nu_3 - \nu_2\nu_3)^{1/2}, \quad \textstyle\sum \nu_i \geqslant 0. \end{aligned} \tag{4.4}$$

The locus $4\nu_2\nu_3 = 1$ (or $4\iota_2\iota_3 = 1$) defining the common boundary of equal maximum resolved shear stress in directions [111] and [$\bar{1}$11] is readily plotted by choosing successive values of ν_1 between $\pm 1/\sqrt{2}$ (the intercepts of the locus with the x-axis $\nu_2 = \nu_3$). The other direction cosines are determined from

$$\begin{aligned} \nu_2 &= (1/\sqrt{2})[(1 - \nu_1^2) \pm \{(1 - \nu_1^2)^2 - \tfrac{1}{4}\}^{1/2}]^{1/2}, \\ \nu_3 &= (1/\sqrt{2})[(1 - \nu_1^2) \mp \{(1 - \nu_1^2)^2 - \tfrac{1}{4}\}^{1/2}]^{1/2}. \end{aligned} \tag{4.5}$$

From equation (4.3) these obviously give a single x-coordinate and a pair of equal positive and negative y-coordinates for each ν_1 in the interval. The complete locus encircling [111] and [$\bar{1}$11] is shown in Fig. 4.7. Loci $4\nu_1\nu_3 = 1$ and $4\nu_1\nu_2 = 1$ are geometrically identical to locus $4\nu_2\nu_3 = 1$ on a [111] stereographic projection and equivalently positioned relative to the axes [111]–[1$\bar{1}$1] ($\nu_1 = \nu_3$) and [111]–[11$\bar{1}$] ($\nu_1 = \nu_2$), respectively, as is the first locus to the x-axis. In Fig. 4.7 only those parts of these curves are drawn which serve to define the initial region of single slip in direction [111] (shown shaded in the figure).

The loci $\nu_i = 0$ ($i = 1, 2, 3$) are great circles opposite the projected lattice axes [100], [010], and [001] (each circle passing through the other pair of poles), which axes are the vertices of an equilateral spherical triangle centered on [111], as seen in Fig. 4.7. The interior circle passes through the intercepts of these great circles with the loci $4\nu_i\nu_j = 1$ and represents the 45° cone (also centered on [111]) of all loading-axis directions for which the resolved-shear-stress factor $\cos\lambda \cos\psi$ (i.e., $(\mathbf{b}\cdot\boldsymbol{\iota})(\mathbf{n}\cdot\boldsymbol{\iota})$, commonly called the "Schmid factor") attains its maximum value 0.5. (The equation of this circle is $x^2 + y^2 = (\sqrt{2} - 1)/(\sqrt{2} + 1)$.)

According to Taylor & Elam's (1926) experimental law, for an initial axis position $\boldsymbol{\iota}_0$ anywhere within the shaded region of Fig. 4.7, single slip in tension is initiated in direction [111] on a plane whose normal $\mathbf{n}$ lies on the perimeter of the projection at the intersection of the radial line through $\boldsymbol{\iota}_0$. As single slip continues, the tension axis rotates along this radial line toward the center [111]. Overshoot occurs when the axis rotates beyond the symmetry line with another ⟨111⟩ system into one of the three unshaded, spade-shaped regions having [111] as apex.

Consider the primary spherical triangle [001]–[011]–[$\bar{1}$11] of Fig. 4.7. Upon comparing this triangle with that of Fig. 4.5, we see that, although the projected shapes are necessarily different, the unshaded region in the [001] corner is approximately the same as that of combined region (1$\bar{1}$2)[$\bar{1}$11] and (2$\bar{1}$3)[$\bar{1}$11] in Fig. 4.5 corresponding to slip in direction [$\bar{1}$11]. This near equality of these regions reflects the point (already noted) made by Taylor (1956) that the maximum resolved shear stress on an arbitrary plane containing a ⟨111⟩ direction differs from the maximum ⟨111⟩ stress on a {110}, {112}, or {123} lattice plane by at most 1.4 percent. If indeed gross single slip in a b.c.c. crystal takes place only on one of the latter three planes, then a tension axis within the shaded region of Fig. 4.7, instead of rotating radially inward from its initial position $\boldsymbol{\iota}_0$, will rotate (almost radially) on the great circle through [111], $\boldsymbol{\iota}_0$, and the nearest {110}, {112}, or {123} plane normal on the perimeter of the [111] projection.

For the compression test we may reinterpret the loci and shaded region of Fig. 4.7 corresponding to a negative Schmid factor $(\mathbf{b}\cdot\boldsymbol{\iota})(\mathbf{n}\cdot\boldsymbol{\iota})$. Thus, loci $4v_iv_j = 1$ and $v_i = 0$ define common boundaries of the opposite slip directions [$\bar{1}\bar{1}\bar{1}$] (the primary system), [1$\bar{1}\bar{1}$], [$\bar{1}$1$\bar{1}$], and [$\bar{1}\bar{1}$1]; and the shaded area is the initial region of compressive single slip in direction [$\bar{1}\bar{1}\bar{1}$]. The slip-plane normal $\mathbf{n}$ is located as before, and the compression axis rotates toward $\mathbf{n}$ as slip proceeds.

4.3 Axial Loading of B.C.C. Crystals

Striking examples of overshooting in b.c.c. metals may be seen from the tension experiments on iron single crystals at 298 K by Keh (1965). In these tests he found by 'two-trace" (p. 16) analyses of the slip lines that "slip takes place on one of the planes containing a ⟨111⟩ zone axis, which bears the maximum resolved shear stress." Thus, for crystals whose axis was oriented in one of the corners [001], [011], or [$\bar{1}$11] of the standard stereographic triangle, "slip always takes place on the {112} planes." (That these are the slip planes in the corner positions may be seen from either Fig. 4.5 or Fig. 4.7.) Keh states that it was not possible to conveniently determine the slip directions because of the shape of the specimen. However, he goes on to say that, using the TEM diffraction

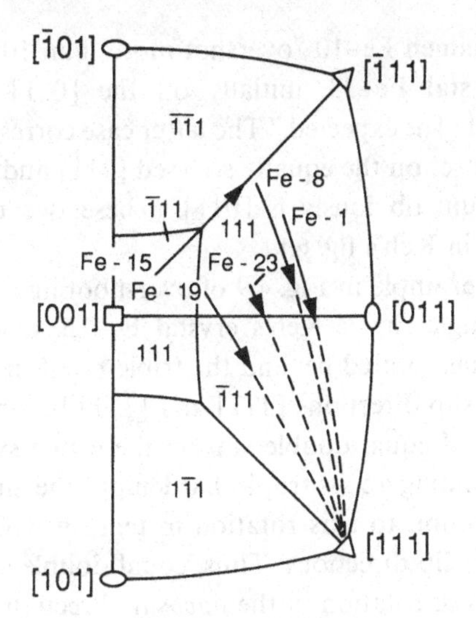

Fig. 4.9. [001] stereographic projection showing b.c.c. crystal slip regions (corresponding to Taylor and Elam's experimental law) and tensile axis rotations of iron crystal specimens of Keh (1965).

contrast method, "the Burgers vectors of the primary dislocations in several crystals were uniquely determined and they agreed in all cases with the assumed ⟨111⟩ slip direction having the maximum resolved shear stress." Although Keh did not measure external shape changes and commonly found "kink bands" of increasing misorientation (with the overall lattice) as testing proceeded, his X-ray determinations of crystal orientations at various stages of deformation nevertheless provide considerable evidence in support of single slip as the dominant mode of deformation well beyond the symmetry line. Apparently considering classical Taylor hardening as the norm, he writes, "Ideally, when the orientation reaches the [001]–[011] line, the [111] and [$\bar{1}$11] slip directions become equally favourable, thus the direction of rotation of the tensile axis should shift from [111] to [011]. This was not observed. Instead, overshooting seems to be a general phenomenon in these crystals" (p. 17).

In Fig. 4.9 are shown, on a partial [001] stereographic projection, the subdivisions of the triangles corresponding to Taylor & Elam's (1926) experimental law, confirmed by Keh (1965), together with all the initial and final crystal-axis positions given in Keh's figure 6. The examples of overshooting and the continuation of single slip beyond the respective symmetry line with another ⟨111⟩-system are self-evident. Keh also mentions that his crystal specimen Fe–10 "overshot more than 10°" and that the tensile axis of crystal Fe–33, initially on the [001]–[011] boundary, "rotates toward [011] as expected." The latter case corresponds to equal double slip at the outset on the equally stressed [111] and [$\bar{1}$11] slip systems, whose vector sum obviously is [011]. (These two crystal specimens were not included in Keh's fig. 6.)

A particularly noteworthy example in Fig. 4.9 of overshooting and the perpetuation of dominant single slip is Keh's crystal Fe–15. Once the tensile axis of this specimen has rotated beyond the triple point in initial single slip, planes containing slip directions [111] and [$\bar{1}\bar{1}$1] become the most highly stressed systems. If equal double slip on these two systems were then initiated, corresponding to isotropic hardening, the limiting direction of their net contribution to axis rotation in tension would be the vector sum [001] of their slip directions. Thus, equal double slip on these systems alone would cause rotation in the *opposite* direction along the [001]–[$\bar{1}$11] axis. Since the result of that rotation would be to return the axis into the region of [$\bar{1}$11] single slip, we conclude that Taylor hardening leads to axis stability and triple slip at the intersection of the three ⟨111⟩ slip regions. What has been observed, however, as shown in Fig. 4.9, is the continuation of single slip in the primary system.

As was remarked earlier, Keh (1965) also conducted tensile tests on iron crystals in the corner orientations of the standard stereographic triangle. In b.c.c metals, according to either the classical experimental law of Taylor & Elam (1926) for iron crystals or the hypothesis that ⟨111⟩ slip occurs on the most highly stressed {110}, {112}, or {123} plane, orientations [001], [011], and [$\bar{1}$11] are, respectively, four-, four-, and sixfold multiple-slip positions (as evident from Figs. 4.5 and 4.9). From his experiments Keh found that "the corner orientations are quite stable. The mean orientation of [001] and [011] crystals remains unchanged after 12% tensile strain, while that of [$\bar{1}$11] changes less than 2° after the same amount of strain" (p. 17). This stability implies (but does not strictly require) equal slip in all critical systems and is consistent with isotropic hardening in these multiple-slip configurations. (Of course, axis stability alone is not direct evidence of isotropic hardening, and critical experiments on b.c.c. crystals parallel to those of Nakada on aluminum (Wonsiewicz & Chin 1970b) have not been performed.)

The most comprehensive study, to my knowledge, of latent hardening in b.c.c metals in single slip is the work of Nakada & Keh (1966), who tested high-purity iron crystals at various temperatures from 77 K to 298 K. Their method of determining the hardening of latent systems is the same

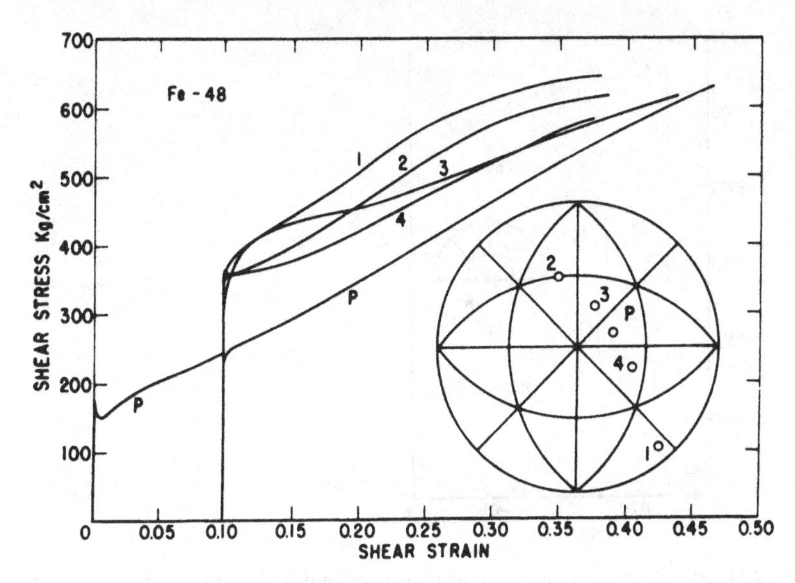

Fig. 4.10. Primary system hardening curve (*P*) in an iron crystal and subsequent resolved shear stress–strain curves in four secondary crystal specimens (from Nakada & Keh 1966, fig. 3). Reprinted with permission from *Acta Met.* 14, Y. Nakada & A. S. Keh, Latent hardening in iron single crystals, Copyright 1966, Pergamon PLC.

as that described in Section 4.1. Although they did not make complete strain measurements, they took considerable care to ensure the integrity of their latent-hardening results (see Nakada & Keh 1966, pp. 964–5) and with reason concluded that "any latent hardening observed is a genuine property of the latent system and is not due to variations in specimen preparation."

Nakada and Keh found the slip system in their parent crystals to be the plane of maximum resolved shear stress containing a $\langle 111 \rangle$ slip direction, but reminded their readers (p. 963) "that at least one of the close packed planes, (110), (211) or (321), was always near the plane which bore the maximum resolved shear stress." They determined the active system in each secondary specimen by two-surface trace analyses of the slip lines, stating that this "was accomplished unambiguously."

Characteristic latent-hardening results of Nakada and Keh are presented in Fig. 4.10 (reproduced from their fig. 3). Here are shown the primary-system curve corresponding to an initial axis orientation near the maximum Schmid-factor position $(\mathbf{b}\cdot\boldsymbol{\iota})(\mathbf{n}\cdot\boldsymbol{\iota}).=0.5$ and active-hardening curves of four previously latent systems after a prestrain $\gamma=0.1$ of the

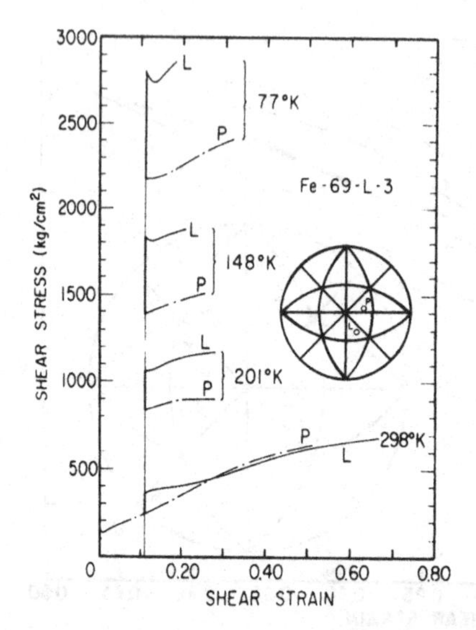

Fig. 4.11. Effect of temperature on primary (*P*) and initially latent (*L*) secondary-system hardening in iron crystals (from Nakada & Keh 1966, fig. 11). Reprinted with permission from *Acta Met.* 14, Y. Nakada & A. S. Keh, Latent hardening in iron single crystals, Copyright 1966, Pergamon PLC.

parent crystal. The initial tensile-axis positions of the secondary specimens may be seen in the figure, and the slip directions are given in Nakada and Keh's table 3.[5] Other results for prestrains from 5 to 30 percent in shear are displayed in their figures 4–9. They also specifically investigated latent hardening in a coplanar system and found, in contrast to the nearly equal coplanar hardening commonly measured in f.c.c crystals, that for b.c.c. iron, coplanar latent hardening was approximately 20 percent greater than active hardening after a 15 percent prestrain in shear at room temperature. However, the subsequent active-hardening curve of the secondary coplanar system essentially merged with the continuation of the primary system curve after an additional 12 percent shear (see their fig. 6). This was distinctly different from the typical behavior of noncoplanar systems (see Fig. 4.10 or Nakada and Keh's fig. 4 or fig. 5) and is closer to that of coplanar systems in f.c.c. crystals.

Nakada and Keh also investigated the effects of temperature and strain rate on the "latent-hardening ratio." They define this as the ratio of the critical shear stress of a previously latent system (at the initiation within a secondary specimen of measurable slip in that system) to the critical

[5] These require interpretation, however, as Nakada and Keh adopted a [100] rather than an [001] stereographic projection. Thus, their primary slip direction is $[11\bar{1}]$ rather than [111].

strength of the primary system at the end of deformation of the parent crystal. With regard to temperature dependence, two kinds of tests were conducted on each of three different parent-crystal and secondary-specimen orientations (Nakada & Keh 1966, p. 967). "First, the parent crystals were prestrained at different temperatures, and then the small latent hardening specimens were restrained at the prestrain temperatures. Secondly, the parent crystals were prestrained at room temperature, and the latent hardening specimens were tested at other temperatures. The latent hardening behavior was essentially the same for both kinds of tests." The latter test is illustrated in Fig. 4.11 (Nakada and Keh's fig. 11). As may be determined from the figure, and as they state in general, "The latent hardening ratio is essentially constant at all temperatures."

To assess strain-rate dependence, Nakada and Keh performed tests on secondary specimens of two different orientations taken from a parent crystal prestrained to 16 percent shear. These specimens were deformed at strain rates ranging widely from 3×10^{-5} to 10^{-1} (see their fig. 13). Nakada and Keh found, "The variation of the latent hardening ratio as a function of the strain rate is not systematic and is rather small" (p. 967).

4.4 Latent-Hardening Theories

The experimental papers that have been cited and discussed (to varying extent) in this chapter are, of course, only a selection from a much larger body of similar studies in the metallurgical and metal physics literature (see Basinski & Basinski (1979) for a comprehensive bibliography). However, they are representative of that literature; and the papers in particular that I have reviewed at greater length are among the most noteworthy and often-cited works from the period (1960s and 1970s) during which the principal qualitative and quantitative features of finite deformation and latent hardening of axially loaded cubic crystals were established. As we have seen from that review, the typical responses of both f.c.c. and b.c.c. crystals are well characterized by the eight-point summary given at the end of Section 4.1, although in the case of b.c.c. metals the evidence for isotropic hardening in high-symmetry orientations is only indirect. This summary may be condensed and restated as follows:

1. In single-slip orientations the hardening of latent slip systems is generally greater than that of the active system (for all initial axis positions that have been tested); and dominant primary slip consistent with overshooting well beyond the crystallographic symmetry line is the norm.

2. In high-symmetry, multiple-slip orientations axis stability, axisymmetric finite deformation, and statistically isotropic dislocation networks are typically observed, consistent with equal slip in critical systems and isotropic hardening of all systems.

In several of the experimental papers reviewed herein, explanations of greater latent than active hardening in single slip are offered from the perspective of dislocation theory. However, dislocation theory is inherently linked to *infinitesimal* strain theory (refer to the standard reference Hirth & Lothe (1982), for example), and those explanations (even if, in essence, correct physically), when not merely descriptive and qualitative, do not appear to be well-founded mathematically within the precise geometry of the finite crystal deformations they presume to address. Be that as it may, theoretical dislocation arguments are outside the scope of this monograph, and we seek instead to characterize and mathematically model the dual (contrasting) anisotropic/isotropic hardening of metal crystals in terms of the gross kinematic mechanism of crystallographic slip.

The one macroscopic feature of crystal deformation that is completely different in single slip than in symmetric multiple slip is the *relative rotation of material and lattice*, present in the former deformation mode but not the latter. Recognizing this, it occurred to me a number of years ago that within the framework of continuum mechanics this relative rotation might account (at least in part) for the observed anisotropy of latent hardening and related overshooting of the load axis in single slip. As we shall find, there is also a mathematical aesthetic (related to the symmetry of the h_{kj} and g_{kj} parameters of Sect. 3.7 and the positive definiteness of the latter) that underlies a particularly simple form incorporating this idea, first presented in Havner & Shalaby (1977). Moreover, and perhaps most notably, this "simple theory" of rotation-dependent anisotropic hardening (given here in Sect. 4.4.2) is a universal theory of overshooting in axially loaded cubic crystals, as established in the series of papers Havner & Shalaby (1977, 1978), Havner, Baker & Vause (1979), Havner & Baker (1979), and Vause & Havner (1979) and reviewed here in Section 4.5.

4.4.1 *General Equations*

Turning again to the equations of slip-systems hardening in Chapter 3, we henceforth consider the (averaged) lattice strains to be very small and geometrically negligible at ordinary pressures. Thus, we shall not distinguish between τ_j and $\tilde{\tau}_j$ or their corresponding critical values, nor between $d\gamma_j$ and $d\tilde{\gamma}_j$. Consequently, from equations $(3.59)_2$ and $(3.49)_2$, the general hardening law of equation (3.54) may be expressed in terms of the effective hardening moduli as

$$d\tau_k^c = \sum_j h_{kj}\, d\gamma_j - \nu_k \sum (\alpha\, d\gamma)_j + \alpha_k \mathscr{M} \sum (\alpha\, d\gamma)_j. \tag{4.6}$$

Since the αs are of the order of stress (from their definition in eq. (3.28)) and $\mathscr{M}$ is the tensor of elastic compliances for any measure, $\mathscr{M}\alpha_k$ is of the order of infinitesimal lattice strain. Consequently, the last term in equation (4.6) may be neglected in comparison with the other terms, and (upon substituting eq. (3.28)) one can write

$$d\tau_k^c = \sum_j h_{kj}\, d\gamma_j - \nu_k (dt - d^*t). \tag{4.7}$$

We now take the current state as reference for strain measure e, whence from equations (A65) (giving the transformation of objective stress rate under change in measure), (3.5)–(3.6), and (3.53) we have

$$\begin{aligned} (dt - d^*t)^{(0)} &= \mathscr{D}\tau - \mathscr{D}^*\tau - m(\sigma D^p + D^p\sigma)\, d\theta, \\ D^p &= \operatorname{sym}\Gamma - \varepsilon = \sum N_j \dot{\gamma}_j. \end{aligned} \tag{4.8}$$

Here θ is a timelike variable as before, m is Hill's measure-dependent parameter defined in equation (A59) (zero for the logarithmic measure of strain), and $\mathscr{D}\tau$, $\mathscr{D}^*\tau$ are, respectively, material and lattice corotational increments in Kirchhoff stress, which differs negligibly from Cauchy stress at ordinary pressures. (Of course, with the current state as reference, $\tau = \sigma$ momentarily in any case.) From standard equations for the corotational derivatives (see eq. (A63), for example) it follows that

$$\mathscr{D}\tau - \mathscr{D}^*\tau = (\tau\Omega - \Omega\tau)\, d\theta, \tag{4.9}$$

with (from eqs. (3.5)–(3.6))

$$\Omega = W - \omega = \sum \Omega_j \dot{\gamma}_j, \quad \Omega_j = \operatorname{skw}(\mathbf{b} \otimes \mathbf{n})_j. \tag{4.10}$$

Ω is the relative spin of material and lattice, henceforward called the *plastic spin.* (In contrast to $W \equiv \operatorname{skw}\Gamma$ and ω, Ω is an objective tensor field.) Furthermore, from $(de - d^*e)^{(0)} = D^p\, d\theta$ and equations (3.19) and (3.53),

$$\nu_j^{(0)} = N_j \equiv \operatorname{sym}(\mathbf{b} \otimes \mathbf{n})_j \tag{4.11}$$

for all measures. Therefore, upon substituting equations $(4.8)_1$, (4.9) (with $\tau = \sigma$), and (4.11) into equation (4.7) evaluated in the reference state, one obtains (Hill & Havner 1982, eq. (7.18))

$$d\tau_k^c = \sum_j h_{kj}\, d\gamma_j - 2\mathrm{tr}(N_k\sigma(\Omega - mD^p))\, d\theta, \tag{4.12}$$

or

$$\dot{\tau}_k^c = \sum_j (h_{kj} + 2m\,\mathrm{tr}(N_k\sigma N_j))\dot{\gamma}_j - 2\mathrm{tr}(N_k\sigma\Omega). \tag{4.13}$$

4.4.2 *The Simple Theory*

From the invariance of τ_k^c, σ, and Ω, it is evident from equation (4.13) that

$$h_{kj} + 2m\,\mathrm{tr}(N_k\sigma N_j) \quad \text{is measure invariant} \tag{4.14}$$

for every pair of systems k, j (Hill & Havner 1982, eq. (7.17)). The simplest possible mathematical theory in the parameters h_{kj} is then obtained by requiring this quantity to be independent of slip systems as well as measure. Denoting this scalar invariant by h (which may be a function of both stress and the history of deformation), we see from statement (4.14) that the result is a theory of equal effective hardening moduli in the logarithmic measure ($m = 0$) which is identically the *simple theory of anisotropic latent hardening* introduced in Havner & Shalaby (1977, eq. (23)), namely

$$d\tau_k^c = \sum_j (h - 2\mathrm{tr}(N_k\sigma\Omega_j))\, d\gamma_j, \quad h > 0, \tag{4.15}$$

or

$$\dot{\tau}_k^c = h\sum \dot{\gamma}_j - 2\mathrm{tr}(N_k\sigma\Omega). \tag{4.16}$$

Obviously, although the corresponding (theoretical) physical hardening moduli

$$H_{kj} = h - 2\mathrm{tr}(N_k\sigma\Omega_j), \quad j = 1,\ldots,n, \quad k = 1,\ldots,N, \tag{4.17}$$

are now both anisotropic and nonsymmetric (as well as explicitly dependent upon the stress state), equation (4.15) gives an incremental net *isotropic* hardening of all systems whenever the relative (or plastic) spin Ω is zero. As this spin can never be zero in single slip but is identically zero in six- or eightfold symmetric multiple slip, the simple theory clearly predicts anisotropic hardening in the former case but isotropic hardening in the latter. That this predicted anisotropy always corresponds to overshooting in single slip in axially loaded f.c.c. and b.c.c. crystals and is qualitatively consistent with the diversity of hardening in latent systems found experimentally will be reviewed and illustrated in Section 4.5.

Consider the invariant second-order work (per unit reference volume) $\frac{1}{2}\mathrm{d}^{\mathrm{p}}t\,\mathrm{d}e$ in a differential strain cycle, postulated in equation (3.80) always to be positive. From equations (3.77) and (3.95) we may write

$$\mathrm{d}^{\mathrm{p}}t\,\mathrm{d}e = \sum\sum(h_{kj} + M_k\mathscr{L}M_j)\,\mathrm{d}\gamma_k\,\mathrm{d}\gamma_j. \tag{4.18}$$

Consequently for the simple theory (with the current state as reference) it follows that

$$\mathrm{d}^{\mathrm{p}}t\,\mathrm{d}e = h(\sum \mathrm{d}\gamma_k)^2 + \sum(M\,\mathrm{d}\gamma)_k\mathscr{L}_0\sum(M\mathrm{d}\gamma)_j > 0 \tag{4.19}$$

since each $\mathrm{d}\gamma_j \geqslant 0$ and the tensor $\mathscr{L}_0$ of elastic moduli in the logarithmic measure is positive definite for infinitesimal lattice strains. Thus, the simple theory guarantees positive second-order work in a differential strain cycle and in general assures positive-definite g_{kj} as well.[6]

4.4.3 *Taylor Hardening*

In the case of Taylor's classical isotropic hardening rule (Taylor & Elam 1923, 1925) we have

$$\mathrm{d}\tau_k^{\mathrm{c}} = H\sum \mathrm{d}\gamma_j \quad \text{for all } k, \tag{4.20}$$

with $H > 0$ the Taylor hardening modulus (identically the slope of the resolved shear stress versus slip curve in single slip). The work-invariant $\mathrm{d}^{\mathrm{p}}t\,\mathrm{d}e$ in the reference state then may be expressed (from eqs. $(3.77)_1$, (3.94), (3.32), and (4.11)):

$$\mathrm{d}^{\mathrm{p}}t\,\mathrm{d}e = H(\sum \mathrm{d}\gamma_j)^2 + \sum(N\,\mathrm{d}\gamma)_k\mathscr{L}\sum(N\,\mathrm{d}\gamma)_j - \sum(N\,\mathrm{d}\gamma)_j\sum(\alpha\,\mathrm{d}\gamma)_k. \tag{4.21}$$

As $\sum(\alpha\,\mathrm{d}\gamma) = \mathrm{d}t - \mathrm{d}^*t$ by definition, it is evident from equations (4.8)–(4.10) that

$$\alpha_k^{(0)} = 2\mathrm{sym}(\sigma\Omega_k) - 2m\,\mathrm{sym}(\sigma N_k). \tag{4.22}$$

Thus, because H also is an invariant, we again choose the logarithmic measure ($m = 0$) and so write for Taylor hardening (using equations $(4.8)_2$

[6] There are exceptional cases of high-symmetry axial load orientations in which the positive-definite rank of matrix g_{kj} for the simple theory (and also for Taylor hardening) may be one less than its order (e.g., rank five in sixfold multiple slip). Pure plastic spin without plastic deformation would then theoretically be permitted (see Havner (1981, sect. 2) or Havner (1985, sect. IV)).

and $(4.10)_1$)

$$\mathrm{d}^{\mathrm{p}}t\,\mathrm{d}e = H\left(\sum \mathrm{d}\gamma_j\right)^2 + D^{\mathrm{p}}\mathscr{L}_0 D^{\mathrm{p}}(\mathrm{d}\theta)^2 - 2\operatorname{tr}(D^{\mathrm{p}}\sigma\Omega)(\mathrm{d}\theta)^2 \tag{4.23}$$

(since taking the trace with symmetric tensor D^{p} annihilates the skew part of $\sigma\Omega$). Because σ is typically of order $10^{-3}\ \mathscr{L}_0$ and the magnitude of incremental plastic rotation $\Omega\,\mathrm{d}\theta$ is expected to be no greater than that of $D^{\mathrm{p}}\,\mathrm{d}\theta$, the last term on the right-hand side of equation (4.23) may be neglected in comparison with the positive-definite second term (except when they both are zero). Consequently, Taylor hardening also ensures $\mathrm{d}^{\mathrm{p}}t\,\mathrm{d}e > 0$.

4.4.4 *The P.A.N. Rule*

The general family of plastic-spin-dependent hardening theories introduced in Havner & Shalaby (1977, eq. (20)) may be written

$$\dot{\tau}_k^{\mathrm{c}} = \sum_j h_{kj}\dot{\gamma}_j - 2\operatorname{tr}(N_k\sigma\Omega), \quad h_{kj} = h_{jk}, \tag{4.24}$$

of which the simple theory (eq. (4.16)) obviously is the simplest example. (The proposed symmetry of the h_{kj} excludes Taylor's rule, as discussed previously, but leads to the general theoretical structure of saddle potential functions and minimum principles presented in Sect. 3.7.) Another demonstrably useful member of this family was introduced in Peirce, Asaro, & Needleman (1982). The effective hardening moduli (logarithmic measure) for the "P.A.N. rule" as we shall call it are

$$h_{kj} = \hbar + \operatorname{tr}(N_k\sigma\Omega_j + N_j\sigma\Omega_k), \tag{4.25}$$

with $\hbar$ a positive hardening parameter.[7] Therefore the P.A.N. rule can be expressed (from $\Omega_k = -\Omega_k^{\mathrm{T}}$)

$$\mathrm{d}\tau_k^{\mathrm{c}} = \sum_j (\hbar - \operatorname{tr}(N_k\sigma\Omega_j) - \operatorname{tr}(\Omega_k\sigma N_j))\mathrm{d}\gamma_j \tag{4.26}$$

or

$$\dot{\tau}_k^{\mathrm{c}} = \hbar\sum \dot{\gamma}_j - \operatorname{tr}(N_k\sigma\Omega) - \operatorname{tr}(\Omega_k\sigma D^{\mathrm{p}}). \tag{4.27}$$

Upon comparing equations (4.27) and (4.16), we see that the P.A.N. rule incorporates only half the plastic-spin dependence of the simple theory as a source of anisotropy, but adds another anisotropic contribution that

[7] Peirce et al. also give a two-parameter form in which $\hbar$ is replaced by $q\hbar + (1-q)\hbar\delta_{kj}$, $0 < q < 1$.

depends upon the plastic strain rate. These changes may prove advantageous as the rates of hardening of some latent systems in single slip according to the simple theory become unreasonably high beyond shears of order one. (The P.A.N. rule has not been thoroughly investigated in single slip but apparently is a theory of limited overshooting for certain crystal orientations in tension (Peirce et al. 1982).)

The P.A.N. rule shares with the simple theory (and of course Taylor's rule by definition) the capability of predicting isotropic hardening in high-symmetry axial load orientations, as first remarked by Havner (1984, p. 483). This may be seen as follows: For axisymmetric deformation in $\langle 111 \rangle$ or $\langle 100 \rangle$ loading of cubic crystals, Ω is identically zero and the load axis is a principal direction of plastic strain rate. Thus, σ and D^{p} are coaxial tensors, and their product annihilates the trace with every skew-symmetric tensor Ω_k. Consequently, $\dot{\tau}_k^c$ in equation (4.27) reduces to the isotropic first term.

With regard to the invariant second-order work $\frac{1}{2}\mathrm{d}^{\mathrm{p}}t\,\mathrm{d}e$, from $\alpha_k^{(0)} = 2\mathrm{sym}(\sigma\Omega_k)$ (logarithmic measure) and equations $(3.77)_1$, (3.95), $(3.49)_2$, (4.11), $(4.8)_2$, and $(4.10)_1$, one obtains for the P.A.N. rule (with the current state as reference)

$$\mathrm{d}^{\mathrm{p}}t\,\mathrm{d}e = \hbar\left(\sum \mathrm{d}\gamma_j\right)^2 + D^{\mathrm{p}}\mathscr{L}_0 D^{\mathrm{p}}(\mathrm{d}\theta)^2 - 2\mathrm{tr}(D^{\mathrm{p}}\sigma\Omega)(\mathrm{d}\theta)^2. \tag{4.28}$$

Here a term of order stress times infinitesimal lattice strain again has been neglected in comparison with terms of order stress or hardening parameter (to say nothing of elastic moduli), as was done in reducing equation (4.6) to the general form, equation (4.7), encompassing all hardening theories at infinitesimal lattice strain. Observing that equation (4.28) differs from equation (4.23) only by the respective hardening moduli $\hbar$ and H, the same argument applies as before and we conclude that $\mathrm{d}^{\mathrm{p}}t\,\mathrm{d}e > 0$ for the P.A.N. rule as well. The corresponding matrix g_{kj} generally will be positive definite (as well as symmetric); however, the qualification regarding the simple theory and Taylor hardening stated in footnote 6 of this chapter also pertains to the P.A.N. rule.

4.4.5 *A Two-Parameter Rule*

A two-parameter modification of Taylor's rule can be generalized from approximate empirical representations of single-slip latent-hardening data for cubic crystals in Nakada & Keh (1966, fig. 9) and Jackson & Basinski (1967, fig. 16). This two-parameter rule may be expressed (Havner

1982b, 1984) as

$$\dot{\tau}_k^c = H_1 \sum_l \dot{\gamma}_l + H_2 \sum_m \dot{\gamma}_m, \quad \mathbf{n}_l = \mathbf{n}_k, \quad \mathbf{n}_m \neq \mathbf{n}_k, \quad H_2 > H_1 > 0. \quad (4.29)$$

In other words, the contribution of incremental slip $d\gamma_j$ to the hardening of another system k equals the increment in self-hardening of the first system (taken to be uniform over all systems) if the two systems are coplanar. If not, the contribution is greater than the self-hardening and is taken as uniform among the other intersecting systems. (Obviously, this hardening rule is not a member of the family represented by eq. (4.24).)

The two-parameter rule predicts isotropic hardening of f.c.c. crystals in the case of equal eightfold slip in [001] axial loading, for then two slip systems in each of the four {111} slip planes are active, and the net hardening modulus equals $2H_1 + 6H_2$. In [$\bar{1}$11] loading, however, in which latent hardening also has been found to be essentially isotropic in equal sixfold slip (as reviewed in Sect. 4.1), the two-parameter rule gives a net hardening modulus of $2H_1 + 4H_2$ in any a, b, or d slip system (that is, planes (111), ($\bar{1}\bar{1}$1), or (1$\bar{1}$1)) but a greater net modulus of $6H_2$ in any c system (plane ($\bar{1}$11)). Consequently, a finitely prestrained secondary specimen subsequently pulled along [001] could not then deform axisymmetrically in eightfold slip since the more hardened systems $c1$, $c\bar{2}$ (see Fig. 4.2) would not be critical, hence would be inactive. This would be in conflict with Nakada's observations on aluminum crystals reported by Wonsiewicz & Chin (1970b). In b.c.c. crystals the hardening always would be anisotropic according to equation (4.29) because only the {112} family of planes is activated in any one of the three corner axis orientations (refer to Figs. 4.5 and 4.7).

The invariant $d^p t \, de$ for the two-parameter rule is, by analogy with equation (4.23) for Taylor hardening (again with the current state as reference),

$$d^p t \, de = H_1 \sum_k \sum_{j(k)} d\gamma_k \, d\gamma_j + H_2 \sum_k \sum_{l(k)} d\gamma_k \, d\gamma_l + D^p \mathscr{L}_0 D^p (d\theta)^2 - 2\mathrm{tr}(D^p \sigma \Omega)(d\theta)^2, \quad (4.30)$$

$$j(k): \quad \mathbf{n}_j = \mathbf{n}_k; \quad l(k): \quad \mathbf{n}_l \neq \mathbf{n}_k; \quad k = 1, \ldots, n,$$

with the union (sum) of all $j(k)$ and $l(k)$ in critical systems equal to n for each system k. Consequently (as before) $d^p t \, de > 0$.

Other (rate-independent) hardening rules that have been proposed in the mechanics literature, although not specifically in the context of finite strain, are briefly reviewed in Havner & Salpekar (1983, sect. 2). The predictions of these theories do not appear to bear any relation to the observed finite-deformation behavior of f.c.c. and b.c.c. crystals, however, and they will not be considered here.

4.5 Investigations of the Simple Theory in Single Slip

Let $\boldsymbol{\iota}$ denote a unit vector in the axial load direction for either the tension or compression case. We may write

$$\boldsymbol{\iota} = \begin{cases} \mathbf{l}/l, & \text{tension;} \\ \mathbf{a}/a, & \text{compression,} \end{cases} \tag{4.31}$$

with $\mathbf{l}$ and $\mathbf{a}$ embedded (axial) line and (end-plane) areal vectors as defined in Chapter 2. From the geometry of single slip,

$$\mathbf{l} = \mathbf{l}_0 + \gamma(\mathbf{n}\cdot\mathbf{l}_0)\mathbf{b}, \quad \mathbf{a} = \mathbf{a}_0 - \gamma(\mathbf{b}\cdot\mathbf{a}_0)\mathbf{n} \tag{4.32}$$

(with of course $\mathbf{a}\cdot\mathbf{l} = \mathbf{a}_0\cdot\mathbf{l}_0 = V_0$, the constant reference volume, since infinitesimal lattice straining is neglected). Define $\mathbf{m} = \mathbf{b} \wedge \mathbf{n}$ and denote the angles that $\mathbf{l}$ and $\mathbf{a}$ make with $\mathbf{m}$ by η and ζ, respectively. Recalling the definitions of angles λ, χ (for $\mathbf{b}$) and ψ, φ (for $\mathbf{n}$) in Chapter 2, we have

$$\begin{aligned} \mathbf{b}\cdot\mathbf{l} &= l\cos\lambda, \quad & \mathbf{n}\cdot\mathbf{l} &= l\cos\psi, \quad & \mathbf{m}\cdot\mathbf{l} &= l\cos\eta, \\ \mathbf{b}\cdot\mathbf{a} &= a\cos\chi, \quad & \mathbf{n}\cdot\mathbf{a} &= a\cos\varphi, \quad & \mathbf{m}\cdot\mathbf{a} &= a\cos\zeta. \end{aligned} \tag{4.33}$$

There follow from equations (4.32) and (4.33), after some algebra (refer to Havner et al. (1979, sect. 3) for details),

$$\cos\lambda/\cos\psi = (\cos\lambda_0/\cos\psi_0) + \gamma, \quad \cos\varphi/\cos\chi = (\cos\varphi_0/\cos\chi_0) - \gamma, \tag{4.34}$$

and

$$\mathbf{l}/l = \mathbf{b}\cos\lambda + \mathbf{p}\cos\psi, \quad \mathbf{a}/a = \mathbf{n}\cos\varphi + \mathbf{q}\cos\chi, \tag{4.35}$$

with

$$\mathbf{p} = \mathbf{n} + (\cos\eta_0/\cos\psi_0)\mathbf{m}, \quad \mathbf{q} = \mathbf{b} + (\cos\zeta_0/\cos\chi_0)\mathbf{m}. \tag{4.36}$$

(Note that $\mathbf{p} = \mathbf{n}$ and $\mathbf{q} = \mathbf{b}$ if the load axis lies in the plane of $\mathbf{b}$ and $\mathbf{n}$.)

Let $\sigma_a(\gamma)$ and $\tau(\gamma)$ denote uniaxial Cauchy stress and resolved shear stress (critical strength) in the active system in single slip. We obviously have

$$\sigma = \sigma_a\boldsymbol{\iota}\otimes\boldsymbol{\iota}, \quad \sigma_a = \begin{cases} \tau/(\cos\lambda\cos\psi) > 0, & \text{tension;} \\ \tau/(\cos\chi\cos\varphi) < 0, & \text{compression.} \end{cases} \tag{4.37}$$

(In compression $\cos\chi < 0$ for all orientations.) Therefore, upon substitution of equations (4.31), (4.34), and (4.35) into equation (4.37), we obtain (Havner, et al. 1979, eq. (4.3))

$$\sigma = \begin{cases} \tau(\gamma)\{2\mathrm{sym}(\mathbf{b}\otimes\mathbf{p}) + \mathbf{b}\otimes\mathbf{b}(c_2+\gamma) + \mathbf{p}\otimes\mathbf{p}/(c_2+\gamma)\}, & \text{tension;} \\ \tau(\gamma)\{2\mathrm{sym}(\mathbf{b}\otimes\mathbf{q}) - \mathbf{n}\otimes\mathbf{n}(c_2+\gamma) - \mathbf{q}\otimes\mathbf{q}/(c_2+\gamma)\}, & \text{compression} \end{cases} \tag{4.38}$$

in which the positive constant c_2 is given by

$$c_2 = \begin{cases} \cos\lambda_0/\cos\psi_0, & \text{tension;} \\ -\cos\varphi_0/\cos\chi_0, & \text{compression.} \end{cases} \tag{4.39}$$

Equation (4.38) for the stress tensor (in single slip) is, of course, independent of latent-hardening theory.

Consider now the simple theory. From equation (4.15), as $\dot{\tau} = H\dot{\gamma}$ in the active system, the theoretical latent-hardening rates are

$$\dot{\tau}_k^c = \{H(\gamma) + 2\mathrm{tr}((N_1 - N_k)\sigma\Omega_1)\}\dot{\gamma}, \quad k = 1,\ldots,N, \tag{4.40}$$

with subscript 1 indicating the active system. Substituting equation (4.38) (using the orthogonality of **b**, **n** and **m**) we have

$$\dot{\tau}_k^c = \{H(\gamma) + \tau(\gamma)[A_k + B_k\gamma + C_k/(c_2+\gamma)]\}\dot{\gamma}, \tag{4.41}$$

where (Havner et al. 1979, eqs. (4.12) and (4.13))

$$\begin{aligned} &\text{in tension: } A_k = c_2B_k + \mathbf{b}N_k\mathbf{b} - \mathbf{p}N_k\mathbf{n}, \\ &\qquad B_k = \tfrac{1}{2} - \mathbf{b}N_k\mathbf{n}, \quad C_k = -\tfrac{1}{2} + \mathbf{b}N_k\mathbf{p}; \end{aligned} \tag{4.42a}$$

$$\begin{aligned} &\text{and in compression: } A_k = c_2B_k + \mathbf{q}N_k\mathbf{b} - \mathbf{n}N_k\mathbf{n}, \\ &\qquad B_k = \tfrac{1}{2} - \mathbf{b}N_k\mathbf{n}, \quad C_k = -\tfrac{1}{2} + \mathbf{q}N_k\mathbf{n}. \end{aligned} \tag{4.42b}$$

4.5.1 *F.C.C. Crystals*

The simple theory was shown to account for the phenomenon of overshooting in f.c.c crystals in tension by Havner & Shalaby (1977, 1978) using active-hardening equations $\tau = \tau_0 + H_0\gamma$ and $\tau = A_0\gamma^{1/2}$, respectively, with τ_0, H_0, and A_0 scalar constants. Because the corresponding integrated forms of equation (4.41) (Havner & Shalaby 1977, eqs. (A4), and 1978, eqs. (19)) are too complicated for algebraic assessment of the net strengths $\tau_k^c - \tau_k$, this was done graphically via computer plots demonstrating that $\tau_k^c > \tau_k$ in all latent systems, to shear $\gamma = 1.5$, for each of five initial positions spanning stereographic triangle $a1$ (selected as the active system). In

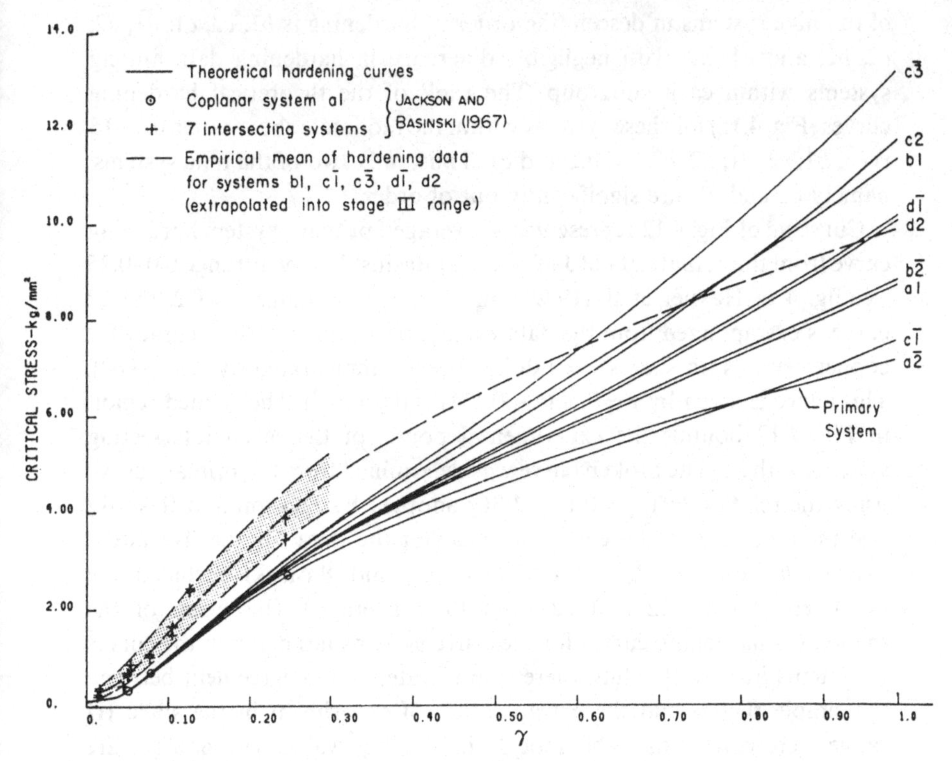

Fig. 4.12. Theoretical and experimental latent hardening in copper crystals (from Havner, Baker & Vause 1979, fig. 5). Reprinted with permission from *J. Mech. Phys. Solids* 27, K. S. Havner, G. S. Baker & R. F. Vause, Theoretical latent hardening in crystals - I, Copyright 1979, Pergamon Press PLC.

addition, the predictions of the simple theory were found to be in qualitative agreement with experimental results of Jackson & Basinski (1967) regarding the diversity of finite distortional latent hardening in copper crystals. Comparisons with these experiments were more fully explored in Havner et al. (1979), and Fig. 4.12 is taken from that work.

Recall from the review in Section 4.1 that the parent crystals of Jackson and Basinski were initially oriented 10° from [011] on the great circle connecting [011] and primary slip direction [$\bar{1}$01] ($a\bar{2}$ the active system). Beginning with six virgin crystals prestrained to shears γ between 5 and 25 percent, Jackson and Basinski obtained hardening data for nine different slip systems (including $a\bar{2}$) from their secondary specimens, several at more than one level of prestrain (see Fig. 4.4 here, table 2 in Havner et al. (1979), or table I in Jackson & Basinski (1967)). As noted in Section 4.1, the rank of the nine systems in descending order of hardening is $b1$, $c\bar{3}$, $c\bar{1}$; $d\bar{1}$, $d2$; $c2$, $b\bar{2}$; and $a1$, $a\bar{2}$, with negligible difference in hardening data among systems within each subgroup. The rank of the theoretical hardening curves (Fig. 4.12) for these systems within the moderate shear range 0.1–0.5 is $c\bar{3}$, $b1$, $c2$, $d\bar{1}$, $d2$, $b\bar{2}$, $a1$, $a\bar{2}$, and $c\bar{1}$. Thus, only two of the nine systems, namely $c2$ and $c\bar{1}$, are significantly out of order.

Curve $a\bar{2}$ of Fig. 4.12 represents the averaged primary-system hardening curve from the actual data of Jackson and Basinski in shear range 0.0–0.25 (see fig. 4 of Havner et al. (1979)). In the stage III range $\gamma > 0.2$, the $a\bar{2}$ curve is extrapolated from the data using a procedure whose accuracy for copper crystals (to shears $\gamma > 1.0$) has been demonstrated by Bell (1964). This curve is given by $\tau = 7.56(\gamma - 0.111)^{1/2}$ (kg/mm^2). The shaded region in Fig. 4.12 bounds the experimental points of the seven intersecting systems with $a\bar{2}$. The broken curve was determined from the primary curve using the relation $(\tau_k^c)_{avg} = 0.17 + 1.36\tau$ adopted by Jackson and Basinski (see their fig. 16) as the empirical mean of the data for the five latent systems $b1$, $c\bar{1}$, $c\bar{3}$, $d\bar{1}$, and $d2$.[8] (Jackson and Basinski excluded the unrelated systems $b\bar{2}$ and $c2$ from their average.) The mean of the theoretical hardening curves for these five systems lies between the curves of systems $b1$ and $d\bar{1}$. Thus, there is moderately good agreement between the simple theory and the extrapolation of experiment in the stage III range. However, the theoretical hardening values obviously are substantially less than the experimental ones in the small-to-moderate strain range of the actual data. Apparently the only latent-hardening experiments on f.c.c. crystals prestrained to levels above $\gamma = 0.25$ are the compression tests of Kocks (1964) on aluminum. Comparisons of theoretical predictions with those experiments are included in the following.

The simple theory was shown to be a theory of overshooting in f.c.c crystals in compression by Vause & Havner (1979). They analyzed seven positions spanning the stereographic triangle of slip system $a2$ (which in compression is the triangle labeled $a\bar{2}$ in Fig. 4.2, as previously remarked). They investigated both linear and parabolic active-system hardening, demonstrating that $\tau_k^c > \tau_k$ in all latent systems well beyond the symmetry line with system $c2$ (triangle $c\bar{2}$ in Fig. 4.2) across which the loading axis rotates in compression. Vause and Havner also made comparisons with

[8] As pointed out by Havner et al. (1979, p. 47), this empirical relation is invalid near the limit $\tau \to \tau_0 \approx 0.1$ kg/mm^2 since it otherwise would define a highly anisotropic virgin crystal with an initial mean critical strength 0.3 kg/mm^2 in the five systems indicated.

the latent-hardening results for aluminum and silver crystals of Kocks (1964) and Ramaswami et al. (1965). It may be recalled from the brief review in Section 4.1 that the initial axis orientation in those experiments was at 45° with both the active slip-plane normal [111] and direction $[\bar{1}01]$. This is the position of maximum Schmid factor $|\cos\varphi_0 \cos\chi_0| = 0.5$ in compression. Kocks investigated latent hardening of systems $b\bar{3}$ and $d\bar{3}$ (triangles $b3$, $d3$ of Fig. 4.2) in both aluminum and silver; and Ramaswami et al. (1965) added latent hardening data for system $d\bar{1}$ (triangle $d1$) in silver. Comparisons of predictions of the simple theory with these data may be found in Vause and Havner's figures 13–17. Here, however, we turn to comparisons for aluminum presented in figures 19 and 20 of Khedro & Havner (1991) (included here as Figs. 4.13 and 4.14), which illustrate the importance of choice of strain measure in single-parameter hardening rules.

Khedro and Havner investigated latent-hardening predictions within a family of measure-dependent, one-parameter hardening rules that includes the simple theory as a particular case (for the logarithmic measure). This family may be expressed as

$$\dot{\tau}_k^c = h_{(m)} \sum \dot{\gamma}_j - 2\,\mathrm{tr}(N_k \sigma(\Omega - mD^p)) \tag{4.43}$$

(with m given by eq. (A59) as before), which clearly is encompassed by the Hill–Havner equation (4.12) and gives the simple theory of equation (4.16) for $m = 0$ and $h_{(m)} = h$. For single slip we have (Khedro & Havner 1991, eq. (10))

$$\dot{\tau}_k^c = \{h_{(m)} - 2\,\mathrm{tr}(N_k \sigma(\Omega_1 - mN_1))\}\dot{\gamma}, \tag{4.44}$$

with subscript 1 denoting the active system as before and σ given by equation (4.38). Khedro and Havner evaluated equation (4.44) for each of the simple theory ($m = 0$), the Green measure ($m = 1$), and the Almansi measure ($m = -1$) for five initial positions spanning stereographic triangle $a\bar{2}$ in both tension and compression. (The evaluations for the simple theory serve to confirm the previous calculations of Havner & Shalaby (1977) and Vause & Havner (1979).)

In Figs 4.13 and 4.14 are shown Khedro and Havner's plots of the theoretical τ_k^c (according to each measure) in latent systems $b\bar{3}$ and $d\bar{3}$ versus τ of active system $a2$ for aluminum crystals in compression, in the maximum Schmid-factor position, together with Kock's (1964) data. The adopted active-system hardening curve is the parabola $\tau = 2\gamma^{1/2}$, in units of kg/mm^2 (in which the data were expressed), which is a good approximation to the resolved shear stress versus slip data reported in

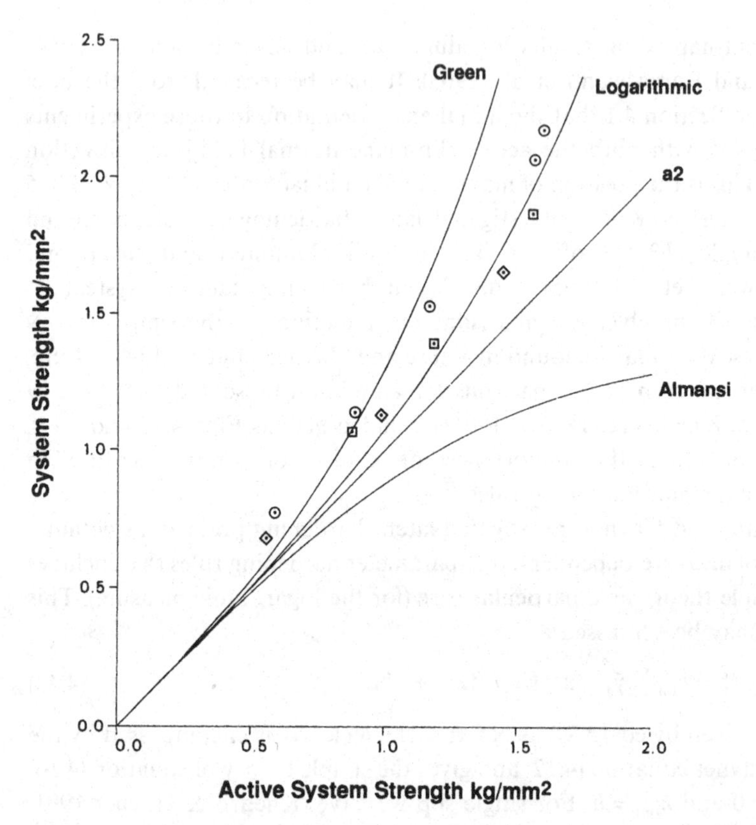

Fig. 4.13. Various theoretical hardening curves and experimental results of Kocks (1964) for latent system $b\bar{3}$ for aluminum crystals in compression (from Khedro & Havner 1991, fig. 19). Reprinted with permission from *Int. J. Plasticity* 7, T. Khedro & K. S. Havner, Investigation of a one-parameter family of hardening rules in single slip in f.c.c. crystals, Copyright 1991, Pergamon Press, PLC.

Kocks et al. (1964) and Kocks (1964), as may be seen from figure 12 of Vause & Havner (1979). (The original comparisons for the simple theory are given in their figs. 13 and 14.) The correlations of both the simple theory and the Green-measure hardening rule with experimental data are seen to be good for system $b\bar{3}$ over the entire range (6–70 percent) of finite prestrains. For system $d\bar{3}$, the quantitative agreement is not as good for either theory, but the results still may be judged satisfactory.[9] The

[9] The simple theory also predicts the "unlimited" overshooting in aluminum from this initial position remarked by Kocks, as may be seen from the net-strength curves in figure 10 of Vause & Havner (1979). The Green-measure rule also predicts unlimited overshooting, as shown by Khedro & Havner (1991).

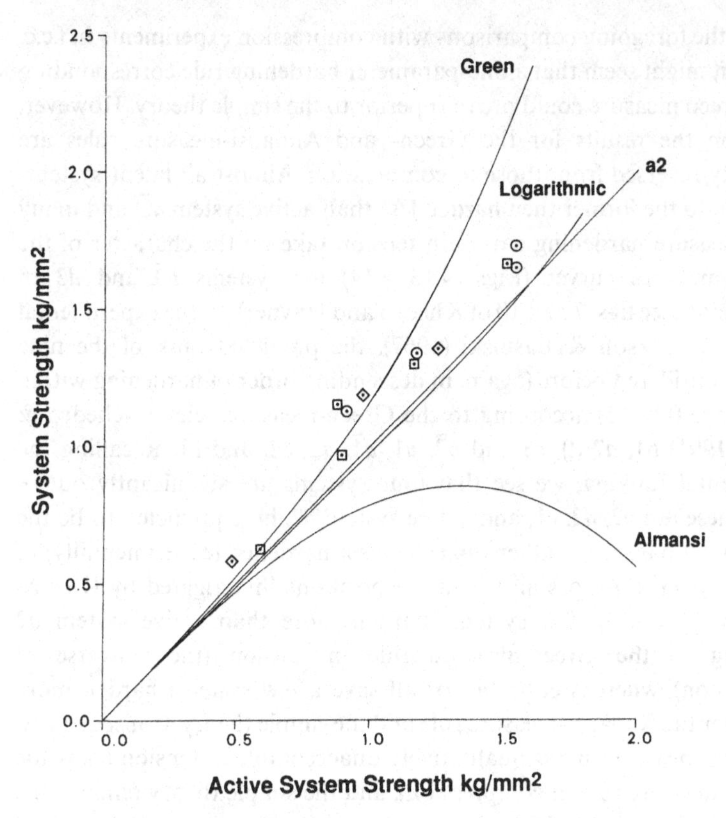

Fig. 4.14. Various theoretical hardening curves and experimental results of Kocks (1964) for latent system $d\bar{3}$ in aluminum crystals in compression (from Khedro & Havner 1991, fig. 20). Reprinted with permission from *Int. J. Plasticity* 7, T. Khedro & K. S. Havner, Investigation of a one-parameter family of hardening rules in single slip in f.c.c. crystals, Copyright 1991, Pergamon Press, PLC.

predictions of the Almansi-measure rule, however, clearly are unsatisfactory in both cases. (The symbols for the various data points are identical to those used in Kocks (1964, fig. 5).)

Comparisons of the three hardening rules with the latent-hardening experiments of Kocks (1964) and Ramaswami et al. (1965) on silver crystals are given in Khedro (1989) (also see Vause & Havner (1979, figs. 15–17) for the simple theory alone). In the much smaller prestrain range (5–17 percent) of those experiments, all theories underpredict the amount of anisotropic hardening in the experimentally investigated systems ($b\bar{3}$, $d\bar{1}$, and $d\bar{3}$), with the Almansi-measure rule farthest from the data and the Green-measure rule a little closer than the simple theory.

From the foregoing comparisons with compression experiments in f.c.c. crystals, it might seem that a one-parameter hardening rule corresponding to the Green measure could prove superior to the simple theory. However, in tension the results for the Green- and Almansi-measure rules are essentially reversed from those in compression. Almost all latent systems according to the former then harden less than active system $a\bar{2}$; and many Green-measure hardening curves in tension take on the character of the Almansi-measure curves (Figs. 4.13–4.14) for systems $b\bar{3}$ and $d\bar{3}$ in compression (see figs. 7 and 10 of Khedro and Havner). In the experimental position of Jackson & Basinski (1967), the predicted rank of the nine systems considered before (again, in descending order of hardening within shear range 0.1–0.5) according to the Green-measure rule is (Khedro & Havner 1991) $b1$, $a\bar{2}$ (!), $c\bar{3}$ and $b\bar{2}$, $a1$, $d\bar{1}$, $d2$, $c2$, and $c\bar{1}$. Recalling the experimental ranking, we see that four systems are significantly out of place. These are $b\bar{2}$, $d2$, $c\bar{1}$, and active system $a\bar{2}$, here predicted to be the second most hardened rather than the least hardened (experimentally) of the nine systems. Across all the initial positions investigated by Khedro and Havner, only a few systems harden more than active system $a\bar{2}$ according to the Green-measure rule in tension (the converse of compression), whereas either all or all save a few systems harden more than $a\bar{2}$ for the Almansi-measure rule and the simple theory. Consequently, the Green-measure rule is qualitatively unacceptable in tension (as is the Almansi-measure rule in compression); and the simple theory remains the best one-parameter hardening rule in the family of equation (4.43), based upon single-slip predictions.[10]

4.5.2 *B.C.C. Crystals*

In b.c.c. metals, as will be recalled from the literature review in Section 4.1, there are historical ambiguities regarding experimental identification of the slip plane. For that reason, in Havner & Baker (1979) parallel analyses were carried out based upon the separate hypotheses of (i) slip on {110}, {112}, and {123} crystallographic planes only and (ii) initial slip on the arbitrary plane of maximum resolved shear stress

[10] Additionally, it may be noted from eq. (4.43) that of this one-parameter family, only the simple theory ($m = 0$) can predict isotropic hardening in high-symmetry multiple slip wherein the relative (plastic) spin Ω is zero. The "special prominence" of the logarithmic measure in this regard also was remarked upon by Hill & Havner (1982, p. 21).

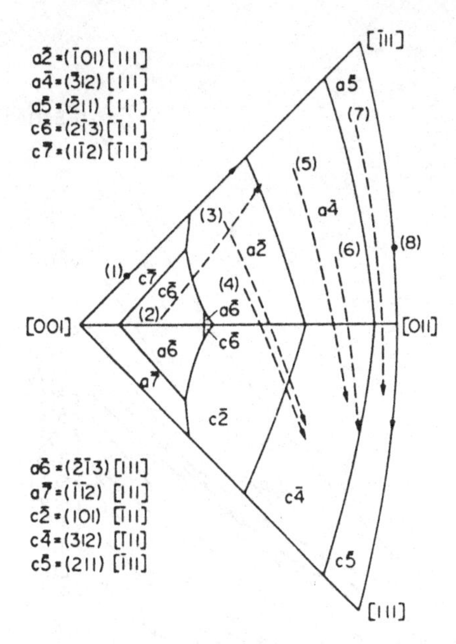

Fig. 4.15. Initial tensile-axis positions and theoretical rotations (to $\gamma = 1.5$) for slip on $\{110\}$, $\{112\}$, and $\{123\}$ planes in b.c.c. crystals (from Havner & Baker 1979, fig. 4). Reprinted with permission from *J. Mech. Phys. Solids* 27, K. S. Havner & G. S. Baker, Theoretical latent hardening in crystals-II, Copyright 1979, Pergamon Press, PLC.

containing a $\langle 111 \rangle$ slip direction. Havner and Baker established that the simple theory accounts for the overshooting phenomenon (and greater latent than active hardening) in b.c.c. metals for both tensile and compressive axial loading according to either specification of the slip plane.

In Fig. 4.15 (taken from Havner and Baker's fig. 4) are shown the theoretical tensile-axis rotations (to $\gamma = 1.5$) from eight different initial positions distributed over stereographic triangle $a\bar{2}$, corresponding to slip on $\{110\}$, $\{112\}$, and $\{123\}$ planes. These rotations may be compared with similar ones in iron crystals from experiments of Fahrenhorst & Schmid (1932, fig. 1) and Keh (1965, fig. 6), the latter rotations also shown in Fig. 4.9 here. For each of the indicated active systems in single slip in Fig. 4.15, using the parabolic curve $\tau = A_0\gamma^{1/2}$, Havner and Baker found that $\tau_k^c > \tau_k$ (according to the simple theory) in all 95 latent systems throughout the deformation. As an example of their results, Fig. 4.16 (from their fig. 7) displays the net-strength curves for those latent systems whose resolved shear stresses successively become the largest, among all 96 systems, corresponding to each of the eight initial positions as the specimen axis

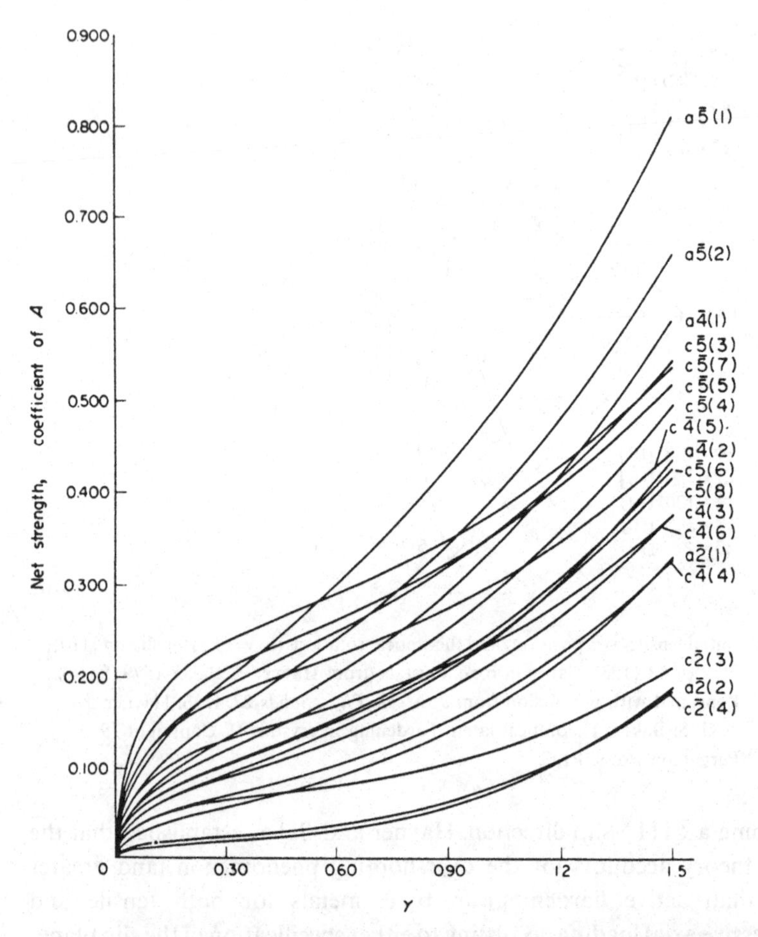

Fig. 4.16. Theoretical net-strength curves (critical strength minus resolved shear stress) of most-highly stressed latent systems in b.c.c. crystals for parabolic active-hardening and the initial axis positions of Figure 4.15 (from Havner & Baker 1979, fig. 7). Reprinted with permission from *J. Mech. Phys. Solids* 27, K. S. Havner & G. S. Baker, Theoretical latent hardening in crystals - II, Copyright 1979, Pergamon Press, PLC.

rotates toward the respective slip direction. For instance, as is evident from Fig. 4.15, for initial positions 3 and 4 in slip-system region $a\bar{2}$ (that is, active system $(\bar{1}01)[111]$) these latent systems are, in sequence, $(101)[\bar{1}11]$, $(312)[\bar{1}11]$, and $(211)[\bar{1}11]$ of triangle $c\bar{2}$. In contrast, for initial positions 1 and 2, $a\bar{2}$ is latent but subsequently becomes the most highly stressed system. (Recall from Fig. 4.9 that position 1 is not a double-slip position, but rather lies on the line that bisects initial-slip region

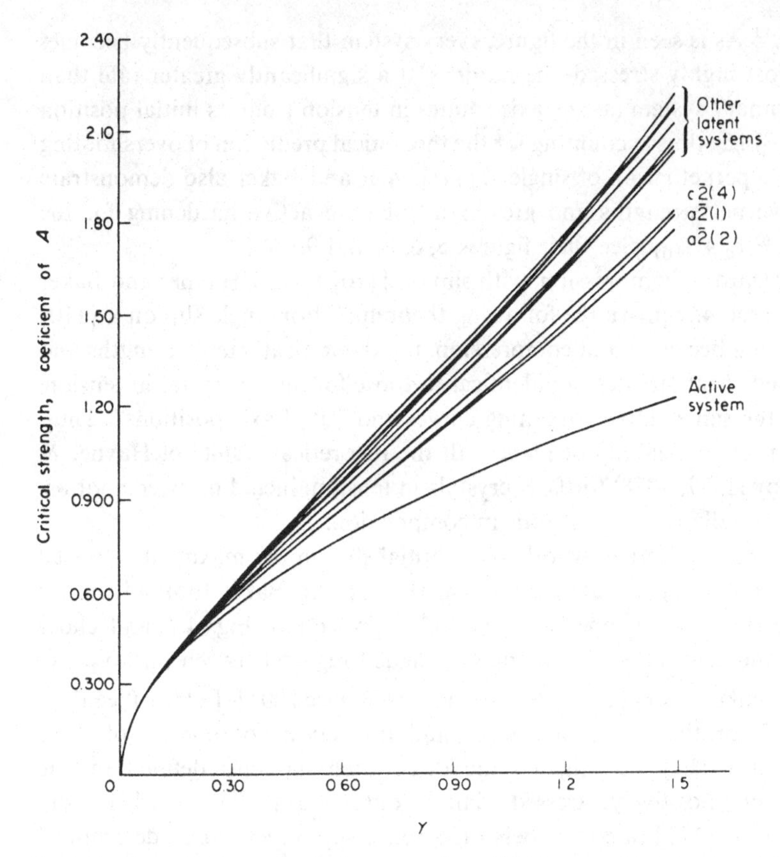

Fig. 4.17. Theoretical latent-hardening curves of most-highly stressed latent systems in b.c.c. crystals for parabolic active-hardening and initial axis positions 1, 2, 4, 6, 8 of Figure 4.15 (from Havner & Baker 1979, fig. 10). Reprinted with permission from *J. Mech. Phys. Solids* 27, K. S. Havner & G. S. Baker, Theoretical latent hardening in crystals - II, Copyright 1979, Pergamon Press, PLC.

$(1\bar{1}2)[\bar{1}11]$. As single slip proceeds, the axis rotates past the triple point into the segment along symmetry line $[001]$–$[\bar{1}11]$ in which systems $(\bar{1}01)[111]$ ($a\bar{2}$) and $(011)[\bar{1}\bar{1}1]$ ($b1$) are most highly stressed; but single slip continues because of the greater hardening of these systems, both theoretically and (apparently) experimentally. (One again may note the behavior of Keh's (1965) crystal specimen Fe–15 discussed in Sect. 4.3 and shown in Fig. 4.9.))

In Fig. 4.17 (from Havner and Baker's fig. 10) are shown critical-strength curves for both the active system (parabolic hardening) and the preceding maximally stressed latent systems for each of initial positions 1, 2, 4, 6, and 8.[11] As is seen in the figure, every system that subsequently becomes the most highly stressed one hardens at a significantly greater rate than the primary system (as the axis rotates in tension from its initial position in Fig. 4.15), thus accounting for the theoretical prediction of overshooting and the perpetuation of single slip. (Havner and Baker also demonstrate positive net strengths and greater latent than active hardening for the case $\tau = \tau_0 + H_0\gamma$. See their figures 5, 6, 8, and 9.)

In regard to compression, with slip on $\{110\}$ planes, Havner and Baker (1979, sect. 4.2) prove the following theorem: " For single slip on a $\{110\}$ plane of a bcc crystal in compression, the theoretical latent strengths and resolved shear stresses are identical to those for an fcc crystal in tension, given the same active hardening curves and initial axis positions." Thus, for slip on this family of planes, all the theoretical results of Havner & Shalaby (1977, 1978) for f.c.c. crystals in tension, including overshooting, apply equally to b.c.c. crystals in compression.

For the alternative hypothesis of initial slip on the maximally stressed plane containing a $\langle 111 \rangle$ direction, Havner and Baker took advantage of the threefold symmetry of the [111] projection, Fig. 4.7, and chose seven initial positions spanning the shaded region between lines $\nu_2 = \nu_3$ (the x-axis), $\nu_1 = \nu_2$ ($y = \sqrt{3}x$), and $\nu_1 = -\nu_2$ (arc $[001]$–$[\bar{1}11]$). (See their table 3 for the direction cosines and projection coordinates of these positions.) They investigated the three latent systems defined by the maximum positively stressed planes containing the other $\langle 111 \rangle$ slip directions ([111] of course being the active slip direction), as determined at every stage of the deformation by the current position $\boldsymbol{\iota}$ of the loading axis. For any $\mathbf{b}_k$, the corresponding continuously changing $\mathbf{n}_k$ is found from

$$\mathbf{b}_k \cdot \mathbf{n}_k = 0, \quad (\boldsymbol{\iota} \wedge \mathbf{b}_k) \cdot \mathbf{n}_k = 0, \quad |\mathbf{n}_k| = 1.$$

Then, from equations (4.35), the loading-axis position after slip γ is given by (Havner & Baker 1979, eq. (5.17))

$$\boldsymbol{\iota} = \begin{cases} [(c_2 + \gamma)\mathbf{b} + \mathbf{n}]/\{1 + (c_2 + \gamma)^2\}^{1/2}, & \text{tension;} \\ [-\mathbf{b} + (c_2 + \gamma)\mathbf{n}]/\{1 + (c_2 + \gamma)^2\}^{1/2}, & \text{compression} \end{cases} \tag{4.45}$$

(with $\mathbf{b}, \mathbf{n}$ defining the primary system, as before). Havner and Baker found that the net strengths of all such latent systems were positive (for each of linear and parabolic active hardening), in both tension and compression, for every initial position throughout the deformation (see their figs. 12–17).

[11] These positions correspond to the maximum possible resolved shear stress on the respective system (that is, a Schmid factor of 0.5).

Thus, as stated at the beginning of this section, they established that the simple theory accounts for overshooting and the perpetuation of single slip in b.c.c. crystals according to either hypothesis about the slip planes.

In regard to latent hardening, as with f.c.c. crystals in tension, the simple theory underpredicts the degree of anisotropy that apparently is developed at small to moderate strains. (Compare the curves in Fig. 4.17 with the high experimental latent-hardening values of Nakada & Keh (1966) at $\gamma = 0.1$ in Fig. 4.10.) It also may be that the theoretical latent-hardening results at shears of order one are unrealistically high. However, data are not available for b.c.c. crystals at such levels of prestrain. (Note that theoretical curves such as those in Fig. 4.17 may not logically be compared with the curves labeled 1 through 4 in Fig. 4.10. Only the initial values of the latter, at $\gamma = 0.1$, are latent-hardening results. The curves themselves represent subsequent *active* hardening of previously latent secondary systems in Nakada and Keh's crystal specimens.)

5

ANALYSIS OF CRYSTALS IN CHANNEL DIE COMPRESSION

Crystal grains within polycrystalline metals typically deform finitely by multiple slip, necessarily compatibly (on a continuum scale) with their adjoining neighbors but also somewhat nonuniformly, even when the macroscopic metal deformation is uniform.[1] The experimental configuration that best represents such constrained deformation of crystals (under multi-axial stress states) while retaining a potential for clarity of interpretation associated with uniform distortion is the channel die compression test. This test was introduced for single crystals by Wever & Schmid (1930), and a similar experimental configuration was used by Kocks & Chandra (1982). However, Wever and Schmid's testing arrangement (see their fig. 1) apparently was not well suited to the achievement of uniform distortions (as evident from the inhomogeneously deformed aluminum and iron crystals shown in their figs. 2a, b).

A superior compression die configuration was designed and adopted by Chin, Nesbitt, & Williams (1966) (see their fig. 3) using Teflon lubricant and carefully machined, electropolished crystals. The remarkably uniform deformation obtained in their test of a Permalloy crystal (4% Mo, 17% Fe, 79% Ni), compressed along (110) to a thickness reduction of 50.5 percent (logarithmic compressive strain $e_L = 0.703$) and constrained along $(\bar{1}1\bar{1})$, may be seen in their figure 9, reproduced here as Fig. 5.1. (Also see fig. 5 of Chin, Thurston, & Nesbitt (1966) showing surface slip traces on the same crystal.) In subsequent experiments on copper and silver crystals (Wonsiewicz & Chin 1970a, b) uniformity of deformation was not always achievable, but a number of examples of essentially uniform distortion of copper crystals are reported by Wonsiewicz & Chin (1970b) and

[1] Under conditions in which macroscopic shear bands form, individual crystal deformations may be highly nonuniform.

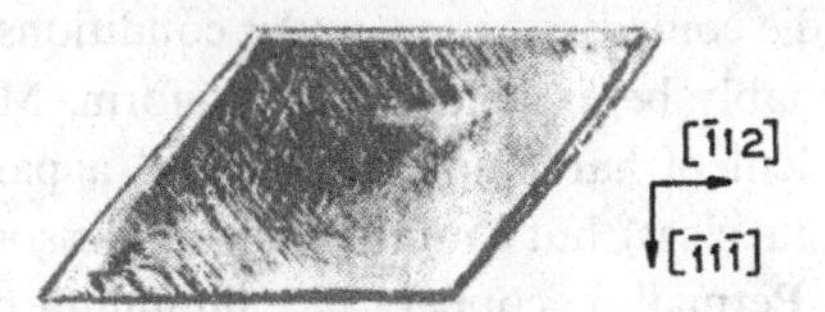

Fig. 5.1. Top view of a Permalloy crystal constrained in direction [$\bar{1}1\bar{1}$] and compressed along (110) (perpendicular to plane) to a thickness reduction of 50.5% (from Chin, Nesbitt & Williams 1966, fig. 9). Reprinted with permission from *Acta Met.* 14, G. Y. Chin, A. Nesbitt & A. J. Williams, Anisotropy of strength in single crystals under plane strain compression, Copyright 1966, Pergamon Press, PLC.

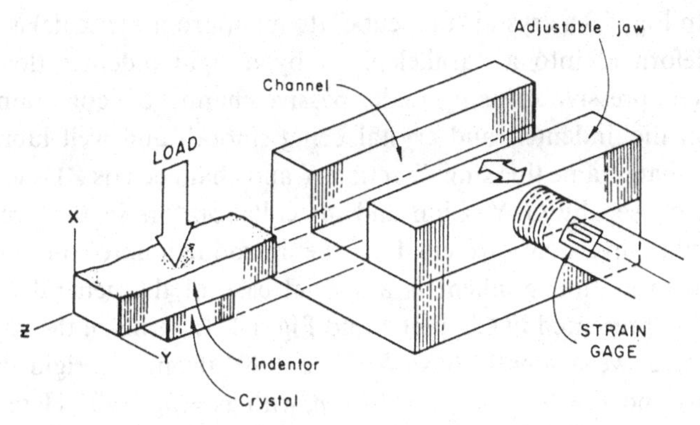

Fig. 5.2. Experimental arrangement of Wonsiewicz, Chin, & Hart (1971, from their fig. 2).

Wonsiewicz, Chin & Hart (1971). In the latter work a further improvement in the channel die testing arrangement was introduced that enabled the experimental determination of the lateral constraint stress. This final experimental configuration is shown in Fig. 5.2, taken from figure 2 of Wonsiewicz et al. (1971).

Numerous examples of uniform distortion of aluminum crystals in channel die compression are found in the experiments of Driver & Skalli (1982), Driver, Skalli, & Wintenberger (1983, 1984), Skalli, Driver, & Wintenberger (1983), and Skalli (1984), who used a testing arrangement similar to that introduced by Chin, Nesbitt, & Williams (1966) (see fig. 16 of Skalli (1984)). The high degree of uniformity of deformation obtained by Driver and his colleagues may be seen in figure 5 of Driver & Skalli (1982), which shows an aluminum crystal compressed to $e_L \approx 1.0$ in the same orientation as that of the Permalloy crystal in Fig. 5.1, and in figure 1b of Driver et al. (1983) and figure 18 of Skalli (1984) for aluminum crystals in other orientations.

The object of this chapter is the presentation of a rational analysis of single crystals in the channel die compression test under conditions where their deformation may reasonably be expected to be uniform. Much of the analysis will be independent of hardening theory (and a part of it independent of particular crystal class), but a number of comparisons with experimental studies of f.c.c. Permalloy, copper, and aluminum crystals will be based upon the hardening rules presented in Chapter 4.

5.1 Deformation Gradient and Stress State

In Fig. 5.3 a crystal "unit cube" (to an appropriate scale) is shown finitely deformed into a parallelepiped by a rigid indenter (load and nominal compressive stress p_x) under passive channel die constraints, the surfaces of die, indenter, and crystal being smooth and well lubricated. The coordinate frame (loading direction X and channel axis Z) is identical with that chosen by G. Y. Chin and his collaborators in their series of experimental investigations cited in the introduction to this chapter; and the deformation gradient is a special case of the general uniform deformation presented in Chapter 1 and Fig. 1.1. In terms of the variables defined there, we evidently have $b=1$, $\delta=90°$ (from the rigid channel constraint), and $\chi_x = 90° - \alpha$, $\chi_y = 90° - \beta$, with $\alpha_0 = \beta_0 = 90°$. Here χ_x and χ_y are the angles of shear in the yz and xz planes, respectively (Fig. 5.3). The compressive spacing stretch in the loading direction X is denoted by $\lambda(<1)$, with $\lambda_{(y)} = 1$ the fixed spacing stretch in the direction Y (unless the crystal contracts laterally during elastic response) and $\lambda_x, \lambda_y, \lambda_z$ the respective fibre stretches. The most convenient variables in which to express the overall deformation are λ, λ_z, χ_x, and χ_y, from which $\lambda_x = \lambda \sec \chi_y$ and $\lambda_y = \sec \chi_x$.

For finite-distortion analysis based upon crystallographic slip, the infinitesimal lattice-strain contribution, hence the volumetric strain, may be neglected in comparison with the large plastic deformations that are our focus; consequently $|A| = 1$ and $\lambda_z = \lambda^{-1}$. The deformation gradient and its inverse then may be expressed (Sue & Havner 1984, eq. (2.1))

$$A = \begin{bmatrix} \lambda & 0 & 0 \\ 0 & 1 & 0 \\ \lambda \tan \chi_y & \tan \chi_x & \lambda^{-1} \end{bmatrix}, \qquad A^{-1} = \begin{bmatrix} \lambda^{-1} & 0 & 0 \\ 0 & 1 & 0 \\ -\lambda \tan \chi_y & -\lambda \tan \chi_x & \lambda \end{bmatrix}, \tag{5.1}$$

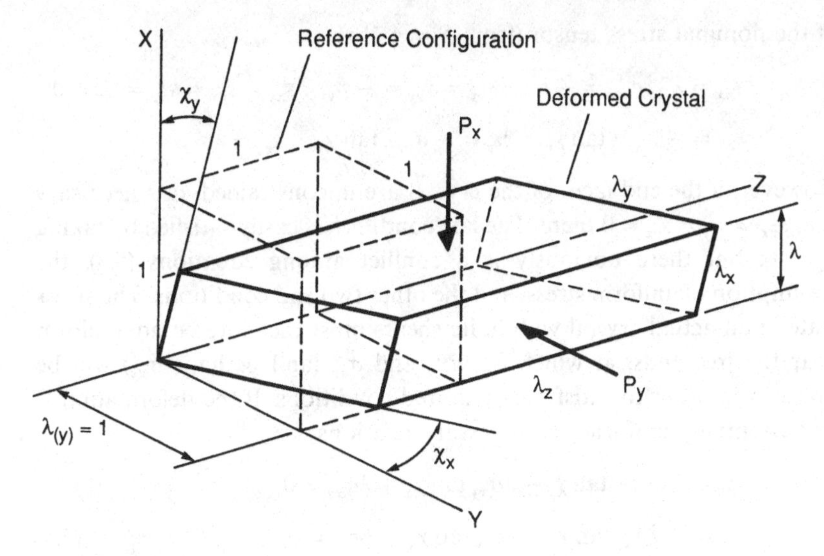

Fig. 5.3. Reference (undeformed) and deformed crystal configurations in a channel die test.

and the components of the velocity gradient $\Gamma = \dot{A}A^{-1}$ are (Sue & Havner 1984, eq. (2.2))

$$\begin{gathered}\Gamma_{xx} = -\Gamma_{zz} = \dot{\lambda}/\lambda < 0, \quad \Gamma_{xy} = \Gamma_{xz} = \Gamma_{yx} = \Gamma_{yy} = \Gamma_{yz} = 0, \\ \Gamma_{zx} = 2(\dot{\lambda}/\lambda)\tan\chi_y + (\tan\chi_y)^{\cdot}, \quad \Gamma_{zy} = (\dot{\lambda}/\lambda)\tan\chi_x + (\tan\chi_x)^{\cdot},\end{gathered} \tag{5.2}$$

from which $d_{xy} = W_{xy} = W_{yx} = d_{yy} = 0$. The additional unknown kinematic variables are the components of lattice spin with respect to the channel frame.

Because Teflon (or other) lubrication and electropolishing of crystal specimens are used to minimize friction in channel die tests, we shall consider it reasonable to adopt in analysis the following ideal frictionless boundary conditions:

$$\sigma_{xy} = \sigma_{xz} = 0 \text{ on the } X\text{-faces}, \quad \sigma_{xy} = \sigma_{yz} = 0 \text{ on the } Y\text{-faces}. \tag{5.3}$$

It also should be noted that if the stress state is theoretically uniform (corresponding to uniform deformation) these conditions *must* hold. For example, a nonzero friction stress σ_{xy} would be positive on face $y = 1$ but negative on face $y = 0$ since both traction components would oppose the downward motion of the crystal; hence σ_{xy} could not be uniform if nonzero. (Similar arguments can be made for σ_{xz} and σ_{yz}.)

Assuming $\sigma_{xy} = \sigma_{xz} = \sigma_{yz} = 0$ everywhere, we obtain for the components of the nominal stress tensor (from $S = A^{-1}\sigma$)

$$s_{xx} = \lambda^{-1}\sigma_{xx} = -p_x, \quad s_{yy} = \sigma_{yy} = -p_y, \quad s_{xy} = s_{yx} = s_{xz} = s_{yz} = 0,$$

$$s_{zx} = -\sigma_{xx}\lambda \tan\chi_y, \quad s_{zy} = -\sigma_{yy}\lambda \tan\chi_x, \quad s_{zz} = \lambda\sigma_{zz}.$$

However, if the end faces of the crystal are unconstrained, it is necessary that $s_{zx} = s_{zy} = s_{zz} = 0$ there. The last condition is easily satisfied by taking $\sigma_{zz} = 0$; but there obviously is a conflict among equations (5.3), the assumption of uniform stress, and the other two end conditions. The stress state in an actual crystal with finite shears must therefore be nonuniform near the free ends, at which σ_{xz}, σ_{yz}, and σ_{zz} (and perhaps σ_{xy}) will be nonzero in order to satisfy the required conditions. If the deformation is still essentially uniform, the necessary relations are

$$\begin{aligned} s_{zx} &= -\lambda\sigma_{xx}\tan\chi_y - \lambda\sigma_{xy}\tan\chi_x + \lambda\sigma_{xz} = 0, \\ s_{zy} &= -\lambda\sigma_{xy}\tan\chi_y - \lambda\sigma_{yy}\tan\chi_x + \lambda\sigma_{yz} = 0, \\ s_{zz} &= -\lambda\sigma_{xz}\tan\chi_y - \lambda\sigma_{yz}\tan\chi_x + \lambda\sigma_{zz} = 0. \end{aligned} \tag{5.4}$$

As remarked by Sue & Havner (1984, p. 420), the additional Cauchy stresses could be relieved by removing material (by spark-cutting techniques, say) to obtain orthogonal end faces at regular deformation intervals throughout an experiment, thus minimizing the conflict; but this apparently has not been done in any channel die test. Alternatively, end effects could be minimized by preparing a crystal specimen that is significantly longer (direction Z) than its width or thickness. (Fortunately, all crystals in channel die compression necessarily increase in length with deformation.)

Here we shall assume that only $\sigma_{xx} = -\lambda p_x$ and $\sigma_{yy} = -p_y$ are nonzero. Strictly, then, we are dealing mathematically with an infinitely long crystal to which equations (5.4) do not apply. Nevertheless, we shall compare theoretical predictions with experimental results for crystals of finite length. As it turns out, $\chi_y = 0$ in all cases that will be investigated. Thus $s_{zx} = s_{zz} = 0$, and only $s_{zy} = \lambda p_y \tan\chi_x$ is nonzero for theoretically uniform distortion of those crystals.

5.2 Initial Elastic Analysis

Let $\boldsymbol{\iota}$ and $\boldsymbol{\kappa}$ denote unit vectors normal to the X (loading) and Y (constraint) faces of the crystal, with $\boldsymbol{\mu}$ a unit vector in the (channel-axis) direction Z. Based upon the preceding arguments, the uniform stress (away from the free ends) may be expressed

$$\sigma = -f\boldsymbol{\iota}\otimes\boldsymbol{\iota} - g\boldsymbol{\kappa}\otimes\boldsymbol{\kappa}, \tag{5.5}$$

where f and g denote the compressive Cauchy stresses $-\sigma_{xx} = \lambda p_x$ and $-\sigma_{yy} = p_y$. Prior to the initiation of significant slip, the crystal behavior will be essentially elastic. The following elementary analysis of that initial response for cubic crystal lattices is taken from Sue & Havner (1984, sect. 3).

The infinitesimal elastic strain is given by $\varepsilon = \mathscr{M}\sigma$, with

$$\begin{aligned}\mathscr{M}_{ijkl} &= s_{12}\delta_{ij}\delta_{kl} + \tfrac{1}{4}s_{44}(\delta_{ik}\delta_{jl} + \delta_{il}\delta_{jk}) \\ &\quad + (s_{11} - s_{12} - \tfrac{1}{2}s_{44})\sum_m a_{im}a_{jm}a_{km}a_{lm}.\end{aligned} \tag{5.6}$$

Here s_{11}, s_{12}, and s_{44} are the elastic compliances of the cubic lattice (in standard notation) and the a_{im} are direction cosines of the channel axes relative to the lattice axes. Upon substituting equation (5.5), one obtains

$$\begin{aligned}\varepsilon_{ij} &= -s_{12}(f+g)\delta_{ij} + \tfrac{1}{2}s_{44}(f\delta_{ix}\delta_{jx} + g\delta_{iy}\delta_{jy}) \\ &\quad - s_A\sum_m a_{im}a_{jm}(f\iota_m^2 + g\kappa_m^2), \quad s_A = s_{11} - s_{12} - \tfrac{1}{2}s_{44}.\end{aligned} \tag{5.7}$$

Assuming the unloaded crystal fits snugly in the (rigid) channel die, and disregarding the compliance effect of the very thin layer of lubricant, one has $\varepsilon_{xy} = 0$ and $\varepsilon_{yy} = 0$ for uniform elastic deformation. Thus

$$\varepsilon_{xy} = -s_A\sum_m \iota_m\kappa_m(f\iota_m^2 + g\kappa_m^2) = 0, \tag{5.8}$$

$$\varepsilon_{yy} = -s_{12}(f+g) - \tfrac{1}{2}s_{44}g - s_A\sum_m \kappa_m^2(f\iota_m^2 + g\kappa_m^2) = 0. \tag{5.9}$$

For a hypothetical elastically isotropic crystal, $s_A = 0$ and equation (5.8) is satisfied identically. For actual anisotropic crystals, however, equation (5.8) serves as a constraint on lattice orientation if both $\sigma_{xy} = 0$ and $\varepsilon_{xy} = 0$ are to be satisfied (given that f and g satisfy eq. (5.9)). It is readily confirmed that the permitted orientations must belong to one of four families. These correspond to either $\boldsymbol{\iota}$ or $\boldsymbol{\kappa}$ a $\{100\}$ or $\{110\}$ direction, with equation (5.8) a zero identity in each case.

The greatest success experimentally in achieving uniform finite deformation in the channel die test has been within the family of (110) loading. Moreover, the only experimental study, to my knowledge, of the lateral constraint stress (Wonsiewicz et al. 1971) is for that family of orientations. Consequently, all evaluations here of single crystals in channel die compression (based upon the work of Sue & Havner (1984), Havner & Sue (1985), Havner & Chidambarrao (1987), Chidambarrao & Havner (1988a, b), and Fuh & Havner (1989)) will be confined to $\boldsymbol{\iota}_0 = (110)$ (initial orientation). As then $\kappa_2 = -\kappa_1$ for all orthogonal $\boldsymbol{\kappa}_0$, equation (5.8) is identically satisfied and equation (5.9) gives (Sue & Havner 1984, eq. (3.6))

$$(g/f)_{\mathrm{E}} = -\frac{s_{12} + s_{\mathrm{A}}\kappa_1^2}{s_{11} - 2s_{\mathrm{A}}\kappa_1^2(2 - 3\kappa_1^2)} \tag{5.10}$$

if positive (the E signifying elastic range). Otherwise the crystal contracts laterally ($\varepsilon_{yy} < 0$) and $g = 0$. This is indeed the case for copper crystals in (and near) the orientation $\boldsymbol{\kappa}_0 = (\bar{1}10)$, as noted by Sue & Havner (1984).

For a small deviation in orientation from one of the exact families satisfying equation (5.8), an ε_{xy} or ε_{yy} differing slightly from zero (in comparison with ε_{xx}) perhaps could be accommodated by the compliance of the thin layer of lubricant between the crystal specimen and the channel walls. For large deviations from $\{100\}$ and $\{110\}$ of both the loading and constraint directions, however, the stress and deformation in the elastic range necessarily will be nonuniform if the crystal fits snugly within a hardened steel die and does not contract laterally under compression by the load piece.

In the finite-deformation range, if slip in the critical systems can laterally spread an unconstrained crystal under compression, then the preceding elastic considerations may be judged to be no longer of significance in determining the constraint stress of a channel die. Correspondingly, uniform finite deformations of crystals may still be possible in orientations that do not have the symmetry required by equations (5.8) and (5.9) in the elastic range. Examples of apparently uniform single-crystal deformations in nonsymmetric orientations in channel die tests may be found in Driver & Skalli (1982), Skalli et al. (1983), and Driver et al. (1984). These cases will not be considered here, however, as a simple yet rational analysis of such orientations that is consistent with the zero shear traction boundary conditions of equations (5.3) is not available. (Such orientations could be treated as nonuniform problems by a numerical approach such as the finite-element method, but the three-dimensional, incremental, multiple-slip, point-to-point calculations would be enormous.)

5.3 Slip-System Inequalities and Kinematic Relations

We shall consider the channel die test to be deformation controlled (in the direction of loading) and signify differentiation with respect to the logarithmic compressive strain $e_L = -\ln\lambda$ by $(\cdot)'$, whence $d_{xx} = -1$. In a critical system

$$\tau_k^c = \mathrm{tr}(N_k\sigma) = m_k f + r_k g, \quad k = 1, \ldots, n, \tag{5.11}$$

where the scalar variables

$$m_k = -\boldsymbol{\iota} N_k \boldsymbol{\iota}, \quad r_k = -\boldsymbol{\kappa} N_k \boldsymbol{\kappa} \tag{5.12}$$

are called the "active and reactive Schmid factors" (following Sue & Havner 1984). Thus, in every critical system

$$(\tau_k^c)' \geqslant m_k f' + r_k g' + m_k' f + g_k' f, \tag{5.13}$$

the last two terms corresponding to lattice rotation with respect to the channel frame. Equations for m_k' and r_k' may be established as follows.

Consider an elemental material areal vector $d\mathbf{a} = \mathbf{n}\,da$. From material differentiation of Nanson's equation one obtains

$$(da)' = (\mathrm{tr}\,D - \mathbf{n}D\mathbf{n})\,da, \quad \mathbf{n}' = (\mathbf{n}D\mathbf{n})\mathbf{n} - \mathbf{n}\Gamma \tag{5.14}$$

(also recall eq. $(2.1)_2$). For finite deformation caused solely by multiple slip, $\mathrm{tr}\,D = \mathrm{tr}\,\Gamma = 0$ (as in eqs. (5.2)) and

$$\Gamma = \sum (\mathbf{b} \otimes \mathbf{n})_j \gamma_j' + \omega, \tag{5.15}$$

with ω now representing the lattice-spin tensor in the channel frame. Therefore, the *lattice-corotational derivative* $\hat{\mathbf{n}} \equiv \mathbf{n}' - \omega\mathbf{n}$ of a material plane unit normal is (Havner & Chidambarrao 1987, eq. (A6))

$$\hat{\mathbf{n}} = (\ln \lambda_{(n)})'\mathbf{n} - \sum (\mathbf{n}\cdot\mathbf{b}_j)\mathbf{n}_j \gamma_j', \tag{5.16}$$

where $\lambda_{(n)}$ is the material spacing stretch in direction $\mathbf{n}$ (whence $\mathbf{n}D\mathbf{n} = (\ln \lambda_{(n)})'$). Substituting $\boldsymbol{\iota}$ and $\boldsymbol{\kappa}$ in turn and using the channel die constraints $\boldsymbol{\iota} D \boldsymbol{\iota} = -1$ and $\boldsymbol{\kappa} D \boldsymbol{\kappa} = 0$, we have

$$\hat{\boldsymbol{\iota}} = -\boldsymbol{\iota} - \sum (\boldsymbol{\iota}\cdot\mathbf{b}_j)\mathbf{n}_j \gamma_j', \quad \hat{\boldsymbol{\kappa}} = -\sum (\boldsymbol{\kappa}\cdot\mathbf{b}_j)\mathbf{n}_j \gamma_j'. \tag{5.17}$$

From equation (5.12), since the effect of infinitesimal lattice strain on slip-system geometry is disregarded, it follows that

$$m_k' = -2\boldsymbol{\iota} N_k \hat{\boldsymbol{\iota}}, \quad r_k' = -2\boldsymbol{\kappa} N_k \hat{\boldsymbol{\kappa}}. \tag{5.18}$$

Thus, the final equations for the rates of change of the active and reactive

Schmid factors are (Havner & Chidambarrao 1987, eqs. (A9)–(A10))

$$m'_k = \sum_j n_{kj}\gamma'_j - 2m_k, \quad r'_k = \sum_j c_{kj}\gamma'_j, \tag{5.19}$$

with

$$n_{kj} = \mathbf{\iota} N_{kj} \mathbf{\iota}, \quad c_{kj} = \mathbf{\kappa} N_{kj} \mathbf{\kappa}, \quad N_{kj} = 2\,\mathrm{sym}\{N_k(\mathbf{n} \otimes \mathbf{b})_j\}. \tag{5.20}$$

(Equivalent expressions under load control are given in Sue & Havner (1984, eqs. (4.6)–(4.8)).)

Upon substitution of equations (5.19) into inequality (5.13), one obtains a system of n inequalities in the critical systems (with the equality necessarily satisfied in a momentarily active system):

$$(\tau^c_k)' - f\sum_j n_{kj}\gamma'_j - g\sum_j c_{kj}\gamma'_j \geqslant m_k f' + r_k g' - 2m_k f, \quad j, k = 1, \ldots, n. \tag{5.21}$$

Following Havner & Chidambarrao (1987), we shall call these inequalities the *consistency conditions* in critical systems. To them must be adjoined the general hardening law

$$(\tau^c_k)' = \sum_j H_{kj}\gamma'_j, \quad j = 1, \ldots, n, \quad k = 1, \ldots, N, \tag{5.22}$$

and the kinematic constraints of the channel die test

$$d_{xx} = \mathbf{\iota} D \mathbf{\iota} = -1, \quad d_{yy} = \mathbf{\kappa} D \mathbf{\kappa} = 0, \quad d_{xy} = \mathbf{\iota} D \mathbf{\kappa} = 0, \tag{5.23}$$

which can be expressed (from $D = \sum N_j\gamma'_j$)

$$\sum m_j\gamma'_j = 1, \quad -\sum r_j\gamma'_j = 0, \quad -\sum s_j\gamma'_j = 0 \tag{5.24}$$

with

$$m_j = -\mathbf{\iota} N_j \mathbf{\iota}, \quad r_j = -\mathbf{\kappa} N_j \mathbf{\kappa}, \quad s_j = -\mathbf{\iota} N_j \mathbf{\kappa}. \tag{5.25}$$

For subsequent comparisons of theoretical predictions with experiment we shall require the lattice spin ω and the crystal shears χ_x and χ_y. Evolutionary equations for the latter are directly obtainable from equations (5.2) for the components of the velocity gradient and may be expressed (Havner & Chidambarrao 1987, eq. (5.15))[2]

$$\begin{aligned}(\tan\chi_x)' &= 2d_{yz} + \tan\chi_x, \quad d_{yz} = \mathbf{\kappa}\sum(N_j\gamma'_j)\mathbf{\mu}, \\ (\tan\chi_y)' &= 2d_{xz} + 2\tan\chi_y, \quad d_{xz} = \mathbf{\iota}\sum(N_j\gamma'_j)\mathbf{\mu}.\end{aligned} \tag{5.26}$$

To obtain equations for the components of lattice spin, we follow Havner & Chidambarrao (1987, appendix) and let $\hat{\mathbf{\iota}}_c, \hat{\mathbf{\kappa}}_c$ denote derivatives of $\mathbf{\iota}, \mathbf{\kappa}$

[2] There are numerous instances in the literature where finite angles of shear are incorrectly calculated from $(\tan\chi_x)' = 2d_{yz}$ and $(\tan\chi_y)' = 2d_{xz}$.

in a lattice-corotational frame momentarily coincident with the channel frame. Then

$$\hat{\boldsymbol{\iota}}_c = -\omega_c \boldsymbol{\iota}_c, \quad \hat{\boldsymbol{\kappa}}_c = -\omega_c \boldsymbol{\kappa}_c, \tag{5.27}$$

with ω_c the matrix array on the channel axes of the lattice-spin tensor (which, of course, is the opposite of the channel spin relative to the momentarily coincident lattice frame), $\boldsymbol{\iota}_c = (1, 0, 0)$, and $\boldsymbol{\kappa}_c = (0, 1, 0)$. Also let ω_x, ω_y, and ω_z denote the components of the axial vector of spin tensor ω_c. We evidently have (from either eqs. (5.27) or a simple sketch of the relatively rotating frames)

$$\omega_x = -\boldsymbol{\mu}\cdot\hat{\boldsymbol{\kappa}}, \quad \omega_y = \boldsymbol{\mu}\cdot\hat{\boldsymbol{\iota}}, \quad \omega_z = \boldsymbol{\iota}\cdot\hat{\boldsymbol{\kappa}} = -\boldsymbol{\kappa}\cdot\hat{\boldsymbol{\iota}}, \tag{5.28}$$

wherein all vectors are expressed in components on the [100][010][001] lattice axes. Thus, upon substitution of equations (5.17) for $\hat{\boldsymbol{\iota}}$ and $\hat{\boldsymbol{\kappa}}$, we obtain the desired results (Havner & Chidambarrao 1987, eq. (A15))

$$\begin{aligned} \omega_x &= \sum (\boldsymbol{\kappa}\cdot\mathbf{b}_j)(\boldsymbol{\mu}\cdot\mathbf{n}_j)\gamma'_j, \quad \omega_y = -\sum (\boldsymbol{\iota}\cdot\mathbf{b}_j)(\boldsymbol{\mu}\cdot\mathbf{n}_j)\gamma'_j, \\ \omega_z &= \sum (\boldsymbol{\iota}\cdot\mathbf{b}_j)(\boldsymbol{\kappa}\cdot\mathbf{n}_j)\gamma'_j = -\sum (\boldsymbol{\kappa}\cdot\mathbf{b}_j)(\boldsymbol{\iota}\cdot\mathbf{n}_j)\gamma'_j. \end{aligned} \tag{5.29}$$

(The last equality merely expresses the continued orthogonality of $\boldsymbol{\iota}$ and $\boldsymbol{\kappa}$, imposed as $\boldsymbol{\iota}D\boldsymbol{\kappa} = 0$ or eq. $(5.24)_3$.)

We now restate the general problem to be solved at each increment of deformation in terms of the net strengths f_k, which are momentarily zero in the n critical systems (presumed already determined). From equations (5.21), (5.22), (5.24), and (3.76) we may write

$$\begin{aligned} &f'_k \geqslant 0, \quad \gamma'_k \geqslant 0, \quad f'_k\gamma'_k = 0, \\ &f'_k = \sum_j (H_{kj} - f n_{kj} - g c_{kj})\gamma'_j - m_k f' - r_k g' + 2m_k f, \\ &\qquad\qquad\qquad\qquad\qquad\qquad j, k = 1, \ldots, n, \\ &\sum m_j\gamma'_j = 1, \quad -\sum r_j\gamma'_j = 0, \quad -\sum s_j\gamma'_j = 0. \end{aligned} \tag{5.30}$$

If the moduli H_{kj} are considered to be completely specified, then the $n + 2$ unknowns are the compressive stress rates f' and g' and the n slip rates γ'_j (some of which may be zero as not all critical systems necessarily are active). Alternatively, f' may be taken as obtained from experiment (the usual choice here) and the H_{kj} left expressed in terms of a single unknown hardening parameter (as in equations (4.15) (the simple theory), (4.20) (Taylor's rule), and (4.26) (the P.A.N. rule)) that must be determined from equations (5.30) (together with g' and the γ'_j as before).

Because the m_k, r_k, s_k, n_{kj}, and c_{kj} may be changing continuously with the deformation due to lattice rotation, a general investigation of mathematical properties of equations (5.30) with regard to issues of existence and uniqueness of solution will not be pursued. Rather, in subsequent sections we shall undertake specific analyses of f.c.c. crystals in (110) channel die compression using these equations. However, some generality of approach still will be possible, as we shall see. It turns out that existence of solution is not an issue for this family of crystal orientations (as indeed it must not be if the theoretical framework has any merit), but nonuniqueness is a problem in particular high-symmetry orientations, dependent in part upon hardening rule.

5.4 Initial Critical Systems in (110) Loading: The Elastoplastic Transition

The complete family of f.c.c. crystal orientations having a (110) lattice direction coincident with loading axis X in the channel die test may be defined by

$$\boldsymbol{\iota\kappa\mu} = (110)(1\bar{1}\bar{b})(\bar{1}1\bar{c}), \quad bc = 2, \quad b \geqslant 0. \tag{5.31}$$

The corresponding range for constraint direction $Y = (1\bar{1}\bar{b})$ is the upper right-hand quadrant of the perimeter of the (110) stereographic projection shown in Fig. 5.4 (from fig. 1 of Chidambarrao & Havner 1988b). From the obvious crystallographic symmetry, all possible crystal responses in initial (110) loading are included by restricting Y to this range. The indicated critical systems in the three subranges and at the singular orientations are the same as those displayed in Skalli et al. (1983, fig. 4), with the exception of $Y = (1\bar{1}0)$. The basis for determination of these initial critical systems is explained in the following.

The "resolving" or Schmid factors m_k, r_k, and s_k in all slip systems, as determined for $\boldsymbol{\iota} = (110)$ and $\boldsymbol{\kappa} = (1\bar{1}\bar{b})$ from equations (5.25) and Table 1, are given in Table 2. From Fig. 5.4 and equation (5.31), the three subranges of constraint direction Y are

$$\begin{aligned} &\text{range I -- between } (00\bar{1}) \text{ and } (1\bar{1}\bar{2}), \quad \infty > b > 2; \\ &\text{range II -- between } (1\bar{1}\bar{2}) \text{ and } (1\bar{1}\bar{1}), \quad 2 > b > 1; \\ &\text{range III -- between } (1\bar{1}\bar{1}) \text{ and } (1\bar{1}0), \quad 1 > b > 0; \end{aligned} \tag{5.32}$$

and the singular orientations are

$$\begin{aligned} &Y = (00\bar{1}), \quad b = \infty; \quad Y = (1\bar{1}\bar{2}), \quad b = 2; \\ &Y = (1\bar{1}\bar{1}), \quad b = 1; \quad Y = (1\bar{1}0), \quad b = 0. \end{aligned} \tag{5.33}$$

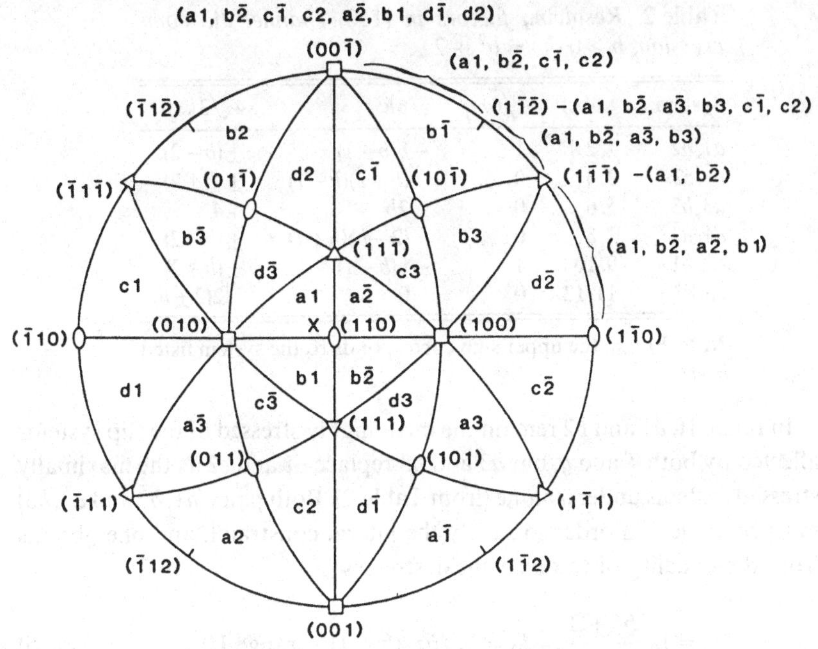

Fig. 5.4. (110) stereographic projection showing f.c.c. crystal slip-system regions in uniaxial compression and identifying initial critical systems in (110) channel die compression (from Chidambarrao & Havner 1988b, fig. 1). Reprinted with permission from *J. Mech. Phys. Solids* 36, D. Chidambarrao & K. S. Havner, On finite deformation of f.c.c. crystals in (110) channel die compression, Copyright 1988, Pergamon Press, PLC.

In range I, it is evident from Table 1 and equation (5.11) that $a1$ and $b\bar{2}$ are the most highly stressed of the slip systems affected by both the f and g stresses, whereas $c\bar{1}$ and $c2$ are the most highly stressed systems affected only by constraint stress g. From the signs of the r_k and s_k factors in these four systems (Table 2), the lateral and shearing constraints in equations (5.24) require that at least three of the systems be active. Consequently, $a1$, $b\bar{2}$, $c\bar{1}$, and $c2$ are the critical systems at the onset of finite plastic deformation in range I, as indicated in Fig. 5.4. Moreover, because the resolved shear stress $m_k f + r_k g$ in each of these critical systems must equal the critical strength τ_0, we have

$$g_0 = \tfrac{1}{2} f_0 \frac{b^2+2}{b^2-1}, \quad f_0 = 2\sqrt{6}\tau_0 \frac{b+1}{b+2} \quad \text{(range I).} \tag{5.34}$$

The subsequent relation between g and f will depend upon hardening theory, as we shall find.

Table 2. *Resolving factors in* (110) *channel die compression;* $b \geqslant 0$, $k^2 = b^2 + 2$

System	Label	$\sqrt{6}m_j$	$\sqrt{6}k^2r_j$	$4\sqrt{3}ks_j$
$a1, b\bar{2}$	1,2	1	$-b(b-1)$	$\pm(b-2)$
$c\bar{1}, c2$	3,4	0	$(b+2)(b-1)$	$\pm(b+2)$
$a\bar{3}, b3$	5,6	0	$2b$	∓ 4
$d1, d\bar{2}$	7,8	0	$(2-b)(b+1)$	$\pm(b-2)$
$a\bar{2}, b1$	9,10	1	$-b(b+1)$	$\pm(b+2)$
$c3, d3$	11,12	0	0	$2(2\pm b)$

Note: For s_j, the upper sign corresponds to the system listed first.

In range II, $a1$ and $b\bar{2}$ remain the most highly stressed of the slip systems affected by both f and g, but $a\bar{3}$ and $b3$ replace $c\bar{1}$ and $c2$ as the maximally stressed systems under g alone (from Table 2). Both pairs ($a1, b\bar{2}$ and $a\bar{3}, b3$) must be critical in order to satisfy the lateral constraint, and one obtains from the equality of resolved shear stresses

$$g_0 = f_0 \frac{b^2+2}{b(b+1)}, \quad f_0 = \tfrac{1}{2}\sqrt{6}\tau_0(b+1) \quad \text{(range II)}. \tag{5.35}$$

In range III, by similar analysis, the critical systems are $a1$, $b\bar{2}$, $a\bar{2}$, and $b1$ (as identified in Fig. 5.4), and there follow

$$f_0 = \sqrt{6}\tau_0, \quad g_0 = 0 \quad \text{(range III)} \tag{5.36}$$

independently of initial orientation.

For the singular orientations we have the following initial values of f_0 and g_0, corresponding to the critical systems indicated in Fig. 5.4 (as dictated by the constraints):

$$\begin{aligned}
&Y = (00\bar{1}), \quad b = \infty: \quad f_0 = \sqrt{6}\tau_0, \quad g_0 = \tfrac{1}{2}\sqrt{6}\tau_0;\\
&Y = (1\bar{1}\bar{2}), \quad b = 2: \quad f_0 = g_0 = \tfrac{3}{4}\sqrt{6}\tau_0;\\
&Y = (1\bar{1}\bar{1}), \quad b = 1: \quad f_0 = \sqrt{6}\tau_0, \quad g_0 < \tfrac{3}{2}f_0;\\
&Y = (1\bar{1}0), \quad b = 0: \quad f_0 = \sqrt{6}\tau_0, \quad g_0 < f_0.
\end{aligned} \tag{5.37}$$

Consider now the relative values of f and g from equations (5.34)–(5.37) in comparison with those from the elastic constraint equation (5.10) evaluated for aluminum and copper, the two metals studied most extensively in channel die compression. Using the following values for the elastic compliances (from Nye 1957, table 10) in units 10^{-2}/GPa,

$$\text{aluminum:}\quad s_{11} = 1.59,\quad s_{12} = -0.58,\quad s_{44} = 3.52;$$
$$\text{copper:}\quad s_{11} = 1.49,\quad s_{12} = -0.63,\quad s_{44} = 1.33,$$

and noting that $\kappa_1^2 = 1/(b^2 + 2)$, we have for the theoretical elastic ratio (eq. (5.10))

$$\text{aluminum:}\quad (g/f)_{\mathrm{E}} = \frac{0.58b^4 + 1.91b^2 + 1.50}{1.59b^4 + 4.72b^2 + 5.54}; \tag{5.38}$$

$$\text{copper:}\quad (g/f)_{\mathrm{E}} = \frac{0.63b^4 + 1.065b^2 - 0.39}{1.49b^4 + 0.14b^2 + 3.05}. \tag{5.39}$$

Observe from the second equation that copper crystals in (110) compression *contract* elastically for $b < 0.556$ (constraint direction Y less than 21.5° from $(1\bar{1}0)$), whence $g_{\mathrm{E}} = 0$ in that part of range III, consistent with equation (5.36). Also, as equations (5.37) give only upper bounds on g_0 for $Y = (1\bar{1}\bar{1})$ and $Y = (1\bar{1}0)$, the elastic values theoretically should govern in those orientations (as first pointed out by Sue & Havner (1984)), from which

$$Y = (1\bar{1}\bar{1})\text{: aluminum, } (g/f)_0 = 0.337\text{; copper, } (g/f)_0 = 0.28;$$
$$Y = (1\bar{1}0)\text{: aluminum, } (g/f)_0 = 0.27\text{ ; copper, } (g/f)_0 = 0.$$

Except in these cases, however, the predictions of the constraint stress from the separate elastic and (initial) finite-plastic-deformation analyses generally differ. For example, throughout range II (Y between $(1\bar{1}\bar{2})$ and $(1\bar{1}\bar{1})$) equation (5.35) requires $g_0 > f_0$ for the initiation of finite multiple slip (independently of specific crystal properties), whereas from equations (5.38) and (5.39) $g_{\mathrm{E}} < f_{\mathrm{E}}$ in aluminum and copper crystals in all (110) compression orientations. We shall reconcile this difference by generalizing the argument of Havner & Sue (1985, sect. 4) as follows.

For an arbitrary orientation in (110) compression, the elastic g/f ratio ordinarily is less (except in range III) than that required to activate the number of slip systems sufficient to satisfy lateral constraint equation $(5.24)_2$. Typically, only systems $a1$ and $b\bar{2}$ become critical initially, and to satisfy that constraint one must include the elastic strain rate ε'_{yy}. Thus, equation $(5.24)_2$ is replaced by (from Table 2)

$$(\varepsilon'_{yy})_0 + (1/\sqrt{6})b(b-1)/(b^2+2)(\gamma'_1 + \gamma'_2)_0 = 0. \tag{5.40}$$

The positive slip rates are now limited in magnitude by the compressive lattice strain rate $-\varepsilon'_{yy}$; and the initial elastoplastic response may be considered infinitesimal. During this interval g/f will increase at a greater rate than its elastic value (because $\Delta\varepsilon_{yy}$ is negative rather than zero) and can attain the ratio required to activate the additional systems after a very small additional deformation.

To make the argument precise, let H_0 designate the average hardening modulus in the active systems during the elastoplastic infinitesimal strain increment. Then, from equations (5.11) and (5.40) and Table 2, as the lattice rotation also is infinitesimal, hence negligible, the additional lattice strain in the constraint direction is

$$\Delta\varepsilon_{yy} = -\frac{1}{6H_0}(b/k^2)(b-1)\{\Delta f - (b/k^2)(b-1)\Delta g\}, \quad k^2 = b^2 + 2, \tag{5.41}$$

when only systems $a\bar{1}$ and $b2$ are active. Also, from equation (5.9) evaluated in (110) loading,

$$\Delta\varepsilon_{yy} = -(s_{12} + s_A/k^2)\Delta f - \{s_{11} - 2s_A(2b^2+1)/k^4\}\Delta g \tag{5.42}$$

(which, of course, gives the ratio of eq. (5.10) for $\Delta\varepsilon_{yy} = 0$). Thus[3]

$$\Delta g = \frac{(b/k^2)(b-1) - 6H_0(s_{12} + s_A/k^2)}{\{(b/k^2)(b-1)\}^2 + 6H_0\{s_{11} - 2s_A(2b^2+1)/k^4\}}\Delta f; \tag{5.43}$$

and Δf is determined by substituting this relation into

$$\Delta g - (g/f)_0 \Delta f = \{(g/f)_0 - (g/f)_E\} f_E, \tag{5.44}$$

with $(g/f)_0$ given by the appropriate one of equations (5.34)–(5.37). As an example, for $Y = (1\bar{1}\bar{2})$ one obtains

$$\Delta f = \frac{\{1 + 27H_0(2s_{11} - s_A)\}(s_{11} + s_{12} - s_A/3)}{\{1 - 27H_0(s_{11} + s_{12} - s_A/3)\}(2s_{11} - s_A)} f_E. \tag{5.45}$$

To estimate the percentage increase above the purely elastic stress f_E, take $H_0 = 0.1$ GPa, a representative order of magnitude. (For aluminum crystals, from fig. 3 of Franciosi et al. (1980), $H_0 \approx 50$ MPa in single slip in the elastoplastic range.) Then, $\Delta f = 0.35 f_E$ for aluminum and $\Delta f = 0.26 f_E$ for copper in this orientation.[4]

[3] This equation also applies to $Y = (00\bar{1})$, $b = \infty$, in which (from Table 2) systems $a\bar{2}$ and $b1$ also are active during the elastoplastic interval. The result is $\Delta g = \Delta f(1 - 6H_0 s_{12})/(1 + 6H_0 s_{11})$, with the elastic ratio $(g/f)_E = -s_{12}/s_{11} (>0)$ from eq. (5.10), and the fully plastic ratio $(g/f)_0 = \frac{1}{2}$ from eq. $(5.37)_1$.

[4] For $Y = (00\bar{1})$, the increases during the elastoplastic transition from four to eight slip systems are (with the same H_0) $\Delta f = 0.27 f_E$ for aluminum and $\Delta f = 0.16$ f_E for copper.

The basic argument also applies in range III. There $\Delta\varepsilon_{yy}$ is positive from equation (5.41) (as $b<1$) during the elastoplastic interval, and g/f decreases (theoretically to zero) from its elastic ratio until systems $a\bar{2}$ and $b1$ become critical and the constraints can be satisfied by finite slipping in at least three of systems $a1$, $b\bar{2}$, $a\bar{2}$, and $b1$. To illustrate, for $Y=(2\bar{2}\bar{1})$, $b=\frac{1}{2}$, we obtain for aluminum (using the same H_0) $\Delta g=-5.43\Delta f$ from equation (5.43). Equation (5.44) then gives (as $g_0=0$ in range III) $\Delta f=0.05\,f_E$. For copper, $g=0$ in the elastic range because the crystal contracts laterally, and an infinitesimal-strain, elastoplastic-transition analysis is not required.

Implicit in the use of $(g/f)_0$ from equations (5.34)–(5.37) in the preceding calculations is the assumption that the hardening of latent systems equals that of active systems $a1$ and $b\bar{2}$ during the small elastoplastic-strain interval between f_E and f_0. Correspondingly, τ_0 is the critical strength at the beginning of *finite* multiple slip in channel die compression, taken as the same in all systems.

To assess the possible significance of any anisotropic hardening during the elastoplastic transition, let us again consider orientation $Y=(1\bar{1}\bar{2})$ and evaluate the increase $\Delta\tau=\tau_0-\tau_E$ in hardening of the active systems for aluminum and copper, using $H_0=0.1$ GPa as before. We have for this orientation $\Delta\tau=(1/\sqrt{6})(\Delta f-\frac{1}{3}\Delta g)$, with $\Delta g=2.82\,\Delta f$ for aluminum and $\Delta g=2.90\,\Delta f$ for copper from equation (5.43). The respective values of τ_E for these metals are $0.358\,f_E$ and $0.339\,f_E$ (from $\tau_E=(1/\sqrt{6})(f_E-\frac{1}{3}g_E)$ and eqs. (5.38) and (5.39) with $b=2$). Thus we obtain $\Delta\tau=0.024\ \tau_E$ for aluminum and $\Delta\tau=0.010\ \tau_E$ for copper (using $\Delta f=0.35\,f_E$ and $0.26\,f_E$, as previously calculated for these metals). Therefore the hardening during the elastoplastic interval is negligible, being only one or two percent or less,[5] and any deviations from equal hardening within this very small increment would be insignificant in magnitude (and probably not measurable) as compared to the critical strengths themselves. Consequently, if Schmid's law of equal critical shear stresses holds at the initiation of slip in the original systems, as has been assumed, it also holds at the onset of finite multiple slipping.

Henceforth all analyses will be in the fully plastic range that begins at the end of the elastoplastic transition interval, at which point the loading and constraint stresses have attained their values f_0 and g_0 given by one

[5] For $Y=(00\bar{1})$ the hardening increases are only 0.25 percent for aluminum and 0.14 percent for copper based upon $H_0=0.1$ GPa. For $Y=(2\bar{2}\bar{1})$, $\Delta\tau=0.012\ \tau_E$ for aluminum whereas $\tau_0=\tau_E$ for copper since $g_E=0$. In addition, $\tau_0\equiv\tau_E$ for $Y=(1\bar{1}\bar{1})$ and $Y=(1\bar{1}0)$ as previously noted.

of equations (5.34)–(5.37), corresponding to critical strength τ_0 in all slip systems. Lattice straining thenceforward will be disregarded in comparison with finite plastic slips, with the subsequent deformation and lattice rotation governed by equations (5.26), (5.29), and (5.30).

5.5 Analysis of Initial Constraint Directions between $(00\bar{1})$ and $(1\bar{1}\bar{2})$ in (110) Loading

The most distinguishing characteristic of the finite deformation of f.c.c. crystals in (110) channel die compression, observed in the experiments of Chin, Nesbitt & Williams (1966), Wonsiewicz & Chin (1970b), Wonsiewicz et al. (1971), Kocks & Chandra (1982), Skalli et al. (1983), and Skalli (1984), is loading-axis stability. The lattice either rotates about the (110) loading direction or is stable relative to the channel axes, depending upon the lattice direction of lateral constraint Y. For each of the three ranges defined in Section 5.4 (eq. (5.32)), which together encompass all orientations save $Y = (00\bar{1})$, $(1\bar{1}\bar{2})$, $(1\bar{1}\bar{1})$, and $(1\bar{1}0)$, Havner & Chidambarrao (1987) and Chidambarrao & Havner (1988b) have shown that loading-axis stability and the appropriate lattice rotation or stability are predicted by a very broad class of hardening theories requiring neither symmetric H_{kj} nor symmetric h_{kj} but permitting both. The singular orientations have been investigated for particular theories by Sue & Havner (1984), Havner & Sue (1985), Fuh & Havner (1986, 1989), and Chidambarrao & Havner (1988a). In the following we illustrate the general analysis for range I, in which the lattice consistently has been observed to rotate about X. We shall find that the constraint stress is dependent upon hardening rule in this range (Havner & Chidambarrao 1987), so experimental knowledge of that stress would serve to distinguish among theories as to their acceptability and usefulness.

5.5.1 *General Solution*

From equations (5.24)–(5.25) and Table 2, the constraints in range I $(\infty > b > 2)$, evaluated for critical systems $a1$, $b\bar{2}$, $c\bar{1}$, and $c2$ (Fig. 5.4), are

$$\begin{gathered}\gamma_1' + \gamma_2' = \sqrt{6}, \quad -b(\gamma_1' + \gamma_2') + (b+2)(\gamma_3' + \gamma_4') = 0, \\ (b-2)(\gamma_1' - \gamma_2') + (b+2)(\gamma_3' - \gamma_4') = 0.\end{gathered} \tag{5.46}$$

Following Havner & Chidambarrao (1987), we take $\gamma_1' - \gamma_2'$ as the primary kinematic unknown, whence

$$\gamma'_1 = \frac{\sqrt{6}}{2}(1+x), \quad \gamma'_2 = \frac{\sqrt{6}}{2}(1-x), \quad x = \frac{1}{\sqrt{6}}(\gamma'_1 - \gamma'_2),$$
$$\gamma'_3 = \frac{\sqrt{6}}{2}\{b-(b-2)x\}/(b+2), \tag{5.47}$$
$$\gamma'_4 = \frac{\sqrt{6}}{2}\{b+(b-2)x\}/(b+2),$$

with $|x| \leqslant 1$. The components of lattice spin relative to the channel axes then may be expressed (from eqs. (5.29) and (5.47) and Table 1)

$$\omega_x = \sqrt{2}(b-1)/(b+2), \quad \omega_y = -(b/k)(b-4)x/(b+2),$$
$$\omega_z = \sqrt{2}(b-1)x/k, \quad k = \sqrt{(b^2+2)}. \tag{5.48}$$

Thus, if $\gamma'_1 = \gamma'_2$ $(x=0)$ the lattice would rotate counterclockwise about the loading axis. Skalli et al. (1983) performed experiments on aluminum crystals in initial orientations $Y = (1\bar{1}\bar{1}6)$, $Z = (\bar{8}8\bar{1})$ and $Y = (1\bar{1}\bar{2}\bar{0})$, $Z = (\bar{1}\bar{0}10\bar{1})$, which lie within the considered range, and found that in each case the lattice rotated counterclockwise about $X = (110)$, and the crystal sheared finitely in the yz plane. Consequently, the loading axis was stable in the (110) lattice direction, with apparently equal slip on systems $a1$ and $b\bar{2}$.

To investigate theoretical predictions of lattice rotation, crystal shearing, and the changing constraint stress in range I, we turn to the consistency conditions, equations (5.21), applicable to every hardening theory. The coefficients n_{kj} and c_{kj} required for their evaluation are readily determined from equations (5.20) and Table 1 and may be found in Havner & Chidambarrao (1987, eqs. (A15)). The resulting inequalities are

$$(\tau^c_1)' - (f/12)(3,5,-1,1)\gamma' - g(b-1)/(6k^2)\{3(b-1), b+1, -(b-1),$$
$$-(b-1)\}\gamma' \geqslant (1/\sqrt{6})f' - (1/\sqrt{6})b(b-1)g'/k^2 - (2/\sqrt{6})f,$$

$$(\tau^c_2)' - (f/12)(5,3,1,-1)\gamma' - g(b-1)/(6k^2)\{b+1, 3(b-1), -(b-1),$$
$$-(b-1)\}\gamma' \geqslant (1/\sqrt{6})f' - (1/\sqrt{6})b(b-1)g'/k^2 - (2/\sqrt{6})f,$$

$$(\tau^c_3)' - (f/12)(-1,1,3,-3)\gamma' - g(b-1)/(6k^2)\{-(b-1), -3(b+1),$$
$$3(b-1), 3(b-1)\}\gamma' \geqslant (1/\sqrt{6})(b+2)(b-1)g'/k^2,$$

$$(\tau^c_4)' - (f/12)(1,-1,-3,3)\gamma' - g(b-1)/(6k^2)\{-3(b+1), -(b-1),$$
$$3(b-1), 3(b-1)\}\gamma' \geqslant (1/\sqrt{6})(b+2)(b-1)g'/k^2, \tag{5.49}$$

where $\gamma' = (\gamma'_1, \gamma'_2, \gamma'_3, \gamma'_4)^T$. Upon substituting equations (5.47), one finds after some algebra that the consistency conditions can be expressed in reduced form as (Havner & Chidambarrao 1987, eqs. (4.8)–(4.9)[6])

$$(\tau_1^c)' + a_1 x \geqslant d_1, \quad (\tau_2^c)' - a_1 x \geqslant d_1, \\ (\tau_3^c)' + a_2 x \geqslant d_2, \quad (\tau_4^c)' - a_2 x \geqslant d_2, \tag{5.50}$$

in which the equality must hold in an active system, and

$$\begin{aligned} \sqrt{6}a_1 &= 2f/(b+2) - (g/k^2)(b-1)(b-2), \quad x = (1/\sqrt{6})(\gamma'_1 - \gamma'_2), \\ \sqrt{6}a_2 &= 2f(b-1)/(b+2) - (g/k^2)(b-1)(b+2), \quad k^2 = b^2 + 2, \\ \sqrt{6}d_1 &= f' - (1/k^2)(b-1)\{bg' - g(b^2+4b-2)/(b+2)\}, \\ \sqrt{6}d_2 &= (1/k^2)(b-1)\{(b+2)g' + g(b^2-8b-2)/(b+2)\}. \end{aligned} \tag{5.51}$$

Seeking to achieve as general an analysis as possible, Havner and Chidambarrao introduced a broad class of hardening theories that follows the particular symmetries of resolved shear stresses in (110) compression (Table 2) but otherwise makes no specification of the hardening moduli. For i, j and k, l arbitrary pairs of equally stressed slip systems (with the resolved stresses typically different between pairs), the proposed symmetries of the moduli may be written (Chidambarrao & Havner 1988b, eqs. (5.6)):

$$H_{ii} = H_{jj}, \quad H_{ij} = H_{ji} \quad \text{within each pair;} \\ H_{ik} = H_{jl}, \quad H_{jk} = H_{il} \quad \text{between pairs.} \tag{5.52}$$

(Note that the symmetry $H_{ik} = H_{ki}$ between unpaired systems is permitted but not required.) Thus, for the critical systems in range I (Fig. 5.4) one has

$$H_{kj} = \begin{array}{c} \begin{matrix} a1 & b\bar{2} & c\bar{1} & c2 \end{matrix} \\ \begin{bmatrix} H_{11} & H_{12} & H_{13} & H_{14} \\ H_{12} & H_{11} & H_{14} & H_{13} \\ H_{31} & H_{32} & H_{33} & H_{34} \\ H_{32} & H_{31} & H_{34} & H_{33} \end{bmatrix} \begin{matrix} a1 \\ b\bar{2}. \\ c\bar{1} \\ c2 \end{matrix} \end{array} \tag{5.53}$$

We shall show subsequently that each of the four hardening theories introduced in Section 4.4 belongs to this class. (Of course, Taylor hardening obviously belongs.)

After substitution of equations (5.53) and (5.47) into the general hardening law of equation (5.22), we find that the $(\tau_k^c)'$ in critical systems

[6] There the systems are numbered 1, 3, 5, and 6 in the same order.

can be expressed in the following simple form (Havner & Chidambarrao 1987, eqs. (4.11)–(4.12)):

$$\begin{aligned} (\tau_1^c)' &= H_1 + h_1 x, \quad (\tau_2^c)' = H_1 - h_1 x, \\ (\tau_3^c)' &= H_2 + h_2 x, \quad (\tau_4^c)' = H_2 - h_2 x, \end{aligned} \tag{5.54}$$

with $x = (\gamma_1' - \gamma_2')/\sqrt{6}$ as before, and

$$\begin{aligned} (2/\sqrt{6})H_1 &= H_{11} + H_{12} + b(H_{13} + H_{14})/(b+2), \\ (2/\sqrt{6})H_2 &= H_{31} + H_{32} + b(H_{33} + H_{34})/(b+2), \\ (2/\sqrt{6})h_1 &= H_{11} - H_{12} - (b-2)(H_{13} - H_{14})/(b+2), \\ (2/\sqrt{6})h_2 &= H_{31} - H_{32} - (b-2)(H_{33} - H_{34})/(b+2). \end{aligned} \tag{5.55}$$

Combining equations (5.50) and (5.54), we obtain the final form of the consistency conditions in range I, applicable to any hardening theory in the broad class of equations (5.52) or (5.53):

$$\begin{aligned} &(a1)\ H_1 + (a_1 + h_1)x \geqslant d_1, \quad (b\bar{2})\ H_1 - (a_1 + h_1)x \geqslant d_1, \\ &(c\bar{1})\ H_2 + (a_2 + h_2)x \geqslant d_2, \quad (c2)\ H_2 - (a_2 + h_2)x \geqslant d_2 \end{aligned} \tag{5.56}$$

(the equality necessarily holding in an active system).

The unique solution is now transparent, with the following proof taken from Havner & Chidambarrao (1987, p. 253). If x were not zero, neither conditions $(a1)$ and $(b\bar{2})$ nor conditions $(c\bar{1})$ and $(c2)$ (eqs. (5.56)) could be equalities simultaneously. Therefore only one slip system from each pair could be active, with either $\gamma_1' = \gamma_4' = 0$ or $\gamma_2' = \gamma_3' = 0$ from equations (5.46) or (5.47), and $|x| = 1$. However, both γ_3' and γ_4' would be positive from equations $(5.47)_{3,4}$ for $|x| = 1$, a contradiction. Thus, $x = 0$ is the unique solution of constraint equations (5.46) (or (5.47)) and inequalities (5.56), independently of specific hardening rule; and

$$\gamma_1' = \gamma_2' = \frac{\sqrt{6}}{2}, \quad \gamma_3' = \gamma_4' = \frac{\sqrt{6}}{2} b/(b+2), \tag{5.57}$$

which is identically the rotation of the lattice about the stable loading axis found experimentally by Skalli et al. (1983) for initial orientations in range I. Thus, the crystal remains in a (110) loading orientation, and the preceding solution applies (as regards theoretical response) until another pair of slip systems becomes critical at a level of finite deformation that depends upon hardening theory.

Consider now the rate of change of constraint stress. From equations

$(5.51)_{3,4}$ and (5.56) (with $x = 0$) one obtains

$$g' = \tfrac{1}{2}\{f' - \sqrt{6}(H_1 - H_2)\}(b^2 + 2)/(b^2 - 1) + 6gb/(b^2 + 3b + 2). \tag{5.58}$$

Thus the evolution of the constraint stress with crystal deformation can be determined for any hardening theory. "Conversely, whenever the constraint stress can be found experimentally, so that g' as well as f' is known, we can obtain information about the relative hardening of systems $a\bar{1}, b\bar{2}$ (H_1) versus systems $c\bar{1}, c2$ (H_2) from this equation" (Havner & Chidambarrao 1987, p. 254). The evolution of this stress in the initial experimental orientation $Y = (1\bar{1}\bar{1}\bar{6})$ of Skalli et al. (1983) will be evaluated in Section 5.5.2 for each of Taylor hardening, the simple theory, and the P.A.N. rule. Unfortunately, Skalli et al. were not able to determine the constraint stress from their experimental configuration, and Wonsiewicz et al. (1971) did not obtain measurements of the constraint stress in range I.

To complete the general analysis of f.c.c. crystals in this range of (110) loading, we develop exact solutions relating crystal shearing, lattice rotation, and finite compressive strain e_L, following Fuh & Havner (1989, pp. 213–14). From equations (5.2) and (5.26),

$$\begin{aligned} (\tan \chi_x)' - \tan \chi_x &= 2w_x, \\ (\tan \chi_y)' - 2 \tan \chi_y &= -2w_y, \end{aligned} \tag{5.59}$$

where $w_x = W_{zy}$, $w_y = W_{xz}$, and $w_z = W_{yx} = 0$ are the components of the axial vector of the total material spin W relative to the channel frame. The components of the plastic spin $\Omega = \sum \Omega_j \gamma'_j$ in range I are (from eqs. (5.57) and Table 1)

$$\Omega_{zy} = -(1/\sqrt{2})(b - 1)/(b + 2), \quad \Omega_{xz} = \Omega_{yx} = 0. \tag{5.60}$$

Thus, from $W = \Omega + \omega$ and equations (5.48) (with $x = 0$) and (5.60), we obtain

$$w_x = \Omega_{zy} + \omega_x = \frac{1}{\sqrt{2}}(b - 1)/(b + 2), \quad w_y = w_z = 0, \tag{5.61}$$

whence $\chi_y = 0$ from equation $(5.59)_2$, and the crystal finitely shears only in the yz (horizontal) plane.

Let ϕ denote the clockwise angle of Y from $(00\bar{1})$. Then ϕ' is the counterclockwise rotation rate of the lattice about $X = (110)$, and from equations $(5.48)_1$ and $(5.61)_1$,

$$2w_x = \omega_x \equiv \phi' \tag{5.62}$$

in range I. Thus, the solution of equation $(5.59)_1$ may be expressed

$$\tan\chi_x = \exp(e_L(\phi))\int_{\phi_0}^{\phi}\exp(-e_L(\varphi))\,d\varphi, \tag{5.63}$$

which is conveniently evaluated in terms of parameter b as follows.

We first require the equation for b'. From equations $(5.17)_2$ and (5.57), the lattice-corotational derivative of $\boldsymbol{\kappa}$ is

$$\hat{\boldsymbol{\kappa}} = (1/k)\{(b-1)/(b+2)\}(b, -b, 2), \quad k^2 = b^2 + 2. \tag{5.64}$$

Equating this to the derivative of $\boldsymbol{\kappa} = (1/k)(1, -1, -b)$, we obtain (Havner & Chidambarrao 1987, eq. (5.9))

$$b' = -(b^2+2)(b-1)/(b+2). \tag{5.65}$$

Upon inverting this equation and integrating, one has (Fuh & Havner 1989, eq. (3.19))

$$e_L = \ln\left\{\frac{(b_0-1)\sqrt{(b^2+2)}}{(b-1)\sqrt{(b_0^2+2)}}\right\}, \quad \infty > b_0 > 2. \tag{5.66}$$

Finally, after substituting this result and

$$\phi = \arctan(\sqrt{2}/b) \tag{5.67}$$

into equation (5.63) and again integrating with respect to parameter b, we find (Fuh & Havner 1989, eq. (3.20))

$$\tan\chi_x = \frac{1}{\sqrt{2}(b-1)}\left(b + 2 - \frac{b_0+2}{\sqrt{(b_0^2+2)}}\sqrt{(b^2+2)}\right). \tag{5.68}$$

5.5.2 *Numerical Analysis of Initial Orientation $Y = (1\bar{1}\bar{1}\bar{6})$*

The longitudinal fibre stretch $\lambda_z = \exp(e_L)$, lattice rotation $\Delta\phi$, and finite shear χ_x at $Y = (1\bar{1}\bar{2})$, developed from initial orientation $Y = (1\bar{1}\bar{1}\bar{6})$ according to equations (5.66)–(5.68), are shown in Fig. 5.5 (from Fuh & Havner 1989, fig. 10). The corresponding values are $e_L = 0.827$, $\lambda_z = 2.29$, $\Delta\phi = 30.2°$, and $\chi_x = 41.6°$. This range-I (initial) orientation was investigated experimentally for an aluminum crystal to a logarithmic compressive strain of 0.6 by Skalli (1984), who reported a shear angle χ_x of 25.4° at that deformation (see his fig. 43). The theoretical value from equation (5.68) is 32.0° (with $b = 2.45$ at $e_L = 0.6$ from eq. (5.66)). Skalli also reports a small shear $\chi_y = 2.4°$ in the xz plane at $e_L = 0.6$, whereas

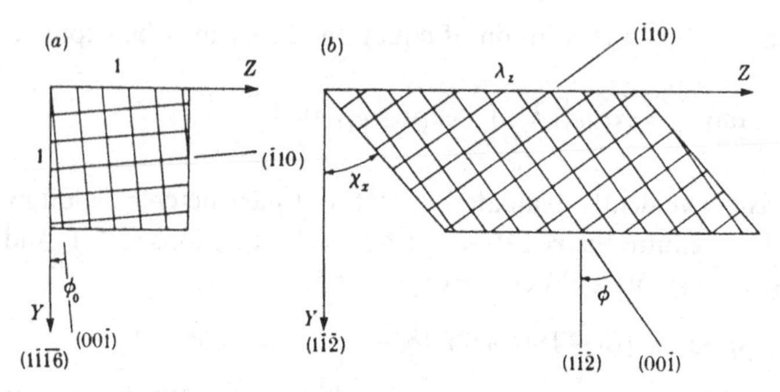

Fig. 5.5. Top view of (a) undeformed configuration ($\phi_0 = 5.05°$) and (b) theoretical deformed configuration ($e_L = 0.827$, $\phi = 35.26°$, $\chi_x = 41.6°$, $\lambda_z = 2.29$) of an f.c.c. crystal in (110) channel die compression (from Fuh & Havner 1989, fig. 10). (Rectangular grid represents underlying lattice).

the theoretical plastic shearing in that plane is zero, as already explained. Perhaps both the lesser yz shear than is predicted and the small xz shear of Skalli's experiment are consequences of some nonuniformity of stress caused by both the free-end conditions and a small amount of longitudinal frictional resistance to the shearing deformation. Whatever the case, the theoretical predictions of the kinematics may be judged reasonably good.

In the following we investigate the development of the constraint stress and latent hardening for this initial orientation according to each of classical Taylor hardening, the simple theory, and the P.A.N. rule. For the assumed biaxial stress state (eq. (5.5)) of the channel die test, the hardening moduli of the latter two theories are (Havner & Chidambarrao 1987, eqs. (5.5)–(5.7))

$$H_{kj} = h - f r_{kj} - g q_{kj} \text{ (simple theory);} \tag{5.69}$$

$$H_{kj} = \hbar - f\,\text{skw}(r_{kj}) - g\,\text{skw}(q_{kj}) \text{ (P.A.N. rule),} \tag{5.70}$$

with

$$\begin{aligned} r_{kj} &= 2\boldsymbol{\iota} N_k \Omega_j \boldsymbol{\iota}, \quad q_{kj} = 2\boldsymbol{\kappa} N_k \Omega_j \boldsymbol{\kappa}, \\ \text{skw}(r_{kj}) &= \tfrac{1}{2}(r_{kj} - r_{jk}), \quad \text{skw}(q_{kj}) = \tfrac{1}{2}(q_{kj} - q_{jk}). \end{aligned} \tag{5.71}$$

In range I of (110) loading, with systems $a1$, $b\bar{2}$, $c\bar{1}$, and $c2$ active, the r_{kj} and q_{kj} are (Havner & Chidambarrao 1987, eqs. (A16))

$$r_{kj} = \frac{1}{24}\begin{bmatrix} 5 & -5 & 1 & -1 \\ -5 & 5 & -1 & 1 \\ 1 & -1 & -3 & 3 \\ -1 & 1 & 3 & -3 \end{bmatrix},$$

$$q_{kj} = \frac{1}{12(b^2+2)} \times \begin{bmatrix} -b^2+6b-3 & 2b^2-2b+1 & -b^2-6b+1 & -2b^2-6b+5 \\ 2b^2-2b+1 & -b^2+6b-3 & -2b^2-6b+5 & -b^2-6b+1 \\ -b^2-6b+1 & 2b^2-2b-3 & -b^2+14b+5 & -2b^2+10b+1 \\ 2b^2-2b-3 & -b^2-6b+1 & -2b^2+10b+1 & -b^2+14b+5 \end{bmatrix} \tag{5.72}$$

It is evident that both matrices (hence their skew-symmetric parts) have the same symmetry as the moduli H_{kj} of equation (5.53). Therefore the simple theory and the P.A.N. rule, as well as Taylor's rule, belong to the broad class of hardening theories to which the general analysis of Section 5.5.1 applies.[7]

Upon substituting equations (5.55), (5.69)–(5.70), and (5.72) (and $H_1 = H_2$ for Taylor hardening) into equation (5.58), we find that the constraint stress g evolves with f and orientation parameter b for the three theories according to (Havner & Chidambarrao 1987, eqs. (5.11))

$$\text{Taylor's rule: } g' = \tfrac{1}{2}f'(b^2+2)/(b^2-1) + 6gb/(b^2+3b+2) \equiv g_{\mathrm{I}}'; \tag{5.73}$$

$$\text{Simple theory: } g' = g_{\mathrm{I}}' - 3gb/(b^2+3b+2); \tag{5.74}$$

$$\text{P.A.N. rule: } g' = g_{\mathrm{I}}' - g/2 \tag{5.75}$$

(with subscript I a mnemonic for the isotropic hardening of Taylor's rule). Thus, for the same $f(e_{\mathrm{L}})$ in each equation, $g(\text{Taylor}) > g(\text{simple theory}) > g(\text{P.A.N. rule})$, as all theoretical predictions begin from the same initial stress rate f_0, g_0 of equation (5.34) in range I.

Havner and Chidambarrao adopted the following form of the stress–strain curve $f(e_{\mathrm{L}})$:

$$f \equiv -\sigma_{xx} = \sigma_0 + \mathscr{L}_{\mathrm{T}} e_{\mathrm{L}} + (\sigma_{\mathrm{T}} - \sigma_0)\tanh(e_{\mathrm{L}}/e_0), \tag{5.76}$$

$$e_0 = (\sigma_{\mathrm{T}} - \sigma_0)/(\mathscr{L}_0 - \mathscr{L}_{\mathrm{T}}). \tag{5.77}$$

This curve has initial slope $\mathscr{L}_0$ and is asymptotic to $\sigma_{\mathrm{T}} + \mathscr{L}_{\mathrm{T}} e_{\mathrm{L}}$. Using data points from Skalli (1984, fig. 45) for a nearby nominal $(110)(00\bar{1})(\bar{1}10)$ orientation (because no stress–strain information specifically for $Y = (1\bar{1}\bar{1}6)$

[7] It also may be noted that the two-parameter rule of Section 4.4.5, the two-parameter form of the P.A.N. rule mentioned in footnote 7 of Chapter 4, and the two-parameter modification of the simple theory investigated by Fuh & Havner (1986, 1989), among other hardening theories, satisfy the general eqs. (5.52) in (110) channel die compression.

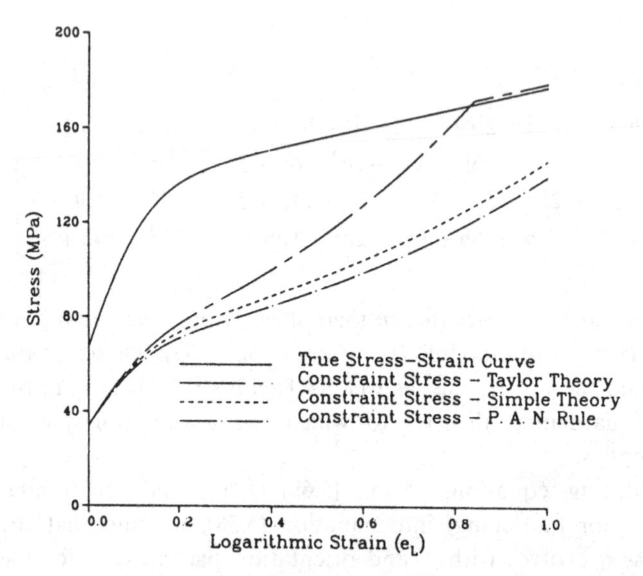

Fig. 5.6. Active stress–strain curve for an aluminum crystal in initial orientation $(110)\,(1\overline{1}\overline{1}6)\,(\overline{8}8\overline{1})$ and predicted constraint stress curves according to various hardening theories (from Havner & Chidambarrao 1987, fig. 3).

was reported) Chidambarrao obtained a best fit of equation (5.76) to that data (see fig. 7 of Chidambarrao (1988)) using a standard nonlinear regression analysis. The result is

$$f(e_{\mathrm{L}}) = 67.1 + 44.1 e_{\mathrm{L}} + 66.5 \tanh(7.64 e_{\mathrm{L}}) \quad (\mathrm{MPa}). \tag{5.78}$$

This curve and the corresponding numerical results for $g(e_{\mathrm{L}})$ according to the three hardening rules considered here are shown in Fig. 5.6 (from fig. 3 of Havner & Chidambarrao (1987)). These and other curves to be presented were obtained using a computer program developed for arbitrary crystal orientations in channel die compression and given in Chidambarrao (1988, appendix A). The program solves numerically the linear complementarity problem with constraints, equations (5.30), for each of the three hardening rules at every deformation step, using strain increments $\delta e_{\mathrm{L}} = 0.01$ with a given $f(e_{\mathrm{L}})$ curve. (The program does not make use of the general analysis in Sect. 5.5.1.)

The curve $g(e_{\mathrm{L}})$ for Taylor hardening in Fig. 5.6 is seen to suffer an abrupt change in slope at $e_{\mathrm{L}} = 0.84$. This strain is the next incremental step after the value $e_{\mathrm{L}} \approx 0.83$ at which the lattice has rotated into position $Y = (1\overline{1}\overline{2})$ (with deformation as shown in Fig. 5.5). For that orientation the exact equations of equal hardening in all slip systems give $f = g$, with systems $a\bar{3}$ and $b3$ as well as the original systems $a1$, $b\bar{2}$, $c\bar{1}$, and $c2$ now critical. (Because of the accumulation of small rounding errors, the incremental numerical procedure has overshot the strain $e_L \approx 0.83$ at $Y = (1\bar{1}\bar{2})$ given by eq. (5.66), with the result that g is slightly greater than f in Fig. 5.6.) In the next solution step the deformation mode changes: Systems $a1$, $b\bar{2}$, $a\bar{3}$, and $b3$ are active; systems $c\bar{1}$ and $c2$ are inactive but remain critical; $g' = f'$ whatever the curve $f(e_L)$; and *the lattice is stable.* Consequently, this solution continues to unlimited deformations, and the subsequent f and g curves coincide for Taylor hardening (but with a slight separation in their computer-generated plots, Fig. 5.6, as already explained).

For both the simple theory and the P.A.N. rule, the predicted anisotropy of hardening (eqs. (5.69) and (5.70)) and different evolution of the constraint stress (eqs. (5.74) and (5.75)) result in continued rotation of the lattice to and beyond $e_L = 1.0$, the limit of the calculations in Havner & Chidambarrao (1987). Thus, from $e_L \approx 0.83$ onward there is a difference in deformation mode between these two theories and Taylor's rule. Unfortunately, the experiment of Skalli (1984) (to $e_L = 0.6$) was not carried sufficiently far to make that kinematic distinction. However, had Skalli used an experimental arrangement similar to that of Wonsiewicz et al. (1971) (see Fig. 5.2 here), permitting determination of the constraint stress during the deformation, that information alone would have served to define the relative merits of these three theories in (110) channel die compression (as is obvious from Fig. 5.6).

The evolving critical strengths according to the simple theory and the P.A.N. rule are displayed in Figs. 5.7 and 5.8, respectively (taken from Havner and Chidambarrao's figs. 7 and 8). The isotropic Taylor hardening curve is shown superimposed on each figure. The net-strength curves may be found in Havner and Chidambarrao's figures 4–6. Those curves demonstrate that systems 1–6 of Table 2 are critical from $e_L \approx 0.83$ for Taylor hardening, but that only the original four systems are critical and active to $e_L \approx 1.0$ (and beyond) for the other two theories. There apparently is no experimental information on latent hardening of crystals in channel die compression.

Figure 5.9, reproducing Havner and Chidambarrao's figure 9, shows the initial yield locus in the positive, biaxial-compression quadrant of f, g stress space, the subsequent yield loci at $Y = (1\bar{1}\bar{2})$ ($e_L \approx 0.83$) according to the three theories, and the corresponding stress paths. Note that the trace of the coincident yield planes for active systems $a1$ and $b\bar{2}$ in this stress subspace is continuously rotating counterclockwise (as well as

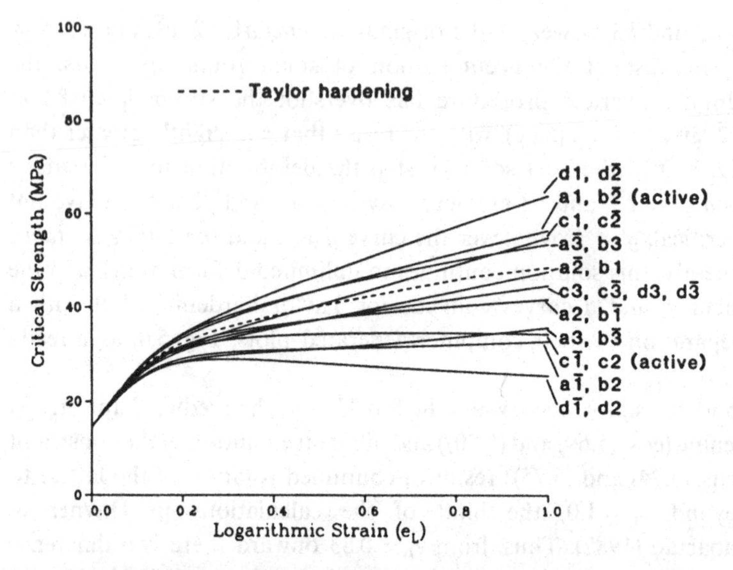

Fig. 5.7. Theoretical critical-strength curves for an aluminum crystal in initial orientation $(110)(1\bar{1}\bar{1}6)(\bar{8}8\bar{1})$ according to the simple theory as compared with Taylor hardening (from Havner & Chidambarrao 1987, fig. 7).

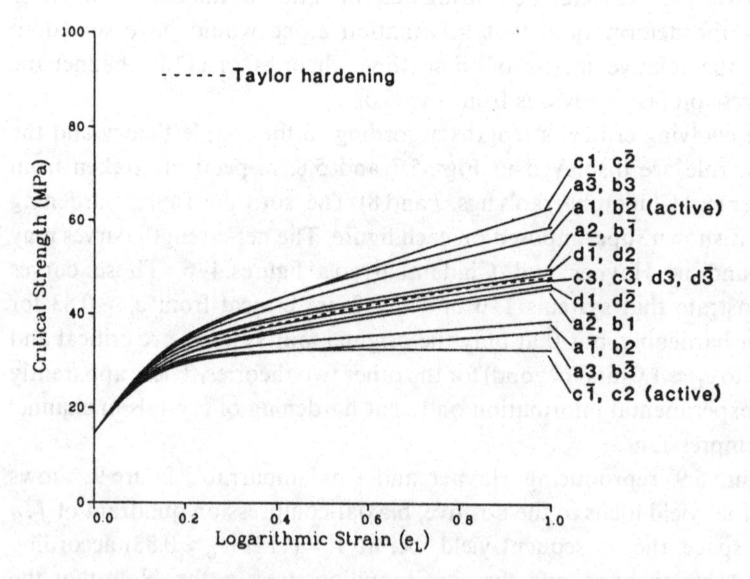

Fig. 5.8. Theoretical critical-strength curves for an aluminum crystal in initial orientation $(110)(1\bar{1}\bar{1}6)(\bar{8}8\bar{1})$ according to the P.A.N. rule as compared with Taylor hardening (from Havner & Chidambarrao 1987, fig. 8).

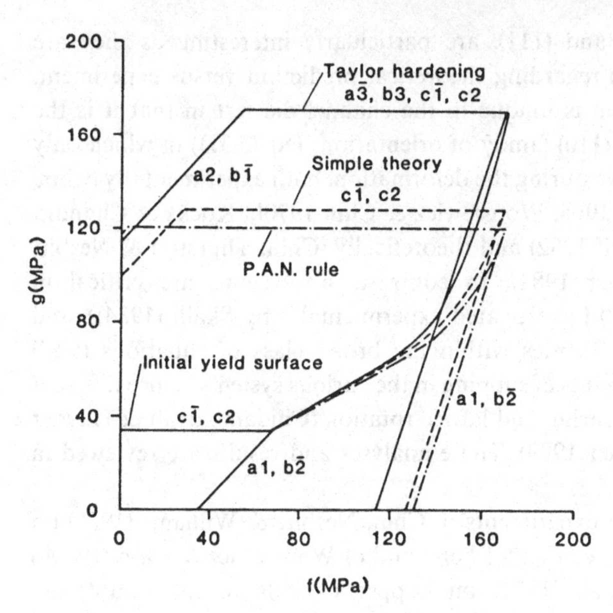

Fig. 5.9. Initial yield locus of an aluminum crystal in orientation $(110)(1\bar{1}\bar{1}\bar{6})(\bar{8}8\bar{1})$ and subsequent yield loci in orientation $(110)(1\bar{1}\bar{2})(\bar{1}1\bar{1})$ for various hardening theories, including the corresponding stress paths (from Havner & Chidambarrao 1987, fig. 9).

moving outward) because of the lattice rotation about the loading axis in range I. This is the reason that the stress path for Taylor hardening in Fig. 5.9 may appear to have passed outside the corresponding yield locus. For subsequent deformation according to Taylor's rule, the path abruptly changes to a straight line emanating from the apex at a 45° angle, and the locus expands isotropically because the lattice no longer rotates. The other theories, however, predict continued lattice rotation and corresponding counterclockwise rotation of the $a1$ and $b\bar{2}$ yield-plane trace to strains beyond $e_L = 1.0$. (It also should be noted that the predicted change in deformation mode at $e_L \approx 0.83$ to lattice stability according to Taylor's rule results in an increased rate of shearing in the yz plane as compared with that corresponding to continued lattice rotation. See fig. 21 of Chidambarrao (1988).)

5.6 Analysis of Two Singular Orientations

Of the four singular constraint directions in (110) channel die compression defined by equations (5.33), the two endpoint orientations of range II, $Y=(1\bar{1}\bar{2})$ and $(1\bar{1}\bar{1})$, are particularly interesting as they are essentially opposites regarding theoretical prediction versus experiment. The latter orientation is unique in the channel die test in that it is the only one within the (110) family of orientations (eq. (5.31)) in which only two systems are active during the deformation, both experimentally (Chin, Nesbitt & Williams 1966; Wonsiewicz & Chin 1970b; Kocks & Chandra 1982; Driver & Skalli 1982) and theoretically (Chin, Thurston & Nesbitt 1966; Sue & Havner 1984). In contrast, six systems are critical in orientation $Y=(1\bar{1}\bar{2})$ (investigated experimentally by Skalli (1984)), and different hardening theories within the broad class of equations (5.52) predict different amounts of slipping in the various systems, hence different results for crystal shearing and lattice rotation (Chidambarrao & Havner 1988a; Fuh & Havner 1989). These analyses and results are reviewed in Sections 5.6.2 and 5.6.3.

In the channel die experiments of Chin, Nesbitt & Williams (1966) on a Permalloy crystal (see Fig. 5.1 here) and of Wonsiewicz & Chin (1970b) and Wonsiewicz et al. (1971) on copper crystals in the double-slip orientation, Chin and his colleagues defined their orientation to be $Y=(\bar{1}1\bar{1})$, which is symmetric with $Y=(1\bar{1}\bar{1})$ relative to the $[00\bar{1}]$–$[110]$ plane (see Fig. 5.4.). Sue & Havner (1984) also presented their analysis in terms of orientation $Y=(\bar{1}1\bar{1})$, and we adopt this choice in the next section in order to make direct use of the equations and hardening curves of that work.

5.6.1 *Orientation* $(110)(\bar{1}1\bar{1})(\bar{1}12)$

As we have seen in Section 5.4, the theoretical elastic value of the ratio g/f holds for constraint orientation $Y=(1\bar{1}\bar{1})$. Thus, it also holds for the equivalent orientation $Y=(\bar{1}1\bar{1})$ in which the critical systems are $a\bar{2}$ and $b1$ (rather than $a1$ and $b\bar{2}$), as is evident from symmetry in Fig. 5.4. For convenience in the following analysis we shall label these systems 1 and 2.

For any metal in this double-slip orientation, the initial (elastic) stress ratio is (Sue & Havner 1984, eq. (3.10))

$$(g/f)_{\mathrm{E}} = \frac{s_{44}/2-(s_{11}+s_{12})}{s_{44}+(s_{11}+2s_{12})}, \tag{5.79}$$

and no elastoplastic transition region is required. As noted in Section 5.4, the ratio is 0.337 for aluminum and 0.28 for copper. From Table 2 (evaluated for $b = 1$ and systems $a1$ and $b\bar{2}$) and equations (5.24)–(5.25), there is no contribution of the slip rates to the constraint $d_{yy} = 0$. Consequently, the elastic constraint, equation (5.79), still applies. The constraint $d_{xy} = 0$ requires

$$\gamma_1' = \gamma_2' = \gamma' = \frac{\sqrt{6}}{2}, \tag{5.80}$$

whence $\gamma = (\sqrt{6}/2)e_{\mathrm{L}}$. Moreover, from equations (5.17) and Table 1,

$$\hat{\mathbf{\imath}} = \hat{\mathbf{k}} = \mathbf{0}.$$

Therefore the lattice does not rotate relative to the channel axes, $m_k' = r_k' = 0$ for all k, and the original consistency conditions, equations (5.13), reduce to

$$(\tau_1^{\mathrm{c}})' = (\tau_2^{\mathrm{c}})' = \frac{1}{\sqrt{6}} f'. \tag{5.81}$$

Thus, $\tau_1^{\mathrm{c}} = \tau_2^{\mathrm{c}} = (1/\sqrt{6}) f(e_{\mathrm{L}})$, and the constraint stress continues to be given by equation (5.79). In addition, from $\omega = 0$, the velocity gradient reduces to $\Gamma = \{(\mathbf{b} \otimes \mathbf{n})_1 + (\mathbf{b} \otimes \mathbf{n})_2\}\gamma'$; whence from Table 1 it follows that $\Gamma_{zx} = 0$ and $\Gamma_{zy} = -\gamma'/\sqrt{3}$. Consequently, from equations (5.2) (or eqs. (5.26)) and (5.80) we obtain

$$\chi_y = 0, \quad \tan \chi_x = -\frac{1}{\sqrt{2}}(\exp e_{\mathrm{L}} - 1), \tag{5.82a}$$

or

$$\tan \chi_x = \frac{1}{\sqrt{2}}(1 - \lambda^{-1}), \tag{5.82b}$$

as first given by Chin, Thurston & Nesbitt (1966, eq. 36). The excellent agreement between this kinematic prediction and experimental results for a Permalloy crystal may be seen in their figure 6. Comparable results were reported by Wonsiewicz & Chin (1970b) for copper crystals and Kocks & Chandra (1982) and Driver & Skalli (1982) for aluminum.

The requirement $\tau_1^{\mathrm{c}} = \tau_2^{\mathrm{c}}$ from the consistency conditions is satisfied identically by every hardening theory within the broad class of equations (5.52); hence it is satisfied by each of the four specific hardening rules in Section 4.4. Because the lattice is stable, systems $a\bar{2}$ and $b1$ remain the most highly stressed for the constant ratio g/f of the elastic constraint, equation (5.79). Furthermore, because all latent systems harden at either the same or a greater rate than the active systems for Taylor's rule and the two-parameter rule, no other systems can become critical according to these two theories, and they predict equal double slip in systems $a\bar{2}$ and $b1$ to unlimited deformation. For the simple theory and the P.A.N. rule further analysis is required, and numerical results for copper crystals obtained by Sue & Havner (1984) are reviewed in the following.

From equations (4.15) and (4.26) (or eqs. (5.69)–(5.71)), the respective predictions for $Y = (\bar{1}1\bar{1})$ according to the latter two theories are (see Sue & Havner (1984, sect. 6) for details)

$$\tau_k^c = \tau_c + \sqrt{2}(g/f)_0\{(N_{yz})_k + 1/(4\sqrt{3})\}\int_0^\gamma \tau_c \, d\gamma \text{ (simple theory)}, \tag{5.83}$$

$$\tau_k^c = \tau_c + (1/\sqrt{2})(g/f)_0(b_y n_z)_k \int_0^\gamma \tau_c \, d\gamma \text{ (P.A.N. rule)}, \tag{5.84}$$

in which $\tau_c(\gamma) = (1/\sqrt{6})f(e_L)$ is the critical strength of the active systems. For copper crystals Sue and Havner found that the equation

$$f(e_L) = 21.8e_L + 27.8 \tanh(5.26e_L) \text{ } (10^3 \text{ psi}) \tag{5.85}$$

is virtually indistinguishable from the experimental curve reported in Wonsiewicz & Chin (1970b, fig. 1, curve 4) for constraint direction $(\bar{1}1\bar{1})$, as may be seen in Sue and Havner's figure 3.[8] Thus we have (from eqs. (5.80) and (5.81))

$$\tau_c(\gamma) = 7.4\gamma + 11.4 \tanh(4.3\gamma) \text{ } (10^3 \text{ psi}) \tag{5.86}$$

and (Sue & Havner 1984, eq. (6.26))

$$\int_0^\gamma \tau_c \, d\gamma = 3.7\gamma^2 + 2.65 \ln(\cosh 4.3\gamma) \text{ } (10^3 \text{ psi}). \tag{5.87}$$

Plots of equations (5.83) and (5.84) for all 24 slip systems, corresponding to equations (5.86)–(5.87) and $(g/f)_0 = 0.28$ for copper, are shown in Figs. 5.10 and 5.11, respectively (taken from Sue & Havner's figs. 5 and 6). In the case of the simple theory it is seen that 16 of the 22 latent systems harden at a greater rate than active systems $a\bar{2}$ and $b1$, whereas for the P.A.N. rule all but six latent systems harden at the same or a lesser rate

[8] Wonsiewicz and Chin present their curve only within the logarithmic strain range 0.035–0.79. It is likely that experimental data would deviate from equation (5.85) at smaller strains, with a nonzero f_0, hence nonzero τ_0, and a brief "easy-glide" stage.

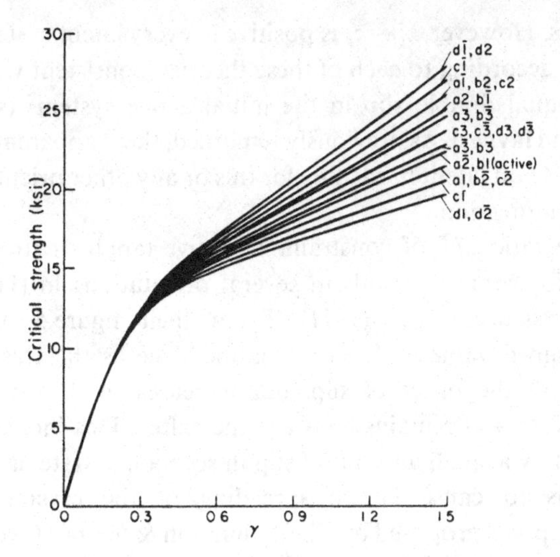

Fig. 5.10. Theoretical critical-strength curves for a copper crystal in orientation $(110)(\bar{1}1\bar{1})(\bar{1}12)$ according to the simple theory (from Sue & Havner 1984, fig. 5). Reprinted with permission from *J. Mech. Phys. Solids* 32, P. L. Sue & K. S. Havner, Theoretical analysis of the channel die compression test-I, Copyright 1984, Pergamon Press, PLC.

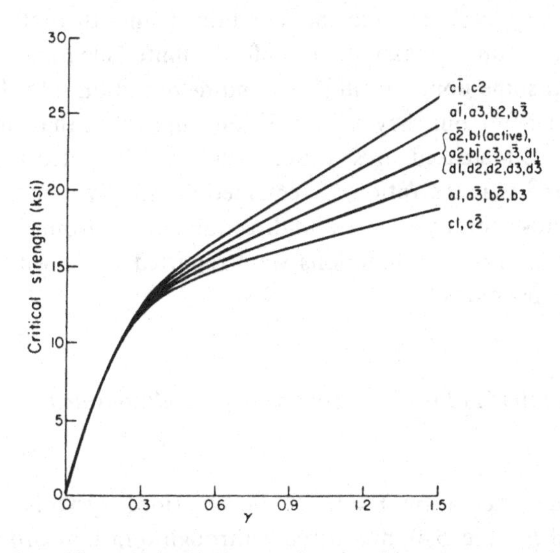

Fig. 5.11. Theoretical critical-strength curves for a copper crystal in orientation $(110)(\bar{1}1\bar{1})(\bar{1}12)$ according to the P.A.N. rule (from Sue & Havner 1984, fig. 6). Reprinted with permission from *J. Mech. Phys. Solids* 32, P. L. Sue & K. S. Havner, Theoretical analysis of the channel die compression test-I, Copyright 1984, Pergamon Press, PLC.

than the active systems. However, $\tau_k^c - \tau_k$ is positive in every latent system to $\gamma = 1.5$ and beyond according to each of these theories, consistent with the perpetuation of equal double slip in the initial active systems (see figs. 7 and 8 of Sue and Havner). As previously remarked, there apparently are no experimental data on latent hardening for this or any other orientation in channel die compression.

With regard to the ratio g/f of constraint to active (applied) stress, this was determined for copper crystals in several orientations in (110) compression by Wonsiewicz et al. (1971). From their figure 5, the experimentally determined value at $Y = (\bar{1}1\bar{1})$ is the same as the elastic constraint ratio 0.28 at the onset of slip, but increases to 0.35 at a logarithmic strain of 0.05 and remains constant thereafter. This increase may have been caused by a small amount of slip in secondary systems $a1$ and $b\bar{2}$, which tends to cause lateral spreading of the crystal if unconstrained. Such slip was proposed by Chin, Thurston & Nesbitt (1966) to explain the slightly greater shearing than that given by equation (5.82b) for a Permalloy crystal in the same orientation. However, this slip is not predicted theoretically because these systems are not critical. Sue & Havner (1984, sect. 6) pointed out that since the virgin crystal specimen of Wonsiewicz et al. (1971) was slightly wider than it was long (6 mm by 5 mm), the inconsistency between free-end conditions and theoretical zero-friction conditions under (assumed) uniform finite shearing χ_x necessarily introduced some nonuniform stress and deformation into the crystal (recall Sect. 5.1 here), but they refrained from suggesting that this is the explanation of the increased stress ratio. It also may be noted that Wonsiewicz et al. found that the ratio g/f remained essentially constant at 0.28 for a crystal whose nominal constraint direction differed from $(\bar{1}1\bar{1})$ by only 2°. (The actual crystal orientations were reported to be within $\pm 1°$ of their nominal positions.)

5.6.2 *Orientation $(110)(1\bar{1}\bar{2})(\bar{1}1\bar{1})$: Consistency Conditions and Hardening*

In the singular orientation $Y = (1\bar{1}\bar{2})$, the six critical systems are $a1$, $b\bar{2}$, $c\bar{1}$, $c2$, $a\bar{3}$, and $b3$ (Fig. 5.4), numbered 1 through 6 in that order. These systems are shown in Fig. 5.12 (from fig. 2 of Chidambarrao & Havner (1988a)) on a crystal pyramid that may be thought of as cut out from the initially rectangular crystal in the channel die. (The heavy lines in the figure represent range I (Sect. 5.5) of the constraint and channel-axis

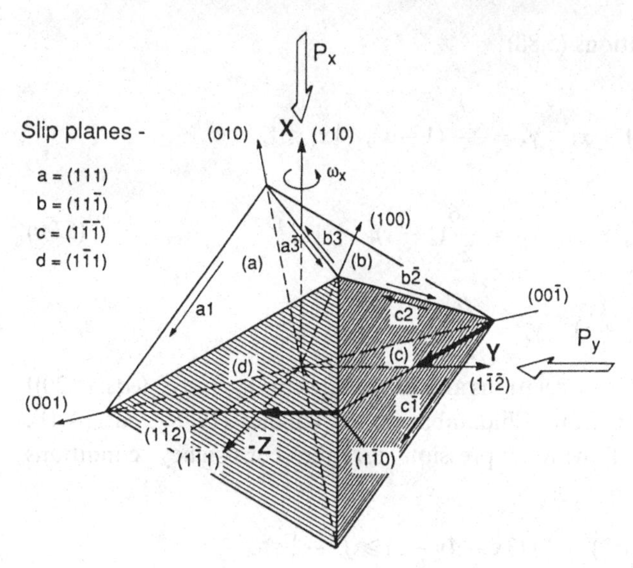

Fig. 5.12. Critical slip systems in initial orientation (110)(1$\bar{1}\bar{2}$)($\bar{1}$1$\bar{1}$) (from Chidambarrao & Havner 1988a, fig. 2). Reprinted with permission from *Int. J. Plasticity* 4, D. Chidambarrao & K. S. Havner, Finite deformation analysis of f.c.c. crystals in (110)(1$\bar{1}\bar{2}$)($\bar{1}$1$\bar{1}$) channel die compression, Copyright 1988, Pergamon Press, PLC.

directions.) At the onset of finite plastic deformation in this orientation, $f_0 = g_0 = (3/2)\sqrt{6}\tau_0$ (from eq. $(5.37)_2$), which state is preceded by an infinitesimal-strain, elastoplastic transition from the initial elastic stage

$$(g/f)_E = -\frac{s_{12} + s_A/6}{s_{11} - s_A/2}, \quad s_A = s_{11} - s_{12} - \tfrac{1}{2}s_{44},$$

as explained in Section 5.4. (For aluminum and copper the respective elastic ratios are 0.369 and 0.508, in contrast to $(g/f)_0 = 1.0$.)

From equations (5.24)–(5.25) and Table 2, the constraints for $\boldsymbol{\kappa}_0 = (1\bar{1}\bar{2})$ reduce to

$$\gamma_1' + \gamma_2' = \sqrt{6}, \quad \gamma_3' + \gamma_6' = \gamma_4' + \gamma_5' = \frac{\sqrt{6}}{4}. \tag{5.88}$$

We define the kinematic variables

$$x = \frac{1}{\sqrt{6}}(\gamma_1' - \gamma_2'), \quad y = \frac{1}{\sqrt{6}}(\gamma_4' - \gamma_5'), \quad z = \frac{1}{\sqrt{6}}(\gamma_3' - \gamma_6'), \tag{5.89}$$

whence from equations (5.88)

$$\gamma_1' = \frac{\sqrt{6}}{2}(1+x), \quad \gamma_2' = \frac{\sqrt{6}}{2}(1-x), \quad |x| \leqslant 1,$$

$$\gamma_4' = \frac{\sqrt{6}}{2}(\tfrac{1}{4}+y), \quad \gamma_5' = \frac{\sqrt{6}}{2}(\tfrac{1}{4}-y), \quad |y| \leqslant \tfrac{1}{4}, \tag{5.90}$$

$$\gamma_3' = \frac{\sqrt{6}}{2}(\tfrac{1}{4}+z), \quad \gamma_6' = \frac{\sqrt{6}}{2}(\tfrac{1}{4}-z), \quad |z| \leqslant \tfrac{1}{4}.$$

Upon substituting these relations and the coefficients n_{kj} and c_{kj}(eqs. (5.20)) for this orientation from Chidambarrao & Havner (1988a, eqs. (A.1)), we obtain the following expression for the consistency conditions (eqs. (5.21)):

$$
\begin{aligned}
&(a1) \quad \sqrt{6}(\tau_1^c)' + \tfrac{1}{6}f(3x-4y-z) \geqslant f' - \tfrac{1}{3}g' + \tfrac{5}{24}f,\\
&(b\bar{2}) \quad \sqrt{6}(\tau_2^c)' + \tfrac{1}{6}f(-3x-y-4z) \geqslant f' - \tfrac{1}{3}g' + \tfrac{5}{24}f,\\
&(c\bar{1}) \quad \sqrt{6}(\tau_3^c)' + \tfrac{1}{6}f(-x+4y+3z) \geqslant \tfrac{2}{3}g' - \tfrac{7}{24}f,\\
&(c2) \quad \sqrt{6}(\tau_4^c)' + \tfrac{1}{6}f(x+3y+4z) \geqslant \tfrac{2}{3}g' - \tfrac{7}{24}f,\\
&(a\bar{3}) \quad \sqrt{6}(\tau_5^c)' + \tfrac{1}{6}f(4x-3y+z) \geqslant \tfrac{2}{3}g' + \tfrac{1}{12}f,\\
&(b3) \quad \sqrt{6}(\tau_6^c)' + \tfrac{1}{6}f(-4x+y-3z) \geqslant \tfrac{2}{3}g' + \tfrac{1}{12}f.
\end{aligned}
\tag{5.91}
$$

(Note that $\sqrt{6}\Sigma(\tau_k^c)' \geqslant 2(f'+g')$ for all theories.)

Consider again the broad class of hardening theories given by equations (5.52). The corresponding matrix of hardening moduli H_{kj} over the six critical systems is

$$
H_{kj} = \begin{array}{c} \begin{matrix} a1 & b\bar{2} & c\bar{1} & c2 & a\bar{3} & b3 \end{matrix} \\ \left[\begin{matrix}
H_{11} & H_{12} & H_{13} & H_{14} & H_{15} & H_{16}\\
H_{12} & H_{11} & H_{14} & H_{13} & H_{16} & H_{15}\\
H_{31} & H_{32} & H_{33} & H_{34} & H_{35} & H_{36}\\
H_{32} & H_{31} & H_{34} & H_{33} & H_{36} & H_{35}\\
H_{51} & H_{52} & H_{53} & H_{54} & H_{55} & H_{56}\\
H_{52} & H_{51} & H_{54} & H_{53} & H_{56} & H_{55}
\end{matrix}\right] \end{array}
\begin{matrix} a1 \\ b\bar{2} \\ c\bar{1} \\ c2' \\ a\bar{3} \\ b3 \end{matrix}
\tag{5.92}
$$

and the general hardening rule (eq. (5.22)) may be expressed (with eqs. (5.90))

$$\begin{aligned}
2\sqrt{6}(\tau_1^c)' &= H_1 + h_1 x + h_2 y + h_3 z, \\
2\sqrt{6}(\tau_2^c)' &= H_1 - h_1 x + h_3 y + h_2 z, \\
2\sqrt{6}(\tau_3^c)' &= H_3 + h_4 x + h_5 y + h_6 z, \\
2\sqrt{6}(\tau_4^c)' &= H_3 - h_4 x + h_6 y + h_5 z, \\
2\sqrt{6}(\tau_5^c)' &= H_5 + h_7 x + h_8 y + h_9 z, \\
2\sqrt{6}(\tau_6^c)' &= H_5 - h_7 x + h_9 y + h_8 z,
\end{aligned} \tag{5.93}$$

where

$$\begin{aligned}
&H_i = \sum_{j=1}^{2} H_{ij} + \frac{1}{4}\sum_{k=3}^{6} H_{ik}, \quad i = 1, 3, 5, \\
&h_1 = H_{11} - H_{12}, \quad h_2 = H_{14} - H_{15}, \quad h_3 = H_{13} - H_{16}, \\
&h_4 = H_{31} - H_{32}, \quad h_5 = H_{34} - H_{35}, \quad h_6 = H_{33} - H_{36}, \\
&h_7 = H_{51} - H_{52}, \quad h_8 = H_{54} - H_{55}, \quad h_9 = H_{53} - H_{56}.
\end{aligned} \tag{5.94}$$

Thus, the consistency conditions for $Y = (1\bar{1}\bar{2})$ finally become

$$\begin{aligned}
&(a1)\ (3H_1 - \tfrac{5}{4}f) + 3(h_1 + f)x + (3h_2 - 4f)y + (3h_3 - f)z \geqslant 6f' - 2g', \\
&(b\bar{2})\ (3H_1 - \tfrac{5}{4}f) - 3(h_1 + f)x + (3h_3 - f)y + (3h_2 - 4f)z \geqslant 6f' - 2g', \\
&(c\bar{1})\ (3H_3 + \tfrac{7}{4}f) + (3h_4 - f)x + (3h_5 + 4f)y + 3(h_6 + f)z \geqslant 4g', \\
&(c2)\ (3H_3 + \tfrac{7}{4}f) - (3h_4 - f)x + 3(h_6 + f)y + (3h_5 + 4f)z \geqslant 4g', \\
&(a\bar{3})\ (3H_5 - \tfrac{1}{2}f) + (3h_7 + 4f)x + 3(h_8 - f)y + (3h_9 + f)z \geqslant 4g', \\
&(b3)\ (3H_5 - \tfrac{1}{2}f) - (3h_7 + 4f)x + (3h_9 + f)y + 3(h_8 - f)z \geqslant 4g',
\end{aligned} \tag{5.95}$$

with the equality necessarily satisfied if a system is active, and the range of kinematically admissible solutions restricted only by $|x| \leqslant 1$, $|y| \leqslant \frac{1}{4}$, and $|z| \leqslant \frac{1}{4}$.

For classical Taylor hardening, all $H_i = 3H$, all $h_i = 0$, and it can be proved (see Chidambarrao & Havner (1988a, sect. IV)) that the solution of the corresponding consistency conditions is unique and given by $x = 0$ and $y = z = -\frac{1}{4}$, from which

$$\begin{aligned}
&\gamma_1' = \gamma_2' = \frac{\sqrt{6}}{2}, \quad \gamma_3' = \gamma_4' = 0, \\
&\gamma_5' = \gamma_6' = \frac{\sqrt{6}}{4}, \quad g' = f' \quad \text{(Taylor's rule).}
\end{aligned} \tag{5.96}$$

This solution is identical with lattice stability, as is immediately evident from equations (5.29) for lattice spin evaluated for the present orientation and critical systems:

$$\omega_x = \frac{1}{2\sqrt{3}}(\gamma_3' + \gamma_4'), \quad \omega_y = \tfrac{1}{6}(-\gamma_1' + \gamma_2' - \gamma_3' + \gamma_4'),$$
$$\omega_z = \frac{1}{3\sqrt{2}}(\gamma_1' - \gamma_2' - 2\gamma_3' + 2\gamma_4'). \tag{5.97}$$

Obviously, $\omega_x = \omega_y = \omega_z = 0$ according to Taylor's rule, and equations (5.96) continue to hold to unlimited deformations. As stated in Section 5.5.2, this also is the solution for Taylor's rule when the constraint direction has rotated into orientation $(1\bar{1}\bar{2})$ from an initial position within range I (in which systems 1 through 4 are active) because of the isotropy of the hardening.

Lattice stability for initial constraint directions in or near $(1\bar{1}\bar{2})$ has not been found experimentally (Skalli 1984), and we return to the general inequalities (5.95). First it may be noted that if some (x, y, z) is a solution to the consistency conditions and constraints, so is $(-x, z, y)$. Thus, if the solution is unique according to a particular hardening theory (as in the case of Taylor's rule), it is of the form $(0, y, y)$. Correspondingly, from equations (5.90) and (5.97), for a unique solution

$$\gamma_1' = \gamma_2' = \frac{\sqrt{6}}{2}, \quad \gamma_3' = \gamma_4' = \frac{\sqrt{6}}{2}(\tfrac{1}{4} + y), \quad \gamma_5' = \gamma_6' = \frac{\sqrt{6}}{2}(\tfrac{1}{4} - y),$$
$$\omega_x = \frac{1}{\sqrt{2}}(\tfrac{1}{4} + y), \quad \omega_y = \omega_z = 0, \quad |y| \leqslant \tfrac{1}{4}. \tag{5.98}$$

Thus, every hardening theory that provides a unique solution to inequalities (5.95) predicts either lattice stability or counterclockwise rotation of the lattice about the (110) loading axis, the latter motion having been found by Skalli (1984) for a nominal $(110)(1\bar{1}\bar{2})(\bar{1}1\bar{1})$ initial orientation of an aluminum crystal. Moreover, for the class of unique solutions, the consistency conditions (eqs. (5.95)) reduce to

$$(a1, b\bar{2}) \quad (3H_1 - \tfrac{5}{4}f) + \{3(h_2 + h_3) - 5f\}y = 6f' - 2g',$$
$$(c\bar{1}, c2) \quad (3H_3 + \tfrac{7}{4}f) + \{3(h_4 + h_5) + 7f\}y \geqslant 4g', \tag{5.99}$$
$$(a3, b\bar{3}) \quad (3H_5 - \tfrac{1}{2}f) + \{3(h_8 + h_9) - 2f\}y \geqslant 4g',$$

with at least one of the last two conditions an equality (corresponding, of course, to the smaller of the two left-hand sides if different for a given theory).

In the case of the simple theory (eq. (5.69)), the consistency conditions (see eqs. (42) of Chidambarrao & Havner (1988a)) give the same unique kinematic result as for Taylor hardening: lattice stability. However, systems $c\bar{1}$ and $c2$ are "inactivated" (that is, $(\tau_k^c)' > \tau_k'$, so the systems are no longer critical after the first strain increment). In addition (Chidambarrao and Havner, eqs. (44)–(45)),

$$g' = f' + \tfrac{1}{2}f, \quad h = \frac{2}{27}(f' + f) \quad \text{(simple theory)}, \tag{5.100}$$

and the hardening among the six systems is anisotropic:

$$\begin{aligned}
&(a1, b\bar{2}) \quad (\tau_1^c)' = (\tau_2^c)' = \frac{\sqrt{6}}{9}f' - \frac{1}{6\sqrt{6}}f,\\
&(c\bar{1}, c2) \quad (\tau_3^c)' = (\tau_4^c)' = \frac{\sqrt{6}}{9}f' + \frac{11}{6\sqrt{6}}f,\\
&(a\bar{3}, b3) \quad (\tau_5^c)' = (\tau_6^c)' = \frac{\sqrt{6}}{9}f' + \frac{1}{3\sqrt{6}}f.
\end{aligned} \tag{5.101}$$

For the P.A.N. rule (eq. (5.70)), the solution of the corresponding consistency conditions (Chidambarrao and Havner, eqs. (65)) is not unique. The final results are

$$\begin{aligned}
&\gamma_1' = \frac{\sqrt{6}}{2}(1 + x), \quad \gamma_2' = \frac{\sqrt{6}}{2}(1 - x), \quad |x| \leqslant \tfrac{1}{6},\\
&\gamma_3' = \frac{\sqrt{6}}{2}(\tfrac{1}{3} - x), \quad \gamma_4' = \frac{\sqrt{6}}{2}(\tfrac{1}{3} + x),\\
&\gamma_5' = \frac{\sqrt{6}}{2}(\tfrac{1}{6} - x), \quad \gamma_6' = \frac{\sqrt{6}}{2}(\tfrac{1}{6} + x),
\end{aligned} \tag{5.102}$$

whence each of systems 1 through 4 and at least one of systems 5 and 6 ($a\bar{3}$ and $b3$) is active. The lattice spin is (from eqs. (69) of Chidambarrao and Havner)

$$\omega_x = \frac{\sqrt{2}}{6}, \quad \omega_y = 0, \quad \omega_z = \frac{x}{2}. \tag{5.103}$$

Thus, the P.A.N. rule predicts counterclockwise rotation of the lattice about the loading direction for the solution $x = 0$. In addition, for all x

$$g' = f' + \tfrac{1}{3}f, \quad \hbar = \tfrac{2}{27}(f' + \tfrac{1}{6}f) \quad \text{(P.A.N. rule)}, \tag{5.104}$$

and the initial anisotropic hardening among the critical systems is uniquely given by (from Chidambarrao and Havner's eqs. (62), (66), and (67))

$$\begin{aligned}
&(a1, b\bar{2}) \quad (\tau_1^c)' = (\tau_2^c)' = \frac{\sqrt{6}}{9}f' + \frac{1}{6\sqrt{6}}f, \\
&(c\bar{1}, c2) \quad (\tau_3^c)' = (\tau_4^c)' = \frac{\sqrt{6}}{9}f' + \frac{1}{12\sqrt{6}}f, \\
&(a\bar{3}, b3) \quad (\tau_5^c)' = (\tau_6^c)' = \frac{\sqrt{6}}{9}f' + \frac{1}{3\sqrt{6}}f.
\end{aligned} \tag{5.105}$$

Chidambarrao and Havner (1988a) present an extensive numerical investigation of an aluminum crystal at $\boldsymbol{\kappa}_0 = (1\bar{1}\bar{2})$ for each of Taylor's rule, the simple theory, and the P.A.N. rule. Their analyses are based upon a best fit of the data points in Skalli (1984, fig. 45) corresponding to a nominal $(110)(1\bar{1}\bar{2})(\bar{1}1\bar{1})$ initial orientation, using equation (5.76) for the general form of the active stress–strain curve. The result is

$$f(e_L) = 58.0 + 44.5e_L + 44.7\tanh(11.48e_L) \quad \text{(MPa)}. \tag{5.106}$$

In the case of the P.A.N. rule, because the initial solution (eqs. (5.102)) of the consistency conditions and constraints is not unique, a number of different discrete values of x were chosen as starting solutions for the computer program mentioned in Section 5.5.2. Only for $x = 0$ (i.e., $\gamma_1' = \gamma_2' = \sqrt{6}/2$) did a numerical solution exist after the initial strain increment $\delta e_L = 0.01$, and that solution corresponded to continued rotation of the lattice about the loading axis. It is likely that a second-order analysis, as in Havner (1984) and Havner & Sue (1985), would confirm analytically that $x = 0$ is the unique solution for a small but finite strain increment δe_L.

In Fig. 5.13 (which reproduces Chidambarrao and Havner's fig. 9) are shown the initial yield locus in the biaxial-compression quadrant of f, g stress space, the stress path from $f_0 = g_0$ to $e_L = 0.72$ for each of the three theories, and the subsequent yield loci at that strain. Because only the P.A.N. rule predicts lattice rotation, the traces of individual yield loci $\tau_k = \tau_k^c$ in this subspace merely translate parallel to their initial positions for the other two theories, as is evident from the figure.

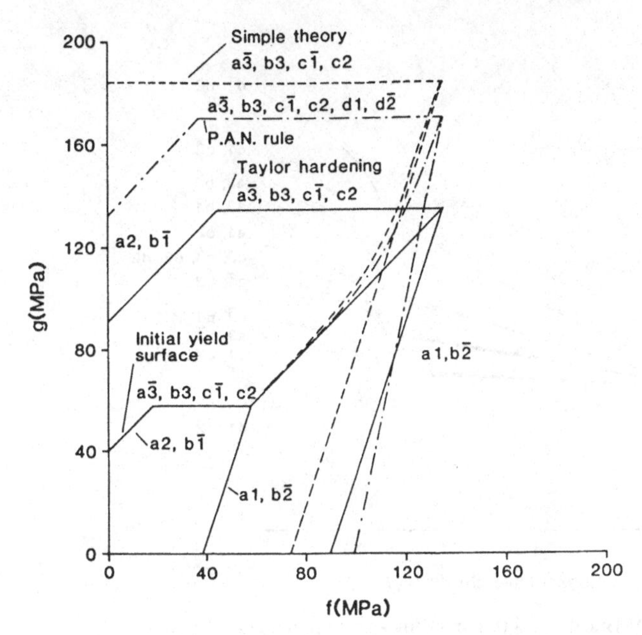

Fig. 5.13. Initial yield locus of an aluminum crystal in orientation $(110)(1\bar{1}\bar{2})(\bar{1}1\bar{1})$ and subsequent yield loci for various hardening theories at a logarithmic strain of 0.72, including the corresponding stress paths (from Chidambarrao & Havner 1988a, fig. 9). Reprinted with permission from *Int. J. Plasticity* 4, D. Chidambarrao & K. S. Havner, Finite deformation analysis of f.c.c. ccrystals in $(110)(1\bar{1}\bar{2})(\bar{1}1\bar{1})$ channel die compression, Copyright 1988, Pergamon Press, PLC.

For the P.A.N. rule, as may be seen in Chidambarrao and Havner's figure 8, systems $d1$ and $d\bar{2}$ (numbered 7 and 8) become critical at $e_L = 0.723$ from the computer calculations; and the solution at the next step ($e_L = 0.73$) is $\gamma_1' = \gamma_2' = \sqrt{6}/2$ (as before), $\gamma_3' = \gamma_4' = \gamma_5' = \gamma_6' = 0$, and $\gamma_7' = \gamma_8' \approx 13/12$. Thenceforward only systems $a1$, $b\bar{2}$, $d1$, and $d\bar{2}$ are critical and active. Equal slip on the latter two systems (as on $a1$ and $b\bar{2}$) contributes to ω_x but not to ω_y and ω_z. Thus, the lattice continues to rotate about the loading axis, although at a different rate with e_L. Moreover, there are finite jumps in parameter $\hbar$ and $g'(e_L)$ at this change in deformation mode (see Chidambarrao and Havner's figs. 3 and 4). Consequently, the theoretical hardening curves for the P.A.N. rule also exhibit a finite change in slope at $e_L = 0.73$, as evident in Fig. 5.14 (from fig. 7 of Chidambarrao and Havner). Hardening curves for the simple theory are given in Chidambarrao and Havner's figure 5. Lattice rotations (zero for Taylor hardening and the simple theory, as already noted) and crystal shears are

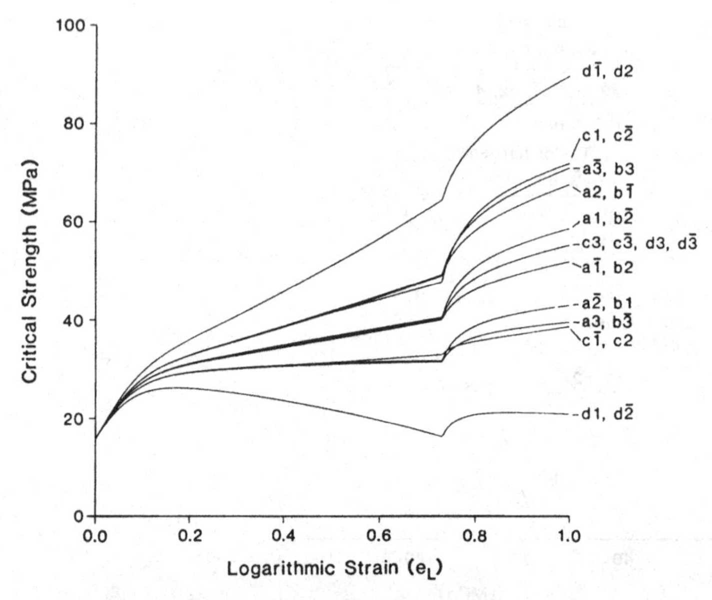

Fig. 5.14. Theoretical critical-strength curves for an aluminum crystal in initial orientation $(110)(1\bar{1}\bar{2})(\bar{1}1\bar{1})$ according to the P.A.N. rule (from Chidambarrao & Havner 1988a, fig. 7). Reprinted with permission from *Int. J. Plasticity* 4, D. Chidambarrao & K.S. Havner, Finite deformation analysis of f.c.c. crystals in $(110)(1\bar{1}\bar{2})(\bar{1}1\bar{1})$ channel die compression, Copyright 1988, Pergamon Press, PLC.

compared with experiment and with the predictions of a very different theoretical approach in the next section.

5.6.3 *Orientation $(110)(1\bar{1}\bar{2})(\bar{1}1\bar{1})$: Minimum Plastic Spin*

The concept of *minimum plastic spin* in crystal mechanics was introduced in Havner (1987b) and extensively explored by Fuh & Havner (1989) for three families of multiple-slip configurations of f.c.c. crystals, including analyses of both axial loading and the channel die test. The basic idea may be expressed as follows:

> It is postulated that, at each stage of the deformation, incremental slip occurs in those critical systems and proportional amounts as is required to make the relative spin of material and lattice a minimum, subject to the loading conditions and constraints, if the solution is otherwise nonunique.

For range I in (110) channel die compression, Fuh and Havner showed that minimum plastic spin, without reference to hardening rule, predicts the identical lattice rotation and crystal shearing (eqs. (5.66)–(5.68)) as are determined by the class of hardening theories given by equations (5.52). This equivalence of minimum-plastic-spin and (initial) hardening-theory solutions also holds in ranges II and III (Fuh and Havner, sect. 3.1), in each of which ranges the lattice is theoretically stable. In the singular orientation $Y=(1\bar{1}\bar{2})$ having three degrees of freedom, however, the general consistency conditions (eqs. (5.95)) permit an infinite variety of solutions, with particular hardening rules giving various specific results for slip rates, as we have seen.

The range of all kinematically admissible solutions for $Y=(1\bar{1}\bar{2})$ may be represented in *plastic-spin space* (Fuh & Havner 1989). For this purpose it is useful to choose orthogonal lattice directions [110], $[3\bar{3}\bar{2}]$, and $[\bar{1}1\bar{3}]$ as a reference frame. Let $\mathbf{a}_\Omega$, with components a_1, a_2, a_3 on the standard lattice axes ([100], [010], and [001]) and components a_x, a_y, a_z on the channel axes, denote the axial vector of plastic spin Ω. We further denote the components of $\mathbf{a}_\Omega$ on the new reference frame by a_1^*, a_2^*, and a_3^*. Then, from $\Omega=\sum\Omega_j\gamma_j'$, Table 1, and the corresponding orthogonal transformation, we have (Fuh and Havner, eqs. (3.39))

$$
\begin{aligned}
a_1^* &= \Omega_{32}^* = \frac{1}{4\sqrt{3}}(\gamma_1'+\gamma_2') - \frac{\sqrt{3}}{4}(\gamma_3'+\gamma_4') + \frac{1}{2\sqrt{3}}(\gamma_5'+\gamma_6'),\\
a_2^* &= \Omega_{13}^* = \frac{11}{4\sqrt{33}}(\gamma_1'-\gamma_2') - \frac{1}{4\sqrt{33}}(\gamma_3'-\gamma_4') + \frac{1}{\sqrt{33}}(\gamma_5'-\gamma_6'),\\
a_3^* &= \Omega_{21}^* = \frac{2}{\sqrt{66}}(\gamma_3'-\gamma_4') + \frac{3}{\sqrt{66}}(\gamma_5'-\gamma_6').
\end{aligned}
\tag{5.107}
$$

Upon substituting the constraints, equations (5.88), and solving for the γ_k', one obtains the following kinematic inequalities (eqs. (3.40) of Fuh and Havner):

$$
\begin{aligned}
\gamma_1' &= \frac{\sqrt{6}}{2} + \frac{2\sqrt{33}}{11}a_2^* - \frac{3\sqrt{66}}{110}a_3^* \geqslant 0,\\
\gamma_2' &= \frac{\sqrt{6}}{2} - \frac{2\sqrt{33}}{11}a_2^* + \frac{3\sqrt{66}}{110}a_3^* \geqslant 0,\\
\gamma_3' &= \frac{\sqrt{6}}{5} - \frac{2\sqrt{3}}{5}a_1^* + \frac{\sqrt{66}}{10}a_3^* \geqslant 0,
\end{aligned}
\tag{5.108}
$$

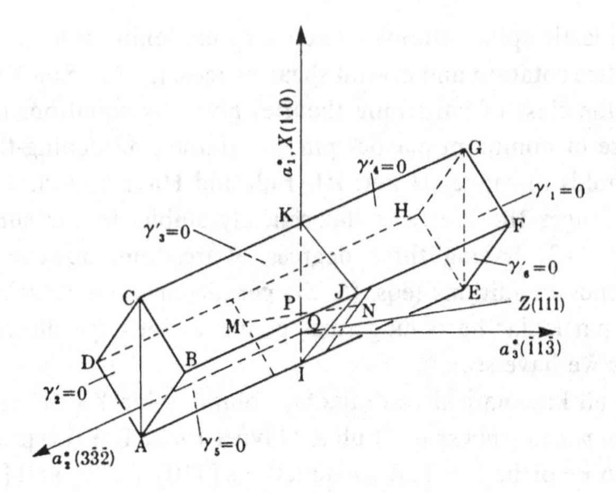

Fig. 5.15. Kinematically admissible solutions parallelepiped in plastic-spin space for orientation $(110)(1\bar{1}\bar{2})(\bar{1}1\bar{1})$ (from Fuh & Havner 1989, fig. 11).

$$\gamma'_4 = \frac{\sqrt{6}}{5} - \frac{2\sqrt{3}}{5} a_1^* - \frac{\sqrt{66}}{10} a_3^* \geqslant 0,$$

$$\gamma'_5 = \frac{\sqrt{6}}{20} + \frac{2\sqrt{3}}{5} a_1^* + \frac{\sqrt{66}}{10} a_3^* \geqslant 0,$$

$$\gamma'_6 = \frac{\sqrt{6}}{20} + \frac{2\sqrt{3}}{5} a_1^* - \frac{\sqrt{66}}{10} a_3^* \geqslant 0.$$

These inequalities define a parallelepiped in plastic-spin space whose long axis is parallel to a_2^*, as shown in Fig. 5.15 (from fig. 11 of Fuh and Havner); and all kinematically admissible solutions are contained within or on this parallelepiped. Points I, J, K are corners of a rhombus that is the intersection of the $a_1^* a_3^*$-plane (lattice plane $(3\bar{3}\bar{2})$) with the parallelepiped. Faces $ABCD$ ($\gamma'_2 = 0$) and $EFGH$ ($\gamma'_1 = 0$) are parallel to axis a_1^* and to each other, but not to plane $a_1^* a_3^*$.

From equations (5.97) and (5.108), the lattice-spin components on the channel axes may be expressed

$$\omega_x = \frac{\sqrt{2}}{5} - \frac{2}{5} a_1^*, \quad \omega_y = -\frac{2}{\sqrt{33}} a_2^* - \frac{8}{5\sqrt{66}} a_3^*,$$

$$\omega_z = \frac{4}{\sqrt{66}} a_2^* - \frac{5}{\sqrt{33}} a_3^*, \tag{5.109}$$

and line IK ($a_2^* = a_3^* = 0$) in Fig. 5.15 is the locus of all solutions corresponding to rotation of the lattice about the loading axis (which also is the set of hardening-theory-dependent unique solutions, eqs. (5.98), to the consistency conditions). Point K, at which $\gamma_3' = \gamma_4' = 0$ and $a_1^* = 1/\sqrt{2}$, is the stable lattice solution $\omega = 0$ of Taylor's rule and the simple theory. Line MN is the locus of (first-order) solutions for the P.A.N. rule given by equations (5.102). This line is parallel to $Z = (\bar{1}1\bar{1})$, as is evident from equations (5.103), with point M in face $ADHE$ ($\gamma_5' = 0$, $x = \frac{1}{6}$) and point N in face $ABEF$ ($\gamma_6' = 0, x = -\frac{1}{6}$). (The other line drawn through point N is parallel to AB and EF.) Point P, the intersection of lines MN and IK (at which $a_1^* = 1/(6\sqrt{2})$), is the initial solution $x = 0$ adopted for the P.A.N. rule, as explained in the preceding section. Moreover, it obviously is the minimum-plastic-spin solution among equations (5.102) (that is, along line MN) but not, of course, the absolute (kinematically admissible) minimum-plastic-spin solution $\mathbf{a}_\Omega = \mathbf{0}$ represented by point O.

As remarked by Fuh and Havner, $Y = (1\bar{1}\bar{2})$ is unique among initial constraint directions in (110) channel die compression in that it is the only orientation for which Taylor hardening, the simple theory, and the P.A.N. rule do not permit (or uniquely predict) the solution corresponding to absolute minimum plastic spin. However, another set of hardening parameters can readily be chosen that will satisfy the reduced consistency conditions, equations (5.99), for the minimum-spin solution ($y = 3/20$) in this orientation. Moreover, this set also will give the minimum-plastic-spin solution in each of ranges I, II, and III (eqs. (5.32)), as that solution is identically the unique one found in the respective range for *any* hardening theory of the general family, equations (5.52) (Havner & Chidambarro 1987; Chidambarrao & Havner 1988b). Therefore, following Fuh and Havner (pp. 219–21), we now turn to an investigation of the consequences of $\mathbf{a}_\Omega = \mathbf{0}$ for lattice rotation and crystal shearing from initial orientation $Y = (1\bar{1}\bar{2})$, with no further consideration of a specific hardening rule associated with that minimum spin.

For active systems 1 to 6 and any subsequent orientation in range II, after a finite rotation about $X = (110)$, the constraints become (from eqs. (5.24) and Table 2)

$$\gamma_1' + \gamma_2' = \sqrt{6}, \quad (b+2)(b-1)(\gamma_3' + \gamma_4') + 2b(\gamma_5' + \gamma_6') = \sqrt{6}b(b-1),$$
$$(2-b)(\gamma_1' - \gamma_2') - (b+2)(\gamma_3' - \gamma_4') + 4(\gamma_5' - \gamma_6') = 0, \qquad (5.110)$$

which reduce to equations (5.88) at $Y = (1\bar{1}\bar{2})$ ($b = 2$). The components of plastic spin on the channel axes are (Fuh and Havner, eqs. (3.47)[9])

$$a_x = \frac{1}{4\sqrt{3}}\{\sqrt{6} - 3(\gamma_3' + \gamma_4') + 2(\gamma_5' + \gamma_6')\},$$

$$a_y = 1/(2k\sqrt{6})\{(b+3)(\gamma_1' - \gamma_2') + (b-1)(\gamma_3' - \gamma_4') + 2b(\gamma_5' - \gamma_6')\},$$

$$a_z = 1/(4k\sqrt{3})\{-(3b-2)(\gamma_1' - \gamma_2') + (b+2)(\gamma_3' - \gamma_4') + 4(\gamma_5' - \gamma_6')\},$$

$$(5.111)$$

with $k = \sqrt{(b^2+2)}$ as before. Thus, for the minimum-plastic-spin solution $\Omega = 0$ one obtains

$$\gamma_1' = \gamma_2' = \frac{\sqrt{6}}{2}, \quad \gamma_3' = \gamma_4' = \frac{\sqrt{6}}{2} b^2/(b^2 + 4b - 2),$$

$$\gamma_5' = \gamma_6' = \frac{\sqrt{6}}{2}(b-1)^2/(b^2 + 4b - 2), \tag{5.112}$$

each of which is positive, as required, for $b > 1$.

The corresponding lattice-spin components on the channel frame are (from eqs. (5.29) and (5.112) and Table 1)

$$\omega_x = \sqrt{2b(b-1)/(b^2+4b-2)}, \quad \omega_y = \omega_z = 0. \tag{5.113}$$

Hence $Y = (1\bar{1}\bar{1})$ $(b = 1)$ is the limiting position at infinite slips. Then, from $\omega_x = \phi'$ and $b = \sqrt{2}\cot\phi$ (eqs. (5.62) and (5.67)) there follow (Fuh and Havner, eqs. (3.50) and (3.51))

$$b' = -b(b-1)(b^2+2)/(b^2+4b-2) \tag{5.114}$$

and

$$e_L = \ln[(b^2+2)/(3b(b-1))], \quad 2 \geqslant b > 1, \tag{5.115}$$

for minimum plastic spin with active systems $a1, b\bar{2}, c\bar{1}, c2, a\bar{3}$, and $b3$. Finally, from equation $(5.59)_1$, with $w_x = \omega_x = \phi'$ (as $\Omega_{zy} \equiv a_x = 0$ in eq. $(5.61)_1$), one has

$$\tan\chi_x = 2\exp(e_L(\phi)) \int_{\phi_0}^{\phi} \exp(-e_L(\varphi))\, d\varphi, \tag{5.116}$$

and upon substituting equations (5.67) and (5.115) and integrating with

[9] I have corrected a misprint in the third one of those equations.

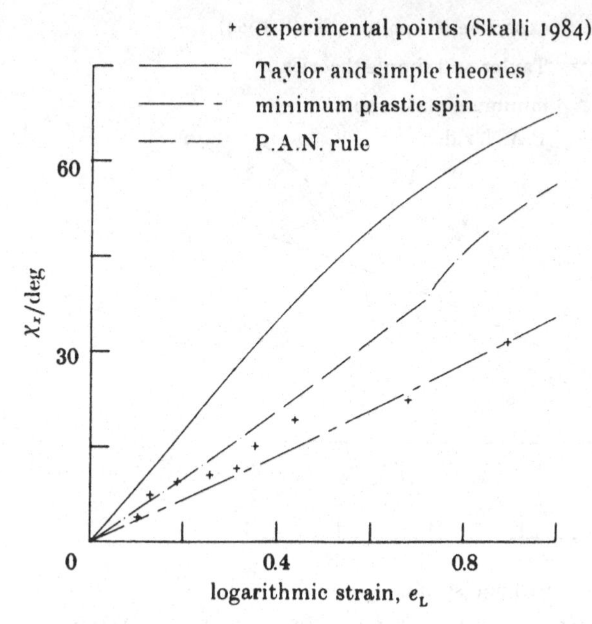

Fig. 5.16. Theoretical predictions of shear angle χ_x against logarithimic compressive strain for initial orientation $(110)(1\bar{1}\bar{2})(\bar{1}1\bar{1})$, and experimental points for the nominal $(110)(1\bar{1}\bar{2})(\bar{1}1\bar{1})$ orientations of Skalli (1984) (from Fuh & Havner 1989, fig. 12).

respect to parameter b we obtain (Fuh and Havner, eq. (3.53))

$$\tan \chi_x = \frac{b^2+2}{b(b-1)}\left\{\frac{\sqrt{2}(b-1)}{b^2+2} + \arctan(\sqrt{2}/b) - \frac{\sqrt{2}}{6} - \arctan\frac{1}{\sqrt{2}}\right\}. \tag{5.117}$$

Comparisons of the minimum-plastic-spin solution with numerical results for Taylor's rule, the simple theory, and the P.A.N. rule (from Chidambarrao & Havner 1988a, b) and the experimental data of Skalli (1984) for each of crystal shear χ_x and the channel-axis orientation are shown in Figs. 5.16 and 5.17 (Fuh and Havner's figs. 12 and 13). In the experiment, the aluminum crystal orientation was only nominally $(110)(1\bar{1}\bar{2})(\bar{1}1\bar{1})$, and there are ambiguities in Skalli (1984) as to its actual initial position. As explained by Fuh and Havner (p. 221), the points $(a)_{\text{proj}}$ and $(b)_{\text{proj}}$ in Fig. 5.17 "are the projections on the (110) plane of the initial channel-axis position as determined from Skalli's table 2 and figure 37b, respectively; and the corresponding unit vectors $\boldsymbol{\mu}_a$ and $\boldsymbol{\mu}_b$ are 2.4° and 1.2° from their projected positions." In addition, the initial loading-axis orientation apparently was about 5° from (110). However, the loading

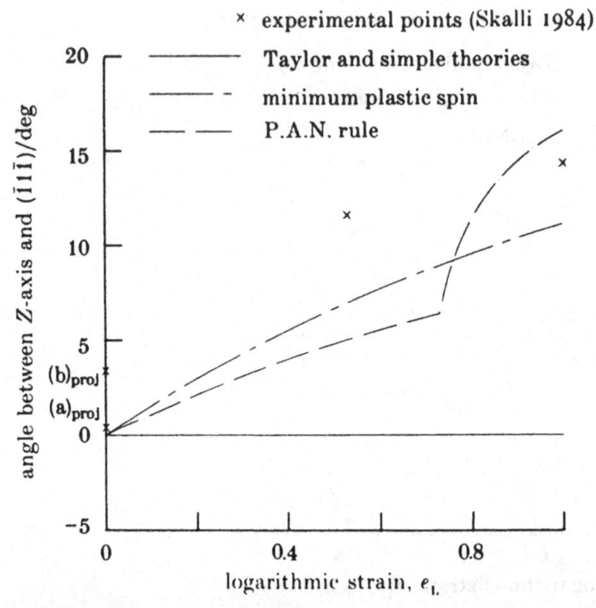

Fig. 5.17. Theoretical predictions of the angle between the channel axis and $(\bar{1}1\bar{1})$ against logarithmic strain for initial orientation $(110)(1\bar{1}\bar{2})(\bar{1}1\bar{1})$, and experimental points for the nominal $(110)(1\bar{1}\bar{2})(\bar{1}1\bar{1})$ orientation of Skalli (1984) (from Fuh & Havner 1989, fig. 13).

axis rotated into (110) and remained there; and the subsequent experimental points in Fig. 5.17 correspond to positions of the channel axis on the perimeter of the (110) projection (Fig. 5.4), as in the theoretically determined cases.

From Figs. 5.16 and 5.17, the predictions of minimum plastic spin as compared with experiment are seen to be quite good over the range of finite deformation. In particular, as pointed out by Fuh and Havner, if $\boldsymbol{\mu}_b$ (from fig. 37b of Skalli) is the actual initial orientation, then the channel-axis rotation relative to the lattice in Fig. 5.17 "is almost identical to that predicted by minimum plastic spin. Direction $\boldsymbol{\mu}_b$ is 3.6° from $(\bar{1}1\bar{1})$ at the outset; and the last experimental point, at $e_L = 1.0$, is 3.2° from the minimum-spin curve" (p. 221). In contrast, the three hardening rules do not give generally acceptable predictions (although the P.A.N. rule is clearly superior to the others for this orientation). It also is worth noting that Toth, Jonas & Neale (1990) investigated the "rate-insensitive limit" of a rate-dependent slip theory[10] and found the result for χ_x (their

[10] Rate-dependent theories are reviewed in Chapter 7.

eq. (31)) to be "identical with that obtained from the Taylor and simple theories" (i.e., the uppermost curve in Fig. 5.16).

5.7 Survey of Other Orientations

Because of the analytical uniqueness of the exact orientation $Y = (1\bar{1}\bar{2})$ in (110) compression and the fact that the comparison experimental position in Skalli (1984) did not precisely coincide with it, Chidambarrao & Havner (1988b) investigated Taylor hardening, the simple theory, and the P.A.N. rule for nearby constraint directions $(3\bar{3}\bar{7})$ (range I) and $(5\bar{5}\bar{9})$ (range II) on either side of $(1\bar{1}\bar{2})$. Results for the former case are found in Chidambarrao and Havner, section 4. Here we briefly review the latter (from their sect. 5.2).

In range II $(2 > b > 1)$, all theories defined by equations (5.52) predict initial lattice stability corresponding to the deformation mode

$$\gamma_1 = \gamma_2 = \frac{\sqrt{6}}{2} e_L, \quad \gamma_5 = \gamma_6 = \frac{\sqrt{6}}{4}(b_0 - 1)e_L \tag{5.118}$$

(Chidambarrao & Havner 1988b, eqs. (5.12)), with plastic spin

$$a_x = \frac{b_0}{2\sqrt{2}}, \quad a_y = a_z = 0 \tag{5.119}$$

(from eqs. (5.111)), which is the minimum-plastic-spin solution among critical systems $a1$, $b\bar{2}$, $a\bar{3}$, and $b3$ (Fuh & Havner 1989, sect. 3.1). For Taylor's rule this solution continues to unlimited deformations, but for each of the simple theory and the P.A.N. rule there subsequently will be a change in mode (because of the anisotropy of hardening for these theories), with associated lattice rotation about the loading axis as shown in Chidambarrao and Havner's figure 20. For the P.A.N. rule there is a double change: first to active systems 1 through 6 (as with $Y = (1\bar{1}\bar{2})$ initially), then to active systems 1, 2, 7, and 8 (again as with $Y = (1\bar{1}\bar{2})$). For $Y = (5\bar{5}\bar{9})$ these changes occur at $e_L \approx 0.3$ and 0.53, respectively. For the simple theory, the mode changes directly from lattice stability to that of active systems 1, 2, 7, and 8 at $e_L \approx 0.7$. For both theories this latter mode is defined by (Fuh & Havner 1989, eqs. (3.63)–(3.64))

$$\begin{gathered} \gamma_1' = \gamma_2' = \frac{\sqrt{6}}{2}, \quad \gamma_7' = \gamma_8' = \frac{\sqrt{6}}{2} b(b-1)/(2+b-b^2), \\ \omega_x = \sqrt{2}(b-1)/(2-b), \quad \omega_y = \omega_z = 0, \quad 2 > b > 1. \end{gathered} \tag{5.120}$$

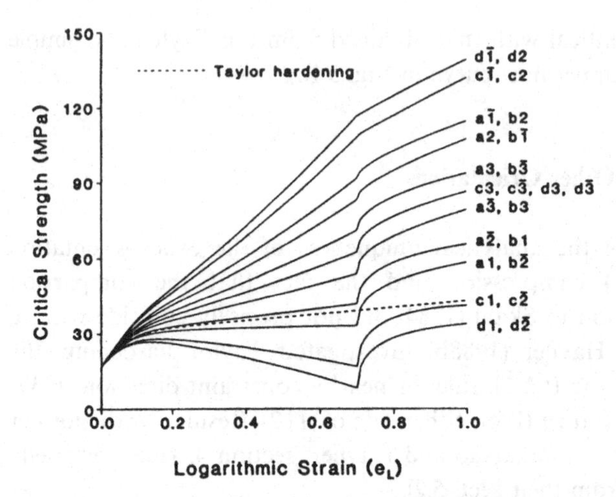

Fig. 5.18. Theoretical critical-strength curves for an aluminum crystal in initial orientation $(110)(5\bar{5}\bar{9})(\bar{9}9\overline{10})$ according to the simple theory (from Chidambarrao & Havner 1988b, fig. 13). Reprinted with permission from *J. Mech. Phys. Solids* 36, D. Chidambarrao & K. S. Havner, On finite deformation of f.c.c. crystals in (110) channel die compression, Copyright 1988, Pergamon Press, PLC.

Experimental lattice orientations in range II apparently rotate about the loading axis (Skalli 1984); but whether there is an initial interval of lattice stability as the general theory predicts cannot be determined from the limited data. With regard to crystal shear χ_x, predicted results to $e_L = 1.0$ differ little among the three theories (see fig. 17 of Chidambarrao and Havner), and each theory gives values approximately twice those obtained experimentally by Skalli (1984) in his nearby nominal $(110)(1\bar{1}\bar{2})(\bar{1}1\bar{1})$ orientation (Fig. 5.16 here).

Critical-strength curves for initial orientation $(110)(5\bar{5}\bar{9})(\bar{9}9\overline{10})$ according to the simple theory and the P.A.N. rule are shown in Figs. 5.18 and 5.19 (figs. 13 and 14 of Chidambarrao & Havner (1988b)), with the Taylor-hardening curve superimposed on each figure. The mode changes at the respective strains previously noted are evident from the finite changes in slope. (All curves were determined from successive incremental computer solutions of eqs. (5.30), with $\delta e_L = 0.01$ and $f(e_L)$ given by eq. (5.106).)

In range III $(1 > b > 0)$, all theories (eqs. (5.52)) predict initial lattice stability (Chidambarrao and Havner, sect. 6). The active systems are $a1, b\bar{2}, a\bar{2}$, and $b1$, with finite slips

$$\gamma_1 = \gamma_2 = \frac{\sqrt{6}}{4}(1 + b_0)e_L, \quad \gamma_9 = \gamma_{10} = \frac{\sqrt{6}}{4}(1 - b_0)e_L. \tag{5.121}$$

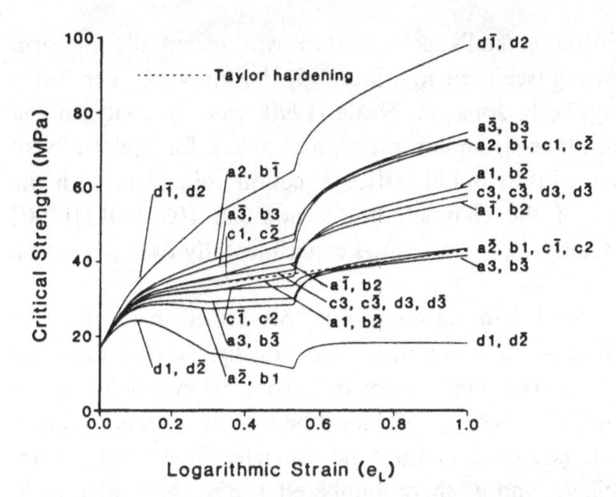

Fig. 5.19. Theoretical critical-strength curves for an aluminum crystal in initial orientation $(110)(5\bar{5}\bar{9})(\bar{9}9\bar{10})$ according to the P.A.N. rule (from Chidambarrao & Havner 1988b, fig. 14). Reprinted with permission from *J. Mech. Phys. Solids* 36, D. Chidambarrao & K. S. Havner, On finite deformation of f.c.c. crystals in (110) channel die compression, Copyright 1988, Pergamon Press, PLC.

Moreover, although plastic spin Ω is not zero, Chidambarrao and Havner show that both $\sigma\Omega$ and σD equal zero in range III (corresponding to eqs. (5.36)). Thus, from the defining equations (4.16) and (4.27) (with $D^p = D$ in the present analysis), the simple theory and the P.A.N. rule, as well as Taylor's rule, predict isotropic hardening $\tau_k^c = f/\sqrt{6}$, $k = 1, \ldots, N$. The foregoing solution also may be extended to $Y = (1\bar{1}0)$ ($b = 0$; see Chidambarrao and Havner, pp. 304–5, or Sue & Havner (1984, sect. 7)). Therefore, in the range $1 > b \geqslant 0$, equation (5.121) applies without limit to the three theories. Thus, all three predict the continued lattice stability that has been observed for copper (Wonsiewicz & Chin 1970b) and aluminum crystals (Skalli et al. 1983) in this range. (Lattice stability also is the minimum-plastic-spin solution for both range III and $Y = (1\bar{1}0)$, as shown by Fuh & Havner (1989, sect. 3.1). In addition, a stable lattice is predicted in each of ranges II and III and orientations $Y = (1\bar{1}\bar{1})$ ($b = 1$) and $(1\bar{1}0)$ by the rate-insensitive limit of the rate-dependent flow rule investigated by Toth, Jonas & Neale (1990, sects. 4.1–4.2), in agreement with minimum-plastic-spin theory.)

Seeking to achieve fully constrained deformation of particular aluminum crystals in channel die compression, Driver & Skalli (1982) devised a tricrystal configuration in which a crystal of orientation $\boldsymbol{\iota}$, $\boldsymbol{\kappa}$, $\boldsymbol{\mu}$ was embedded between two other crystals of orientations $-\boldsymbol{\iota}$, $-\boldsymbol{\kappa}$, $\boldsymbol{\mu}$, with the result that the central crystal's deformation was essentially uniform with negligible shearing (see their figs. 4 and 5).[11] Fuh & Havner (1989, sect. 2), followed by Toth, Jonas & Neale (1990, sect. 3), analyzed the corresponding pure-plane-strain-compression problem for the family of f.c.c. crystals having a [100] initial lattice direction coincident with the longitudinal axis Z of the channel die. A nominal $[021][0\bar{1}2][100]$ orientation in this family was investigated experimentally by Driver et al. (1983).

As the (central) crystal deformation is fully prescribed in the channel frame, with all strain-rate components identically zero except for $-d_{xx} = d_{zz} = 1$ at finite straining, it turns out (from the symmetry of this family of orientations) that eight systems must be critical in order to satisfy these five constraints (see fig. 2 of Fuh and Havner). These systems are $a2$, $a\bar{3}$, $b2$, $b\bar{3}$, $c2$, $c\bar{3}$, $d2$, and $d\bar{3}$ (here numbered 1–8 in that order, as in sect. 2 of Fuh and Havner), with equal biaxial stress state $f_0 = g_0 = \sqrt{6}\tau_0$. Consequently, the unknown slip rates again may be expressed in terms of the three components of plastic spin Ω. The resulting kinematic inequalities are (Fuh and Havner, eqs. (2.19))

$$
\begin{aligned}
\frac{4}{\sqrt{6}}\gamma'_1 &= -a_1 + a_2 + \sin\theta(\cos\theta + \sin\theta) \geqslant 0,\\
\frac{4}{\sqrt{6}}\gamma'_2 &= a_1 - a_3 + \cos\theta(\cos\theta + \sin\theta) \geqslant 0,\\
\frac{4}{\sqrt{6}}\gamma'_3 &= a_1 - a_2 - \sin\theta(\cos\theta - \sin\theta) \geqslant 0,\\
\frac{4}{\sqrt{6}}\gamma'_4 &= -a_1 - a_3 + \cos\theta(\cos\theta - \sin\theta) \geqslant 0,\\
\frac{4}{\sqrt{6}}\gamma'_5 &= -a_1 - a_2 + \sin\theta(\cos\theta + \sin\theta) \geqslant 0,\\
\frac{4}{\sqrt{6}}\gamma'_6 &= a_1 + a_3 + \cos\theta(\cos\theta + \sin\theta) \geqslant 0,\\
\frac{4}{\sqrt{6}}\gamma'_7 &= a_1 + a_2 - \sin\theta(\cos\theta - \sin\theta) \geqslant 0,\\
\frac{4}{\sqrt{6}}\gamma'_8 &= -a_1 + a_3 + \cos\theta(\cos\theta - \sin\theta) \geqslant 0,
\end{aligned}
\tag{5.122}
$$

[11] A comparable result was obtained by Chin, Nesbitt, & Williams (1966, fig. 10) on a Permalloy crystal, orientation $(110)(\bar{1}1\bar{1})(\bar{1}12)$, constrained between two rectangular polycrystalline blocks.

where a_1, a_2, a_3 (as before) are the components of $\mathbf{a}_\Omega$ on the lattice axes [100][010][001] and θ is the counterclockwise orientation of loading axis X from [010] on a [100] stereographic projection (see fig. 1 of Fuh and Havner). From symmetry, it is only necessary to consider the range $0 \leqslant \theta \leqslant 45°$ to encompass all orientations in the family $Z = [100]$. (Note that $\sum \gamma'_k = \sqrt{6}$.)

Similarly as for orientation $(110)(1\bar{1}\bar{2})(\bar{1}1\bar{1})$ (Fig. 5.15), the domain of all kinematically admissible solutions can be displayed in plastic-spin space as a polyhedron bounded by planes $\gamma'_k = 0$ (eqs. (5.122)). Four representative polyhedra are shown in figures 3–6 of Fuh & Havner (1989), corresponding to different ranges of θ, and Fig. 5.20 reproduces figure 5 of that work, with

$$b = 1 - \sin 2\theta, \quad c = \sin\theta(\cos\theta - \sin\theta), \quad e = \sin 2\theta - \cos 2\theta. \tag{5.123}$$

(For $\theta = 0$ ($X = [010]$) the hexahedron of Fig. 5.20 degenerates into the line segment from -1 to $+1$ on the a_3-axis. For $\theta = 45°$ ($X = [011]$) the hexahedron degenerates to a point at the origin. This position is crystallographically equivalent to the stable-lattice orientation $Y = (1\bar{1}0)$ in (110) compression of a single crystal that deforms rectangularly without the necessity of additional constraints, as was found experimentally for copper by Wonsiewicz & Chin (1970b) and shown analytically by Sue & Havner (1984, sect. 7).)

The solution of the consistency conditions (eqs. (2.31) of Fuh and Havner) for each of Taylor hardening and the P.A.N. rule is represented by segment FE of the a_1-axis in Fig. 5.20, and by a comparable line segment for the other ranges of θ (see figs. 3, 4, and 6 in Fuh and Havner). Because $\omega = -\Omega$ in pure plane-strain compression, this solution set corresponds to clockwise rotation of the lattice about the longitudinal channel axis $Z = [100]$. It also corresponds to isotropic hardening for both theories and may be continued to unlimited deformations for any a_1 spin within the interval. For the simple theory, all kinematically admissible solutions are permitted by the first-order analysis, but only the solutions (segment FE) associated with rotation about Z give isotropic hardening and may be automatically extended to finite strain. Thus, all three hardening rules permit the minimum-plastic-spin solution (point F) defined by

$$a_1 = \theta' = \sin\theta(\cos\theta - \sin\theta), \quad a_2 = a_3 = 0. \tag{5.124}$$

The minimum-spin result also represents the rigid/plastic limit of the

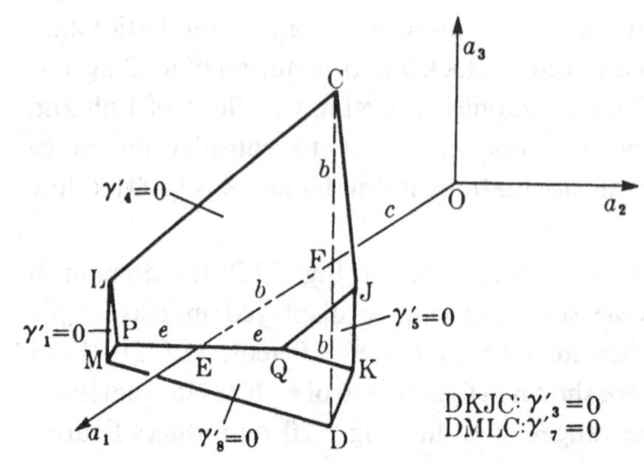

Fig. 5.20. Kinematically admissible solutions hexahedron in plastic-spin space for pure plane-strain compression, with channel axis $Z = [100]$ and θ the counterclockwise orientation of loading axis X from [010]: range $22.5° \leqslant \theta < \arctan(\frac{1}{2})$, evaluated at $\theta = 25°$ (from Fuh & Havner 1989, fig. 5).

unique, elastoplastic solution determined for prescribed deformation by the g_{kj} parameters of Chapter 3 (see eqs. (3.58)), corresponding to a slight modification of the simple theory (Fuh and Havner, sect. 2.4). (With the current state as reference, the general parameters may be expressed

$$g_{kj} = N_k \cdot \mathscr{L}_0 \cdot N_j + H_{kj} - 2\mathrm{tr}(N_j \sigma \Omega_k) \tag{5.125}$$

to first order in infinitesimal lattice strain (from Havner 1981, eq. (2.25)).)

This same family of orientations in pure plane-strain compression of f.c.c. crystals was investigated by Toth, Jonas & Neale (1990) using a rate-dependent theory. For $0 < \theta < 45°$ their equations require a shearing stress σ_{xy} which is zero only in the rate-insensitive limit.[12] They found that for any value of their rate-sensitivity parameter the solution corresponds to lattice rotation about the channel axis $Z = [100]$ (i.e., segment FE in Fig. 5.20) but not to minimum plastic spin. However, from their figure 2, the slip rates in the four pairs of systems $a2$ and $c2$, $a\bar{3}$ and $c\bar{3}$, $b2$ and $d2$, and $b\bar{3}$ and $d\bar{3}$ (which they number 5, 6, 2, and 3, respectively in the figure) differ between the rate-dependent and minimum-spin solutions by only a few percent for a rate-sensitivity parameter 0.001 over the range of θ, with the results exactly coincident at $\theta = 0$ and 45° for all such parameters.

In the experiment of Driver et al. (1983) on an aluminum crystal in a

[12] Recall from equations (5.3) and the subsequent discussion that a uniform stress σ_{xy} cannot be justified by the boundary conditions in channel die compression, whether or not ideally frictionless.

nominal $[021][0\bar{1}2][100]$ initial orientation, the loading and channel-axis directions were approximately 4° and 7°, respectively, from the [021] and [100] lattice directions (as determined from fig. 4 of Driver et al.). In the resulting deformation the lattice essentially rotated about the channel axis, with a rotation at $e_L = 0.42$ of approximately 5.7° (as calculated from the same figure). Also, Driver et al. (1984, p. 510) state that the "experimentally observed slip traces" in their 1983 paper "correspond to planes *a* and *c*" (that is, (111) and $(1\bar{1}\bar{1})$).

Fuh & Havner (1989, sect. 3) solved numerically the differential equations (5.122) and (5.124) corresponding to minimum plastic spin for (exact) initial orientation $[021][0\bar{1}2][100]$ ($\tan\theta = 0.5$). The theoretical lattice rotation at $e_L = 0.42$ is approximately 4.6°, with dominant slip on planes *a* and *c* (in particular, systems $a\bar{3}$ and $c\bar{3}$) throughout the deformation to $e_L = 1.0$ (see their fig. 8). Systems *b*2 and *d*2 (3 and 7) are inactive, as is evident from both the equations and Fig. 5.20 at minimum spin. (For $\tan\theta = 0.5$ planes *JKQ* and *LMP* degenerate to points.) Results from Toth, Jonas & Neale (1990, sect. 3.2) are similar, with the exception that their rate-dependent solution gives small equal slips on systems *b*2 and *d*2. In addition, for their rate-sensitivity parameter they obtained $\Delta\theta \approx 5.3°$ (see their fig. 3), which is closer than the minimum-plastic-spin solution to the single available experimental point of Driver et al. (1983).

6

THEORETICAL CONNECTIONS BETWEEN CRYSTAL AND AGGREGATE BEHAVIOR

In this chapter we return to the general theoretical framework of Chapter 3 and extend it to the analysis of characteristics of overall response of macroscopically uniform polycrystalline solids. The objective is the presentation of a rigorous theoretical connection between single-crystal elastoplasticity and macroscopic crystalline aggregate behavior. The development is based upon the original analysis of Hill (1972) and other basic contributions in Hill & Rice (1973), Havner (1974, 1982a, 1986), and Hill (1984, 1985). Central to an understanding of the crystal-to-aggregate transition is the well-known "averaging theorem" introduced by Bishop & Hill (1951a) but only given its final form and initial proof at finite strain in Hill's (1972) seminal work.

6.1 Crystalline Aggregate Model: The Averaging Theorem

At the beginning of Chapter 3, the scale of a crystal material point in a continuum model was defined to have linear dimension of order 10^{-3} mm: greater than 10^3 lattice spacings but at least an order of magnitude smaller than normal grain sizes in polycrystalline metals. Consider now the choice of physical size of a representative "macroelement" that defines a continuum point at the level of ordinary stress and strain analysis (that is, in structural and mechanical components or materials-forming operations.)

The wall thickness of thin-walled metal tubes used in combined stress tests (say, axial loading and torsion) often is in the range 1–2 mm and 10 to 30 grains (see, for example, Mair & Pugh (1964) or Ronay (1968)). Therefore, a representative material volume corresponding to overall aggregate behavior may reasonably be identified as a "unit cube" in the unstressed reference state, of typical linear dimension 1 mm for moderately fine-grained metals, containing a minimum of 1000 crystal grains (Havner 1971). The basic concept of a unit cube in the analysis of polycrystalline behavior was introduced by Bishop & Hill (1951a).

In this and the following sections we wish to establish general characteristics of macroscopic constitutive relations from individual crystal equations. Correspondingly, we consider quasistatic, macroscopically homogeneous finite-deformation processes with negligible body forces. Thus

$$\operatorname{Div} S = \mathbf{0}, \quad \operatorname{Div} \dot{S} = \mathbf{0} \quad \text{in } V_0 \tag{6.1}$$

and

$$\mathbf{n}_0 S = \mathbf{t}_\mathrm{N}, \quad \mathbf{n}_0 \dot{S} = \dot{\mathbf{t}}_\mathrm{N} \quad \text{on } \partial V_0, \tag{6.2}$$

where V_0 denotes the reference-state unit cube and S and $\mathbf{t}_\mathrm{N}$ are nominal stress tensor and traction vector as before ($\mathbf{n}_0 S$ and $\mathbf{n}_0 \dot{S}$ necessarily are continuous across interfaces between grains). Let $\langle\cdot\rangle$ denote volume average over the macroelement, with $\int dV_0 = 1$ on the appropriate scale (approximately $1\,\mathrm{mm}^3$ as defined here). Then, for A, S, and their tensor product Kirchhoff stress τ (from the Green–Gauss transformation):

$$\langle A \rangle = \int \mathbf{x} \otimes \mathbf{n}_0 \, da_0, \quad \langle S \rangle = \int \mathbf{x}_0 \otimes \mathbf{t}_\mathrm{N} \, da_0, \tag{6.3}$$

$$\langle \tau \rangle = \langle AS \rangle = \int \mathbf{x} \otimes \mathbf{t}_\mathrm{N} \, da_0. \tag{6.4}$$

According to the *averaging theorem at finite strain* (Hill 1972, p. 139), "the product average [in eq. (6.4)] decomposes into the product of averages." That is,

$$\langle AS \rangle = \langle A \rangle \langle S \rangle. \tag{6.5}$$

Alternative proofs of this theorem by Hill (1972) and Havner (1974, 1978) are presented in the following.

Consider first Hill's (1972) boundary data for the macroelement. He supposes that either

$$\mathbf{x} = \langle A \rangle \mathbf{x}_0 \quad \text{or} \quad \mathbf{t}_\mathrm{N} = \mathbf{n}_0 \langle S \rangle \text{ on } \partial V_0. \tag{6.6}$$

In other words, either the macroelement (not necessarily a unit cube in Hill's specification) is embedded in a uniformly strained continuum with deformation gradient $\bar{A} = \langle A \rangle$, or the nominal traction vector on the

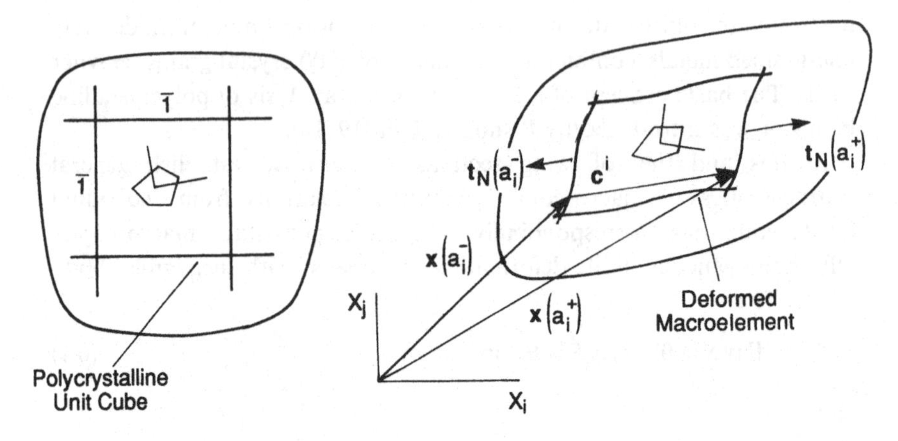

Fig. 6.1. Macroscopically uniform crystalline aggregate model.

element surface is in equilibrium with a uniform nominal stress field $\bar{S} = \langle S \rangle$. It is immediately evident from equations (6.3) and (6.4) that either of equations (6.6) results in the decomposition of equation (6.5).

We now turn to the aggregate model introduced in Havner (1971) and generalized to finite deformation in Havner (1974), namely, an extended array of identical, polycrystalline unit cubes in the undeformed state (each containing on the order 1000 crystal grains, as previously discussed). Consistent with this model, macroscopically uniform stress and deformation fields satisfy

$$\mathbf{x}(a_i^+) = \mathbf{x}(a_i^-) + \mathbf{c}^i, \quad \mathbf{t}_{\mathrm{N}}(a_i^+) = -\,\mathbf{t}_{\mathrm{N}}(a_i^-) \tag{6.7}$$

(Havner 1974, eqs. (4.1)). Here a_i^+ and a_i^- are the pair of reference-cube material faces that were normal to Cartesian axis X_i before deformation (Fig. 6.1); $\mathbf{c}^i$ is independent of position over the faces because a_i^+ and a_i^- are identically deformed and rotated (by definition of the model); and $\mathbf{x}(a_i^+)$, $\mathbf{t}_{\mathrm{N}}(a_i^-)$, and so forth indicate point dependence over the respective face. (In Fig. 6.1, which also depicts the rotation and distortion of an individual crystal grain together with its lattice rotation, the macroelement dimension $\bar{1}$ may be considered of order 1 mm.)

From equations (6.3) and (6.7) we find

$$\langle A_{ik} \rangle = \int x_i n_k^0 \, \mathrm{d}a_0 = \int (x_i^+ - x_i^-) \mathrm{d}a_k^+ = c_i^k, \tag{6.8}$$

$$\langle s_{kj} \rangle = \int x_k^0 t_j^{\mathrm{N}} \, \mathrm{d}a_0 = \int t_j^{\mathrm{N}} \, \mathrm{d}a_k^+ \tag{6.9}$$

(Havner 1974, eqs. (4.3)), with da_k^+ a differential reference area of the (positive) kth unit face of the reference cube. In addition, from equations (6.4) and (6.7) (Havner 1982a, eq. (3.11))

$$\langle \tau_{ij} \rangle = \langle A_{ik} s_{kj} \rangle = \int x_i t_j^{\mathrm{N}} \, da_0 = c_i^k \int t_j^{\mathrm{N}} \, da_k^+ . \tag{6.10}$$

Thus, from equations (6.8)–(6.10), the averaging theorem follows:

$$\langle A_{ik} s_{kj} \rangle = \langle A_{ik} \rangle \langle s_{kj} \rangle, \tag{6.11}$$

which is identically equation (3.8) of Hill (1972).

Equations (6.7) are presumed to hold throughout a macroscopically homogeneous deformation of the aggregate model. Therefore, from rate equations analogous to the relations that led to equation (6.11), we obtain (taking the trace of the respective tensor products)

$$\langle \dot{w} \rangle \equiv \langle s_{ij} \dot{A}_{ji} \rangle = \langle s_{ij} \rangle \langle \dot{A}_{ji} \rangle, \tag{6.12}$$

the total work-rate per unit reference volume (with dw henceforth replacing the notation ρ_0dw of Chapter 3 for simplicity), and

$$\langle \dot{s}_{ij} \dot{A}_{ji} \rangle = \langle \dot{s}_{ij} \rangle \langle \dot{A}_{ji} \rangle, \tag{6.13}$$

in which the stress and deformation rates (or incremental changes) may be unrelated. These equations were first derived by Hill (1972), based upon the boundary data of equations (6.6). The most general boundary conditions for which the averaging theorem and its extensions, equations (6.12) and (6.13), hold precisely may be found in Hill (1984). Henceforth we shall take equations (6.11)–(6.13) to be the defining relations for macroscopically uniform deformation in polycrystalline solids without regard to the details of boundary data.

6.2 Macroscopic Variables and Plastic Potentials

The volume averages $\langle S \rangle$ and $\langle A \rangle$ are appropriate macroscopic measures of stress and deformation for any specification of boundary conditions that result in equations (6.11)–(6.13). (This is particularly clear from equations (6.8) and (6.9), corresponding to the aggregate model of equations (6.7), where $\langle A_{ik} \rangle$ and $\langle s_{kj} \rangle$ are seen to equal the natural, operational definitions of overall deformation gradient and nominal stress for the macroelement (namely, current projected length on direction i per unit reference length of element line in direction k, and force component in direction j per unit reference area of element face k).) Thus, denoting macroscopic tensor quantities for the polycrystalline element by a bar above the corresponding symbol, one has

$$\bar{S} = \langle S \rangle, \quad \bar{A} = \langle A \rangle \tag{6.14}$$

(Hill 1972, eqs. (4.1)).

Other macroscopic tensors are defined in the standard way from their respective connections with the operationally determined quantities $\bar{S}$, $\bar{A}$ and in general are not themselves volume averages. An exception noted by Hill (1972, p. 141) is macroscopic Kirchhoff stress, which from equations (6.10), (6.11), and (6.14) is given by

$$\bar{\tau} = \bar{A}\bar{S} = \langle \tau \rangle. \tag{6.15}$$

The basic (macroscopic) kinematic tensors are defined from

$$\bar{\Gamma} = (\bar{A})^{\cdot} \bar{A}^{-1}, \quad \bar{\Lambda}^2 = \bar{A}^{\mathrm{T}} \bar{A}, \quad \bar{R} = \bar{A}\bar{\Lambda}^{-1}, \tag{6.16}$$

and arbitrary (macroscopic) conjugate stress and strain measures are given by

$$\mathrm{d}\bar{e} = \bar{\mathscr{K}}\,\mathrm{d}\bar{A}, \quad \bar{S} = \bar{t}\bar{\mathscr{K}}, \quad \bar{\mathscr{K}} = \frac{\partial \bar{e}}{\partial \bar{A}^{\mathrm{T}}} \tag{6.17}$$

(analogous to eqs. (3.16) and (3.102) in local variables), with

$$\mathrm{d}\bar{w} \equiv \bar{t}\,\mathrm{d}\bar{e} = \mathrm{tr}(\bar{S}\,\mathrm{d}\bar{A}) = \mathrm{tr}\langle S\,\mathrm{d}A \rangle = \langle t\,\mathrm{d}e \rangle \equiv \langle \mathrm{d}w \rangle \tag{6.18}$$

from equations (6.17), (3.16), (3.102), and (6.12).

Consider now the issue of macroscopic plastic potentials. The original derivation of such potentials from individual crystal relations was made by Hill & Rice (1973). Here we shall follow the approach in Havner (1986, 1987a), which makes use of results from Hill (1984).

Recall from equations (3.104) and (3.105) that, locally,

$$\begin{aligned} \mathrm{d}^{\mathrm{p}}S &= \mathscr{C}\,\mathrm{d}A - \mathrm{d}S = \frac{\partial}{\partial A^{\mathrm{T}}} \sum (\tilde{\tau}\,\mathrm{d}\tilde{\gamma})_j, \\ \mathscr{C} &= \frac{\partial S}{\partial A^{\mathrm{T}}} = \mathscr{K}^{\mathrm{T}} \mathscr{L} \mathscr{K} + \mathscr{T} = \mathscr{C}^{\mathrm{T}} \end{aligned} \tag{6.19}$$

(the gradients evaluated at fixed slips), based solely upon the Green-elasticity of the crystal lattice and the basic kinematics of Section 3.1. Assuming the invertibility of $\mathscr{C}$ (except perhaps at a finite number of singular points within a crystal grain), we also may write

$$\mathrm{d}^{\mathrm{p}}A = \mathrm{d}A - \mathscr{C}^{-1}\,\mathrm{d}S = \frac{\partial}{\partial S^{\mathrm{T}}}\sum(\tilde{\tau}\,\mathrm{d}\tilde{\gamma})_j, \tag{6.20}$$

whence $\tilde{\tau}_j$ serves as a plastic potential for the contribution of $\mathrm{d}\tilde{\gamma}_j$ to plastic stress decrement or plastic deformation increment in the respective nine-dimensional deformation-gradient or nominal-stress space. Our immediate objective is the establishment of counterparts of equations $(6.19)_1$ and (6.20) in macroscopic variables $\bar{S}$ and $\bar{A}$.

As remarked in Sect. 3.6, the symmetric plastic decrement in stress $\mathrm{d}^{\mathrm{p}}t$, whose plastic potential is given by equation $(3.52)_1$, transforms as t under change in strain measure e; hence $\mathrm{d}^{\mathrm{p}}t\delta e$ is a scalar invariant for arbitrary δe (also see Hill & Rice 1972, appendix). Consequently, $\mathrm{d}^{\mathrm{p}}S$ and $\mathrm{d}^{\mathrm{p}}t$ are related like S and t (eq. (3.102)) by the kinematic transformation tensor $\mathscr{K}$:

$$\mathrm{d}^{\mathrm{p}}S = \mathscr{K}^{\mathrm{T}}\,\mathrm{d}^{\mathrm{p}}t, \quad \mathscr{K} = \frac{\partial e}{\partial A^{\mathrm{T}}}. \tag{6.21}$$

This connection can be formally developed from the basic definitions and relationships of equations $(3.104)_1$, (3.105), (3.103), $(3.47)_1$, and $(3.16)_1$, or more directly from the plastic potential equations $(3.104)_2$ (or $(6.19)_1$) and $(3.52)_1$ and the chain rule

$$\frac{\partial}{\partial A^{\mathrm{T}}}(\cdot) = \mathscr{K}^{\mathrm{T}}\frac{\partial}{\partial e}(\cdot) \tag{6.22}$$

(equivalently, $\frac{\partial}{\partial A_{ji}}(\cdot) = \mathscr{K}_{klij}\frac{\partial}{\partial e_{kl}}(\cdot)$) at fixed slips. Then

$$\mathrm{tr}(\mathrm{d}^{\mathrm{p}}S\delta A) = \mathrm{d}^{\mathrm{p}}t\delta e \tag{6.23}$$

from $\delta e = \mathscr{K}\delta A$, which further illustrates the invariance of $\mathrm{d}^{\mathrm{p}}t\delta e$ with change in measure (as well as the point that δe need not be elastic for this invariance to hold).

Let $\mathscr{A}$ denote the influence tensor of elastic heterogeneity in the current state (introduced for finite strain by Hill (1984)), which may be defined

$$\mathscr{A} = \frac{\partial A}{\partial \bar{A}^{\mathrm{T}}}, \quad \text{or} \quad \mathscr{A}_{ijkl} = \frac{\partial A_{ij}}{\partial \bar{A}_{lk}}, \tag{6.24}$$

the gradient taken at fixed slips throughout the polycrystalline macroelement.[1] Thus (Hill, eqs. (3.3))

$$\delta A = \mathscr{A}\delta\bar{A}, \quad \langle\delta A\rangle = \delta\bar{A}, \tag{6.25}$$

so $\langle\mathscr{A}\rangle = \mathscr{I}$ (i.e., $\langle\mathscr{A}_{ijkl}\rangle = \delta_{il}\delta_{jk}$), with δA now representing a purely elastic response to $\delta\bar{A}$, and $\mathscr{C}\delta A$ defining the local nominal stress increment. The *macroscopic* (elastic) nominal stress change is (from eqs. (6.14) and (6.25))

$$\delta\bar{S} = \bar{\mathscr{C}}\delta\bar{A}, \quad \bar{\mathscr{C}} = \langle\mathscr{C}\mathscr{A}\rangle \tag{6.26}$$

(Hill's eqs. $(3.4)_1$ and $(3.6)_1$). Alternatively, we can use equation (6.13), expressible as

$$\operatorname{tr}\langle \mathrm{d}_1 S\, \mathrm{d}_2 A\rangle = \operatorname{tr}\langle \mathrm{d}_1 S\rangle\langle \mathrm{d}_2 A\rangle \tag{6.27}$$

for arbitrary (unrelated) increments, to establish the diagonal symmetry $\bar{\mathscr{C}} = \bar{\mathscr{C}}^{\mathrm{T}}$ of the macroscopic tensor of nominal elastic moduli, which symmetry is not obvious from equation $(6.26)_2$. Thus, for linked elastic changes (with the scalar product understood),

$$\langle\delta A\mathscr{C}\delta A\rangle = \delta\bar{A}\langle\mathscr{A}^{\mathrm{T}}\mathscr{C}\mathscr{A}\rangle\delta\bar{A} = \langle\delta A\mathscr{C}\rangle\langle\delta A\rangle = \delta\bar{A}\bar{\mathscr{C}}\delta\bar{A}$$

from equations (6.25) and (6.27) and the basic definition $\bar{\mathscr{C}}\delta\bar{A} = \langle\mathscr{C}\delta A\rangle$; whence we may write (Hill 1984, eq. $(3.8)_1$)

$$\bar{\mathscr{C}} = \langle\mathscr{A}^{\mathrm{T}}\mathscr{C}\mathscr{A}\rangle \tag{6.28}$$

and the symmetry is proved. (In other words, the averaging theorem in its incremental form gives the decomposition $\langle\mathscr{A}^{\mathrm{T}}\mathscr{C}\mathscr{A}\rangle = \langle\mathscr{A}^{\mathrm{T}}\rangle\langle\mathscr{C}\mathscr{A}\rangle = \langle\mathscr{C}\mathscr{A}\rangle$ from $\langle\mathscr{A}^{\mathrm{T}}\rangle = \mathscr{I}$.)

Macroscopic plastic potentials may now be established. We define plastic decrement in $\bar{S}$ and plastic increment in $\bar{A}$ by

$$\mathrm{d}^{\mathrm{p}}\bar{S} = \bar{\mathscr{C}}\,\mathrm{d}\bar{A} - \mathrm{d}\bar{S}, \quad \mathrm{d}^{\mathrm{p}}\bar{A} = \mathrm{d}\bar{A} - \bar{\mathscr{C}}^{-1}\,\mathrm{d}\bar{S}, \tag{6.29}$$

consistent with local equations $(6.19)_1$ and (6.20) (assuming the overall, elastically weighted volume average of $\mathscr{C}$ from eqs. $(6.26)_2$ or (6.28) is invertible). Taking the volume average of the scalar invariant $\operatorname{tr}(\mathrm{d}^{\mathrm{p}}S\delta A)$, we have (using $\mathscr{C} = \mathscr{C}^{\mathrm{T}}$)

$$\operatorname{tr}\langle\mathrm{d}^{\mathrm{p}}S\delta A\rangle = \operatorname{tr}\langle(\mathrm{d}A\mathscr{C} - \mathrm{d}S)\delta A\rangle.$$

Both $\mathscr{C}\delta A$ and $\mathrm{d}S$ are statically admissible nominal stress increments whereas δA and $\mathrm{d}A$ are kinematically admissible, with $\langle\mathrm{d}A\rangle = \mathrm{d}\bar{A}$.

[1] Henceforth this will be the case for all gradients and no longer will be remarked.

Therefore, from equations $(6.25)_1$, (6.27), and $(6.26)_2$,

$$\langle \mathrm{d^p}S\mathscr{A}\rangle\delta\bar{A} = (\mathrm{d}\bar{A}\langle\mathscr{C}\mathscr{A}\rangle - \mathrm{d}\bar{S})\delta\bar{A} = (\mathrm{d}\bar{A}\bar{\mathscr{C}} - \mathrm{d}\bar{S})\delta\bar{A} \tag{6.30}$$

(the trace understood as before). Hence, with equation $(6.29)_1$, there follows

$$\mathrm{d^p}\bar{S} = \langle\mathscr{A}^{\mathrm{T}}\,\mathrm{d^p}S\rangle, \tag{6.31}$$

which is equivalent to equation $(3.13)_1$ of Hill (1984). Thus we see that the plastic decrement in macroscopic nominal stress, as operationally defined by equation $(6.29)_1$, is the elastically weighted volume average of the local plastic stress decrement. Finally, upon substituting equation $(6.19)_2$ and using (from eq. (6.24) and the chain rule)

$$\frac{\partial}{\partial A^{\mathrm{T}}}(\cdot)\mathscr{A} = \frac{\partial}{\partial \bar{A}^{\mathrm{T}}}(\cdot), \tag{6.32}$$

we obtain (Havner 1986, eq. (3.15))

$$\mathrm{d^p}\bar{S} = \frac{\partial}{\partial \bar{A}^{\mathrm{T}}}\langle\textstyle\sum(\tilde{\tau}\,\mathrm{d}\tilde{\gamma})_j\rangle = \frac{\partial\langle \mathrm{d}w_{\mathrm{p}}\rangle}{\partial \bar{A}^{\mathrm{T}}}. \tag{6.33}$$

In addition, from equations (6.29) and

$$\bar{\mathscr{C}}^{-1} = \frac{\partial \bar{A}}{\partial \bar{S}^{\mathrm{T}}} \tag{6.34}$$

we have (Havner 1987a, eq. (58))

$$\mathrm{d^p}\bar{A} = \frac{\partial}{\partial \bar{S}^{\mathrm{T}}}\langle\textstyle\sum(\tilde{\tau}\,\mathrm{d}\tilde{\gamma})_j\rangle = \frac{\partial\langle \mathrm{d}w_{\mathrm{p}}\rangle}{\partial \bar{S}^{\mathrm{T}}}. \tag{6.35}$$

Thus, the simple volume average of the local plastic work increment (per unit reference volume)

$$\mathrm{d}w_{\mathrm{p}} = \textstyle\sum(\tilde{\tau}\,\mathrm{d}\tilde{\gamma})_j \tag{6.36}$$

is a *macroscopic plastic potential* for $\mathrm{d^p}\bar{S}$ or $\mathrm{d^p}\bar{A}$ in the respective $\bar{A}$ or $\bar{S}$ nine-dimensional space. Consequently, the plastic potentials that are an intrinsic part of ordinary (macroscopic) plasticity theory can be founded on the polycrystalline averaging theorem and the Green-elasticity of the local crystal lattice.

6.3 Macroscopic Plastic Potentials and Generalized Normality in Arbitrary Conjugate Variables

Equation (6.33) may be readily transformed into a plastic-potential equation in arbitrary macromeasures $\bar{t}, \bar{e}$ through macroscopic counterparts of equations (6.21) and (6.22), namely,

$$\mathrm{d^p}\bar{S} = \bar{\mathscr{K}}\,\mathrm{d^p}\bar{t}, \quad \bar{\mathscr{K}} = \frac{\partial \bar{e}}{\partial \bar{A}^{\mathrm{T}}}, \tag{6.37}$$

and

$$\frac{\partial}{\partial \bar{A}^{\mathrm{T}}}(\cdot) = \bar{\mathscr{K}}^{\mathrm{T}} \frac{\partial}{\partial \bar{e}}(\cdot). \tag{6.38}$$

Thus, analogous to equations $(3.47)_1$ and $(3.52)_1$,

$$\mathrm{d^p}\bar{t} = \bar{\mathscr{L}}\,\mathrm{d}\bar{e} - \mathrm{d}\bar{t} = \frac{\partial}{\partial \bar{e}}\langle \textstyle\sum(\tilde{\tau}\,\mathrm{d}\tilde{\gamma})_j \rangle, \quad \bar{\mathscr{L}} = \frac{\partial \bar{t}}{\partial \bar{e}} \tag{6.39}$$

(Havner 1986, eq. (3.19); see also Hill & Rice (1973) for a comparable relation). Therefore $\mathrm{d}\bar{w}_{\mathrm{p}} = \langle \mathrm{d}w_{\mathrm{p}} \rangle$ is a plastic potential for macrostress decrement $\mathrm{d^p}\bar{t}$ in six-dimensional macrostrain space.

The dual plastic potential in macrostress space follows from equation (6.39) by defining $\mathrm{d^p}\bar{e}$ in a manner consistent with the relation between local variables $\mathrm{d^p}t$ and $\mathrm{d^p}e$ (i.e., $\mathrm{d^p}t = \mathscr{L}\,\mathrm{d^p}e$). Thus

$$\mathrm{d^p}\bar{e} = \bar{\mathscr{M}}\,\mathrm{d^p}\bar{t}, \quad \bar{\mathscr{M}} = \frac{\partial \bar{e}}{\partial \bar{t}} = \bar{\mathscr{L}}^{-1}, \tag{6.40}$$

so (from eq. (6.39) and the chain rule)

$$\mathrm{d^p}\bar{e} = \mathrm{d}\bar{e} - \bar{\mathscr{M}}\,\mathrm{d}\bar{t} = \frac{\partial}{\partial \bar{t}}\langle \textstyle\sum(\tilde{\tau}\,\mathrm{d}\tilde{\gamma})_j \rangle \tag{6.41}$$

(Havner 1986, eq. (3.21)), in which the symmetry of $\bar{\mathscr{L}}$ (whence $\bar{\mathscr{M}}$) has momentarily been assumed (again see Hill & Rice (1973) for a comparable relation). That symmetry is easily shown as follows.

From equations $(6.29)_1$, $(6.37)_1$, $(6.39)_1$, and $(6.17)_1$ and the differential of equation $(6.17)_2$ we have (analogous to eqs. $(3.105)_1$ and $(3.103)_2$ in local variables)

$$\bar{\mathscr{C}} = \bar{\mathscr{K}}^{\mathrm{T}}\bar{\mathscr{L}}\bar{\mathscr{K}} + \bar{\mathscr{T}}, \quad \bar{\mathscr{T}} = \bar{t}\frac{\partial^2 \bar{e}}{\partial \bar{A}^{\mathrm{T}} \partial \bar{A}^{\mathrm{T}}} = \bar{\mathscr{T}}^{\mathrm{T}}. \tag{6.42}$$

The symmetry of $\bar{\mathscr{L}}$ is evident from the symmetry of $\bar{\mathscr{C}}$ given by equation (6.28). In addition, the connection between $\bar{\mathscr{L}}$ and $\mathscr{L}$ for measure e is readily obtained from equations (6.28), (6.42), and $(3.105)_1$. The result is

$$\bar{\mathscr{K}}^{\mathrm{T}}\bar{\mathscr{L}}\bar{\mathscr{K}}+\bar{\mathscr{T}}=\langle\mathscr{K}\mathscr{A})^{\mathrm{T}}\mathscr{L}(\mathscr{K}\mathscr{A})\rangle+\langle\mathscr{A}^{\mathrm{T}}\mathscr{T}\mathscr{A}\rangle, \tag{6.43}$$

which does not appear to have been given before in this generality. Upon taking the transpose of both sides, one can see the symmetry of $\bar{\mathscr{L}}$ from the symmetry of $\mathscr{L}, \mathscr{T}$, and $\bar{\mathscr{T}}$.

For the Green measure $e = E$: $\mathrm{d}E = \mathrm{sym}(A^{\mathrm{T}}\,\mathrm{d}A)$, $t = T = SA^{-\mathrm{T}}$ (contravariant Kirchhoff stress), and $\mathscr{K}$ equals Hill's (1984) transformation tensor $\mathscr{G}$, which is linear in A, whence $\bar{\mathscr{G}} = \langle\mathscr{G}\rangle$ from equation $(6.14)_2$. This simple result doesn't apply to other measures, however. In the case of Almansi strain, for example, $\mathrm{d}e_{\mathrm{A}} = \mathrm{sym}(A^{-1}\,\mathrm{d}AC^{-1})$, with $C \equiv A^{\mathrm{T}}A$, and one obtains (Havner 1986, eq. (2.14))

$$\mathscr{K}^{\mathrm{A}}_{ijkl}=\tfrac{1}{2}(A^{-1}_{il}C^{-1}_{jk}+A^{-1}_{jl}C^{-1}_{ik}), \tag{6.44}$$

whose volume average cannot be expected to equal the same expression in macroscopic variables $\bar{A}^{-1}$ and $\bar{C}^{-1} = \bar{A}^{-1}\bar{A}^{-\mathrm{T}}$.

For Green strain $E = \frac{1}{2}(C - I)$, it is worthwhile writing out equation (6.43) in Cartesian-component form. Thus, upon substituting

$$\mathscr{G}_{ijkl}=\tfrac{1}{2}(A_{li}\delta_{jk}+A_{lj}\delta_{ik}),\quad \mathscr{T}_{ijkl}=T_{ik}\delta_{jl} \tag{6.45}$$

(Hill 1984, eqs. (5.3) and $(5.5)_2$), we obtain

$$\langle A_{jm}\rangle\bar{\mathscr{L}}^{\mathrm{G}}_{imkn}\langle A_{ln}\rangle+\bar{T}_{ik}\delta_{jl}=\langle\mathscr{A}_{qpij}A_{qu}\mathscr{L}^{\mathrm{G}}_{purv}A_{sv}\mathscr{A}_{srkl}\rangle+\langle\mathscr{A}_{qpij}T_{pr}\mathscr{A}_{qrkl}\rangle, \tag{6.46}$$

which is equivalent to combining equations (2.15), (3.16), and (3.17) of Hill (1985). We note in passing that the symmetry of $\bar{T}$ is a consequence of

$$\langle A\rangle\bar{T}\langle A\rangle^{\mathrm{T}}\equiv\bar{\tau}=\langle A\rangle\langle S\rangle=\langle\tau\rangle \tag{6.47}$$

(from eq. (6.15)), with, of course, $\tau = (\rho_0/\rho)\sigma$ necessarily symmetric from local balance of moments or moment of momentum.

We now turn to the crystal normality laws, equations (3.61), which as may be recalled are based upon both the Green-elasticity of the lattice and the assumption that the critical strengths $\tilde{\tau}^{\mathrm{c}}_k$ depend only upon the history of the invariant slips $\tilde{\gamma}_j$. From equation (6.23) (with $\mathrm{d}^{\mathrm{p}}S = \mathscr{C}\,\mathrm{d}^{\mathrm{p}}A$), these local normality relations can be expressed in the various equivalent forms

$$\mathrm{d}^{\mathrm{p}}t\delta e=\mathrm{tr}(\mathrm{d}^{\mathrm{p}}S\delta A)=\mathrm{tr}(\mathrm{d}^{\mathrm{p}}A\delta S)=\mathrm{d}^{\mathrm{p}}e\delta t\leqslant 0. \tag{6.48}$$

From equations (6.27) (the averaging theorem in unrelated changes), (6.25), and (6.31),

$$\operatorname{tr}\langle \mathrm{d}^{\mathrm{p}}S\delta A\rangle = \operatorname{tr}(\mathrm{d}^{\mathrm{p}}\bar{S}\delta\bar{A}) = \operatorname{tr}(\mathrm{d}^{\mathrm{p}}\bar{A}\delta S) = \operatorname{tr}\langle \mathrm{d}^{\mathrm{p}}A\delta S\rangle. \tag{6.49}$$

Thus we have the dual inequalities (using eqs. $(6.17)_1$ and $(6.37)_1$)

$$\operatorname{tr}(\mathrm{d}^{\mathrm{p}}\bar{S}\delta\bar{A}) = \mathrm{d}^{\mathrm{p}}\bar{t}\delta\bar{e} \leqslant 0, \quad \operatorname{tr}(\mathrm{d}^{\mathrm{p}}\bar{A}\delta\bar{S}) = \mathrm{d}^{\mathrm{p}}\bar{e}\delta\bar{t} \leqslant 0, \tag{6.50}$$

which are fully equivalent forms from equations (6.29).[2] These inequalities are the *generalized normality relations* in the respective deformation $(\bar{A}, \bar{e})$ or stress $(\bar{S}, \bar{t})$ space (Hill 1972; Havner 1982a, 1987a). They are to be distinguished from the common notion in macroscopic plasticity theory that the Eulerian plastic strain rate $\bar{D}^{\mathrm{p}}$ (say) is normal to the yield surface in Cauchy $(\bar{\sigma})$ or Kirchhoff $(\bar{\tau})$ stress space (or lies within the outward cone of normals at a vertex). However, the difference between the classical, postulated version of normality and equations (6.50), which as we have seen follow from the local crystal normality laws and the averaging theorem, should be negligible in the typical case of infinitesimal lattice strains (when the latter equations are expressed with the current state as reference).

6.4 Work- and Energy-Related Macroscopic Equations

The existence and precise forms of macroscopic plastic potentials and normality laws, based upon local lattice potentials and slip-system critical strengths, are perhaps the most important results for polycrystalline elastoplastic theory at finite strain that can be achieved through use of the averaging theorem. However, other connections between individual-crystal behavior and aggregate response can be established with the aid of this theorem; and in the following we investigate various of these relations, several of which suggest basic constitutive inequalities in macrovariables. The development generally follows that in Havner (1986, sect. 3.2), which makes extensive use of results in Hill (1984).

[2] These equivalent relations also are a direct consequence of the definitions $\mathrm{d}^{\mathrm{p}}t$ and $\mathrm{d}^{\mathrm{p}}e$ and of Hill's invariant bilinear form (eq. (A37) in Appendix), which "has the notable property that its macroscopic value is an unweighted average over the representative volume" (Hill 1972, p. 142). That is, $\langle \delta t\,\mathrm{d}e - \mathrm{d}t\delta e\rangle = \operatorname{tr}\langle \delta S\,\mathrm{d}A - \mathrm{d}S\delta A\rangle = \operatorname{tr}(\delta\bar{S}\,\mathrm{d}\bar{A} - \mathrm{d}\bar{S}\delta\bar{A}) = \delta\bar{t}\,\mathrm{d}\bar{e} - \mathrm{d}\bar{t}\delta\bar{e}$ (Hill 1972, eqs. (4.11) and (4.12); see also Havner 1982a, sect. 3.3), which scalar invariant equals $\delta\bar{t}\mathrm{d}^{\mathrm{p}}\bar{e}$ and $\mathrm{d}^{\mathrm{p}}\bar{t}\delta\bar{e}$ from the defining equations of the respective macroscopic increments.

Let δ now represent the local *imagined* elastic response to a macroscopic deformation increment $\mathrm{d}\bar{A}$ (based upon $\mathrm{Div}(\mathscr{C}\delta A) = \mathbf{0}$ in V_0) and let δS represent the (imagined) elastic response to a corresponding macroscopic nominal stress increment $\mathrm{d}\bar{S}$, with $\mathrm{d}A$ and $\mathrm{d}S$ denoting actual elastoplastic increments as before. (As δA and δS are virtual, $\delta S \neq \mathscr{C}\delta A$ except when $\mathrm{d}\bar{S} = \bar{\mathscr{C}}\,\mathrm{d}\bar{A}$, that is, when the aggregate incremental response is purely elastic.) Again using tensor $\mathscr{A}$ (eq. (6.24)) and adopting another elastic influence tensor $\mathscr{B}$ also introduced by Hill (1984), we have

$$\delta A = \mathscr{A}\,\mathrm{d}\bar{A} = \mathscr{A}\langle \mathrm{d}A\rangle, \quad \delta S = \mathscr{B}\,\mathrm{d}\bar{S} = \mathscr{B}\langle \mathrm{d}S\rangle, \tag{6.51}$$

with

$$\langle \mathscr{A}\rangle = \mathscr{I} = \langle \mathscr{B}\rangle, \quad \mathscr{B}\bar{\mathscr{C}} = \mathscr{C}\mathscr{A}. \tag{6.52}$$

(Hill's eqs. $(3.3)_3$ and $(3.5)_1$), so $\mathscr{B} = \partial S/\partial \bar{S}^{\mathrm{T}}$ at fixed slips. Also, let d^{r} designate local residual change under a differential cycle of overall load (equivalently, macroscopic nominal stress) and let d^{s} denote local self-straining change under a differential cycle of overall deformation gradient, based upon the presumption of elastic response during the removal of the load or deformation increment. Then[3]

$$\begin{aligned} &\mathrm{d}^{\mathrm{r}}S = \mathrm{d}S - \delta S, \quad \mathrm{d}^{\mathrm{r}}A = \mathrm{d}A - \mathscr{C}^{-1}\delta S, \\ &\mathrm{d}^{\mathrm{s}}S = \mathscr{C}\delta A - \mathrm{d}S, \quad \mathrm{d}^{\mathrm{s}}A = \mathrm{d}A - \delta A, \end{aligned} \tag{6.53}$$

and there follow (by making use of eqs. (6.14) and (6.51)–(6.52))

$$\langle \mathrm{d}^{\mathrm{r}}S\rangle = 0, \quad \langle \mathrm{d}^{\mathrm{r}}A\rangle = \mathrm{d}^{\mathrm{p}}\bar{A}, \quad \langle \mathrm{d}^{\mathrm{s}}S\rangle = \mathrm{d}^{\mathrm{p}}\bar{S}, \quad \langle \mathrm{d}^{\mathrm{s}}A\rangle = 0, \tag{6.54}$$

with

$$\mathrm{d}^{\mathrm{p}}\bar{S} = \bar{\mathscr{C}}\,\mathrm{d}^{\mathrm{p}}\bar{A} = \langle \mathscr{A}^{\mathrm{T}}\,\mathrm{d}^{\mathrm{p}}S\rangle \tag{6.55}$$

as before. We also have, from the various defining equations $(6.19)_1$, $(6.20)_1$, and (6.53),

$$\mathrm{d}^{\mathrm{p}}S - \mathrm{d}^{\mathrm{s}}S = \mathscr{C}\,\mathrm{d}^{\mathrm{s}}A, \quad \mathscr{C}(\mathrm{d}^{\mathrm{r}}A - \mathrm{d}^{\mathrm{p}}A) = \mathrm{d}^{\mathrm{r}}S. \tag{6.56}$$

(Recall once more that $\mathrm{d}^{\mathrm{p}}S$ is the local stress decrement after a *local* cycle of $\mathrm{d}A$, and $\mathrm{d}^{\mathrm{p}}A$ is the local deformation increment after a local cycle of $\mathrm{d}S$, whereas $\mathrm{d}^{\mathrm{s}}S$ is the local stress decrement after a *macroscopic* cycle of $\mathrm{d}\bar{A}$, and $\mathrm{d}^{\mathrm{r}}A$ is the local deformation increment after a macroscopic cycle of $\mathrm{d}\bar{S}$ (all with elastic unloading). As expressed in equations (6.54), the macroscopic stress decrement and deformation increment are simple

[3] With the signs chosen, positive d^{s} corresponds to an increment in deformation but a decrement in nominal stress, whereas positive d^{r} represents a positive increment in each variable.

volume averages of the latter changes ($\mathrm{d}^{\mathrm{s}}S$ and $\mathrm{d}^{\mathrm{r}}A$) rather than of the locally defined plastic increments $\mathrm{d}^{\mathrm{p}}S$ and $\mathrm{d}^{\mathrm{p}}A$.)

Consider first the scalar product $\mathrm{tr}\langle \mathrm{d}S\,\mathrm{d}^{\mathrm{p}}A\rangle$, which is not invariant under change in strain measure (see Fig. 3.3 and footnote 7 in Chapter 3). As $\mathrm{d}^{\mathrm{p}}A = \mathrm{d}^{\mathrm{r}}A - \mathscr{C}^{-1}\,\mathrm{d}^{\mathrm{r}}S$ from equation $(6.56)_2$, we have (with eq. $(6.53)_1$)

$$\mathrm{tr}\langle \mathrm{d}S\,\mathrm{d}^{\mathrm{p}}A\rangle = \mathrm{tr}\langle \mathrm{d}S\,\mathrm{d}^{\mathrm{r}}A\rangle - \langle \mathrm{d}^{\mathrm{r}}S\mathscr{C}^{-1}\mathrm{d}^{\mathrm{r}}S\rangle - \langle \delta S\mathscr{C}^{-1}\mathrm{d}^{\mathrm{r}}S\rangle.$$

Recognizing that $\mathrm{d}^{\mathrm{r}}A$ and $\mathscr{C}^{-1}\delta S$ are kinematically admissible whereas $\mathrm{d}S$ and $\mathrm{d}^{\mathrm{r}}S$ are statically admissible,[4] one obtains from the averaging theorem (in the form eq. (6.27)) and equations (6.54) the result

$$\mathrm{tr}(\mathrm{d}\bar{S}\,\mathrm{d}^{\mathrm{p}}\bar{A}) = \mathrm{tr}\langle \mathrm{d}S\,\mathrm{d}^{\mathrm{p}}A\rangle + \langle \mathrm{d}^{\mathrm{r}}S\mathscr{C}^{-1}\,\mathrm{d}^{\mathrm{r}}S\rangle, \tag{6.57}$$

which is equivalent to equation $(4.10)_1$ in Hill (1984). This relation is the finite-strain generalization of a result within infinitesimal-strain theory first given by Mandel (1966) (see also Hill (1971) and Havner & Varadarajan (1973)), namely, $\mathrm{d}\bar{\sigma}\,\mathrm{d}^{\mathrm{p}}\bar{\varepsilon} = \langle \mathrm{d}\sigma\,\mathrm{d}^{\mathrm{p}}\varepsilon\rangle + \langle \mathrm{d}^{\mathrm{r}}\sigma\mathscr{M}_0\,\mathrm{d}^{\mathrm{r}}\sigma\rangle$. In that limited context, with $\mathscr{M}_0$ positive definite, the obvious conclusion is that "microstability" $\mathrm{d}\sigma\,\mathrm{d}^{\mathrm{p}}\varepsilon > 0$ guarantees "macrostability" $\mathrm{d}\bar{\sigma}\mathrm{d}^{\mathrm{p}}\bar{\varepsilon} > 0$, the latter of which inequalities is commonly known as Drucker's postulate (Drucker 1950, 1951). At finite strain, however, both $\mathrm{tr}(\mathrm{d}\bar{S}\mathrm{d}^{\mathrm{p}}\bar{A})$ and the local $\mathrm{tr}(\mathrm{d}S\,\mathrm{d}^{\mathrm{p}}A)$ can be negative after a sufficiently large deformation (in dominant tensile loading with a falling load/extension curve, for example). In addition, $\mathscr{C}$ is generally not positive definite.

As $\mathrm{tr}\langle \mathrm{d}S\,\mathrm{d}^{\mathrm{p}}A\rangle$ is not an invariant, consider the similar scalar product in other conjugate variables. From equations $(3.47)_2$, $(3.16)_1$, (3.103), and $(6.20)_1$ we have

$$\mathrm{tr}(\mathrm{d}S\,\mathrm{d}^{\mathrm{p}}A) = \mathrm{d}t\,\mathrm{d}^{\mathrm{p}}e + \mathrm{d}t\mathscr{M}\,\mathrm{d}t - \mathrm{d}S\mathscr{C}^{-1}\,\mathrm{d}S + \mathrm{d}A\mathscr{T}\,\mathrm{d}A, \tag{6.58}$$

with an analogous relation in macrovariables. Thus, upon substituting equation (6.58) into equation (6.57) and noting that

$$\langle \mathrm{d}^{\mathrm{r}}S\mathscr{C}^{-1}\,\mathrm{d}^{\mathrm{r}}S\rangle = \langle \mathrm{d}S\mathscr{C}^{-1}\mathrm{d}S\rangle - \mathrm{d}\bar{S}\bar{\mathscr{C}}^{-1}\,\mathrm{d}\bar{S} \tag{6.59}$$

from equations $(6.53)_1$, $(6.53)_2$, $(6.54)_2$, and $(6.29)_2$, one finds

$$\begin{aligned} \mathrm{d}\bar{t}\,\mathrm{d}^{\mathrm{p}}\bar{e} = \langle \mathrm{d}t\,\mathrm{d}^{\mathrm{p}}e\rangle + \langle \mathrm{d}t\mathscr{M}\,\mathrm{d}t\rangle - \mathrm{d}\bar{t}\bar{\mathscr{M}}\,\mathrm{d}\bar{t} \\ + \langle \mathrm{d}A\mathscr{T}\,\mathrm{d}A\rangle - \mathrm{d}\bar{A}\bar{\mathscr{T}}\,\mathrm{d}\bar{A} \end{aligned} \tag{6.60}$$

[4] It should be evident and clearly understood from equations (6.56) that the locally defined plastic increments $\mathrm{d}^{\mathrm{p}}A$ and $\mathrm{d}^{\mathrm{p}}S$ are not (respectively) kinematically and statically admissible fields, although they have the pseudo-elastic connection $\mathrm{d}^{\mathrm{p}}S = \mathscr{C}\,\mathrm{d}^{\mathrm{p}}A$.

for arbitrary (work-conjugate) measures t, e. (This result also may be obtained by successive substitution of eqs. $(6.41)_1$, (6.17), (6.14), (6.13), (3.103), $(3.16)_1$, and $(3.47)_2$ into $\mathrm{d}\bar{t}\,\mathrm{d}^{\mathrm{p}}\bar{e}$.) Within infinitesimal-strain theory the last two terms do not appear, $\mathrm{d}t = \mathrm{d}\sigma$ (rotational effects being neglected), $\langle \mathrm{d}t\mathcal{M}\mathrm{d}t\rangle - \mathrm{d}\bar{t}\bar{\mathcal{M}}\,\mathrm{d}\bar{t} = \langle \mathrm{d}^{\mathrm{r}}\sigma\mathcal{M}_0\mathrm{d}^{\mathrm{r}}\sigma\rangle$, and we again have Mandel's (1966) equation, as previously noted.

As remarked in Section 3.6, a local constitutive inequality of the form $\mathrm{d}t\,\mathrm{d}^{\mathrm{p}}e > 0$ is meaningful only if the measure is specified. In Havner (1986 eq. (2.40)) it is postulated that

$$(\mathrm{d}t\,\mathrm{d}^{\mathrm{p}}e)_{\mathrm{L}} > 0 \quad \text{(logarithmic measure)}, \tag{6.61}$$

which from equation $(3.77)_2$ is identically satisfied by the simple theory (eq. (4.15)), as the inequality then reduces to $h(\sum \mathrm{d}\tilde{\gamma}_j)^2 > 0$. Taking inequality (6.61) to be a reasonable hypothesis, we see from equation (6.60) that the corresponding inequality in macroscopic logarithmic strain, $(\mathrm{d}\bar{t}\mathrm{d}^{\mathrm{p}}\bar{e})_{\mathrm{L}} > 0$, hinges upon whether $\langle \mathrm{d}t\mathcal{M}\,\mathrm{d}t\rangle_{\mathrm{L}} - (\mathrm{d}\bar{t}\bar{\mathcal{M}}\,\mathrm{d}\bar{t})_{\mathrm{L}} + \langle \mathrm{d}A\mathcal{T}\,\mathrm{d}A\rangle_{\mathrm{L}} - (\mathrm{d}\bar{A}\bar{\mathcal{T}}\,\mathrm{d}\bar{A})_{\mathrm{L}}$ is positive. This remains an open question and may not be answerable analytically.

We now turn to the work-related scalar product $\mathrm{tr}\langle \mathrm{d}^{\mathrm{p}}S\,\mathrm{d}A\rangle$, which is invariant under change in strain measure from the local invariance $\mathrm{tr}(\mathrm{d}^{\mathrm{p}}S\,\mathrm{d}A) = \mathrm{d}^{\mathrm{p}}t\,\mathrm{d}e$. From equations $(6.56)_1$ and $(6.53)_4$ we have

$$\mathrm{tr}\langle \mathrm{d}^{\mathrm{p}}S\,\mathrm{d}A\rangle = \langle \mathrm{d}^{\mathrm{s}}A\mathscr{C}\,\mathrm{d}^{\mathrm{s}}A\rangle + \langle \mathrm{d}^{\mathrm{s}}A\mathscr{C}\delta A\rangle + \mathrm{tr}\langle \mathrm{d}^{\mathrm{s}}S\,\mathrm{d}A\rangle.$$

As $\mathrm{d}^{\mathrm{s}}S$ and $\mathscr{C}\delta A$ are statically admissible whereas $\mathrm{d}A$ and $\mathrm{d}^{\mathrm{s}}A$ are kinematically admissible, application of the averaging theorem in incremental form (eq. (6.27)) and of equations $(6.54)_{3,4}$ gives

$$\mathrm{tr}(\mathrm{d}^{\mathrm{p}}\bar{S}\,\mathrm{d}\bar{A}) = \mathrm{tr}\langle \mathrm{d}^{\mathrm{p}}S\,\mathrm{d}A\rangle - \langle \mathrm{d}^{\mathrm{s}}A\mathscr{C}\,\mathrm{d}^{\mathrm{s}}A\rangle, \tag{6.62}$$

which is equivalent to equation $(4.10)_2$ of Hill (1984). There follows from the invariance, for any conjugate measures,

$$\mathrm{d}^{\mathrm{p}}\bar{t}\,\mathrm{d}\bar{e} = \langle \mathrm{d}^{\mathrm{p}}t\,\mathrm{d}e\rangle - \langle \mathrm{d}^{\mathrm{s}}A\mathscr{C}\,\mathrm{d}^{\mathrm{s}}A\rangle \tag{6.63}$$

(Hill 1985, eq. (4.13); Havner 1986, eq. (3.34)), with

$$\langle \mathrm{d}^{\mathrm{s}}A\mathscr{C}\,\mathrm{d}^{\mathrm{s}}A\rangle = \langle \mathrm{d}^{\mathrm{s}}e\mathscr{L}\,\mathrm{d}^{\mathrm{s}}e\rangle + \langle \mathrm{d}^{\mathrm{s}}A\mathscr{T}\,\mathrm{d}^{\mathrm{s}}A\rangle, \quad \mathrm{d}^{\mathrm{s}}e = \mathscr{K}\,\mathrm{d}^{\mathrm{s}}A, \tag{6.64}$$

from the definition of $\mathscr{C}$ (eq. $(3.105)_1$). Equation (6.62) or (6.63) is the finite strain generalization of an equation originally given for infinitesimal-strain theory by Hill (1971), which can be expressed $\mathrm{d}^{\mathrm{p}}\bar{\sigma}\,\mathrm{d}\bar{\varepsilon} = \langle \mathrm{d}^{\mathrm{p}}\sigma\,\mathrm{d}\varepsilon\rangle - \langle \mathrm{d}^{\mathrm{s}}\varepsilon\mathscr{L}_0\mathrm{d}^{\mathrm{s}}\varepsilon\rangle$; whence $\mathrm{d}^{\mathrm{p}}\bar{\sigma}\,\mathrm{d}\bar{\varepsilon} < \langle \mathrm{d}^{\mathrm{p}}\sigma\,\mathrm{d}\varepsilon\rangle$ in that context.

As briefly discussed in Section 3.6, the physical reason for the invariance of $\mathrm{d^p}t\,\mathrm{d}e$ ($\mathrm{tr}(\mathrm{d^p}S\,\mathrm{d}A)$) is that this scalar product equals twice the net work (to second order) in a local differential cycle of A, hence of e (see Fig. 3.3). The volume average $\langle \mathrm{d^p}t\,\mathrm{d}e\rangle$ of this invariant would be twice the aggregate net work if every crystal material point ("freed from its surroundings," as expressed by Hill (1984, p. 491)) underwent a local differential cycle of strain, rather than ending at $e+d^{\mathrm{s}}e$ after a cycle of $\bar{e}$. In contrast, the macroinvariant $\mathrm{d^p}\bar{t}\,\mathrm{d}\bar{e}$ ($\mathrm{tr}(\mathrm{d^p}\bar{S}\,\mathrm{d}\bar{A})$) equals twice the actual net work in the polycrystalline volume in a differential cycle of the macroscopic deformation gradient $\bar{A}$, hence of $\bar{e}$. Thus, the fictitious work $\frac{1}{2}\langle \mathrm{d^p}t\,\mathrm{d}e\rangle$ exceeds the actual cyclic work $\mathrm{d^s}\bar{w}=\frac{1}{2}\mathrm{d^p}\bar{t}\,\mathrm{d}\bar{e}$ by $\frac{1}{2}\langle \mathrm{d^s}A\mathscr{C}\,\mathrm{d^s}A\rangle$ (Hill's eq. $(4.14)_2$), which quadratic form may be but is not necessarily positive.

The fictitious work is positive from the fundamental postulate $\mathrm{d^p}t\,\mathrm{d}e>0$ (eq. (3.80)), which was shown to be satisfied by all four specific-hardening theories introduced in Section 4.5 (including classical Taylor hardening). However, we see from equation (6.63) that positive work in a macrostrain cycle is not guaranteed by this basic crystal constitutive inequality. Nevertheless, it is reasonable to expect the positiveness of $\mathrm{d^p}\bar{t}\,\mathrm{d}\bar{e}$; and the term $\langle \mathrm{d^s}A\mathscr{C}\,\mathrm{d^s}A\rangle$ should remain small whatever its sign because $\langle \mathrm{d^s}A\rangle=0$. (Of course, the volume average cannot be decomposed into a product of averages since $\mathscr{C}\,\mathrm{d^s}A$ is not statically admissible.)

An alternative equation for $\mathrm{tr}(\mathrm{d^p}\bar{S}\,\mathrm{d}\bar{A})=\mathrm{d^p}\bar{t}\,\mathrm{d}\bar{e}$ can be obtained as follows. From equations (6.28) and $(6.51)_1$,

$$\mathrm{d}\bar{A}\bar{\mathscr{C}}\,\mathrm{d}\bar{A}=\langle \delta A\mathscr{C}\delta A\rangle \tag{6.65}$$

(equivalent to eq. (3.18) of Hill (1985)); and from equations (6.13), (6.14), and $(6.20)_1$,

$$\mathrm{tr}(\mathrm{d}\bar{S}\,\mathrm{d}\bar{A})=\langle \mathrm{d}S\mathscr{C}^{-1}\,\mathrm{d}S\rangle+\mathrm{tr}\langle \mathrm{d}S\,\mathrm{d^p}A\rangle. \tag{6.66}$$

Upon substituting $\delta A=\mathscr{C}^{-1}(\mathrm{d}S+\mathrm{d^s}S)$ (from eq. $(6.53)_3$) into equation (6.65) and using

$$\mathrm{d^s}S=\mathscr{C}(\mathrm{d^p}A-\mathrm{d^s}A) \tag{6.67}$$

(from $\mathrm{d^p}S=\mathscr{C}\,\mathrm{d^p}A$ and eq. $(6.56)_1$) together with $\mathrm{tr}\langle \mathrm{d}S\,\mathrm{d^s}A\rangle=0$ (from the averaging theorem and eq. $(6.54)_4$), one finds that

$$\mathrm{d}\bar{A}\bar{\mathscr{C}}\,\mathrm{d}\bar{A}=\langle \mathrm{d}S\mathscr{C}^{-1}\,\mathrm{d}S\rangle+2\mathrm{tr}\langle \mathrm{d}S\,\mathrm{d^p}A\rangle+\langle \mathrm{d^s}S\mathscr{C}^{-1}\,\mathrm{d^s}S\rangle. \tag{6.68}$$

Thus, subtracting equation (6.66) (using the basic definition of $\mathrm{d^p}\bar{S}$, eq. $(6.29)_1$), we have

$$\mathrm{tr}(\mathrm{d^p}\bar{S}\,\mathrm{d}\bar{A})=\mathrm{tr}\langle \mathrm{d}S\,\mathrm{d^p}A\rangle+\langle \mathrm{d^s}S\mathscr{C}^{-1}\,\mathrm{d^s}S\rangle \tag{6.69}$$

(Havner 1986, eq. (3.35)). Equations (6.68) and (6.69) are the finite-strain counterparts of equations for infinitesimal-strain theory in Havner & Varadarajan (1973). (Although $\mathrm{tr}(\mathrm{d^p}\bar{S}\,\mathrm{d}\bar{A}) = \mathrm{d^p}\bar{t}\,\mathrm{d}\bar{e}$ for all conjugate measures, the separate volume averages on the right-hand side of equation (6.69) are not invariants (see eq. (6.58), for example).) There now follows from equations (6.57) and (6.69) (Havner 1986, eq. (3.36)):

$$\mathrm{tr}(\mathrm{d^p}\bar{S}\,\mathrm{d}\bar{A} - \mathrm{d}\bar{S}\,\mathrm{d^p}\bar{A}) = \langle \mathrm{d^s}S\mathscr{C}^{-1}\mathrm{d^s}S\rangle - \langle \mathrm{d^r}S\mathscr{C}^{-1}\mathrm{d^r}S\rangle. \tag{6.70}$$

We recall $\langle \mathrm{d^s}S\rangle = \mathrm{d^p}\bar{S}$ and $\langle \mathrm{d^r}S\rangle = 0$, and expect $\mathrm{d^p}\bar{S}\,\mathrm{d}\bar{A} > \mathrm{d}\bar{S}\,\mathrm{d^p}\bar{A}$ (that is, the second-order work in a macrodeformation cycle should be greater than the second-order work in a macroload cycle for the same increments $\mathrm{d}\bar{A}$ and $\mathrm{d}\bar{S}$). However, neither volume average in equation (6.70) is decomposable because neither $\mathscr{C}^{-1}\mathrm{d^s}S$ nor $\mathscr{C}^{-1}\mathrm{d^r}S$ is kinematically admissible.

Consider now the first-order work $\mathrm{tr}(S\,\mathrm{d^p}A)$ in a differential cycle of local nominal stress. This generally does not equal the first-order work $t\,\mathrm{d^p}e$ in a cycle of any other stress measure because S and t do not cycle simultaneously (see eq. $(3.103)_1$) except when the cycle is purely elastic; and neither $\mathrm{tr}(S\,\mathrm{d^p}A)$ nor $t\,\mathrm{d^p}e$ equals incremental plastic work $\mathrm{d}w_\mathrm{p} = \sum(\tilde{\tau}\,\mathrm{d}\tilde{\gamma})_j$, a scalar invariant. Rather, one finds from equations $(6.20)_1$, $(3.105)_1$, $(3.16)_1$, and $(3.47)_2$ that $\mathrm{d^p}A$ and $\mathrm{d^p}e$ are connected by

$$\mathscr{C}\,\mathrm{d^p}A = \mathscr{K}^\mathrm{T}\mathscr{L}\,\mathrm{d^p}e \tag{6.71}$$

while $S = t\mathscr{K}$ (eq. (3.102)). Thus

$$\mathrm{tr}(S\,\mathrm{d^p}A) = t\mathscr{K}\mathscr{C}^{-1}\mathscr{K}^\mathrm{T}\mathscr{L}\,\mathrm{d^p}e. \tag{6.72}$$

In addition, from equations $(3.48)_2$, $(3.49)_2$, and (3.26),

$$t\,\mathrm{d^p}e = \mathrm{d}w_\mathrm{p} - t\mathscr{M}\sum(\alpha\,\mathrm{d}\tilde{\gamma})_j. \tag{6.73}$$

However, because α (eq. (3.29) or (3.32)) is of the order of stress whereas $t\mathscr{M}$ is of the order of lattice strain, the second term on the right-hand side in equation (6.73) is of the order of lattice strain times incremental plastic work. Therefore, it typically will be at least two orders of magnitude smaller than the first term, from which $t\,\mathrm{d^p}e \approx \mathrm{d}w_\mathrm{p}$ for infinitesimal lattice strain. (Then also $\mathrm{tr}(S\,\mathrm{d^p}A) \approx \mathrm{d}w_\mathrm{p}$.)

The macroscopic first-order work in a differential cycle of $\bar{S}$ is, from the averaging theorem and equation $(6.56)_2$,

$$\mathrm{tr}(\bar{S}\,\mathrm{d^p}\bar{A}) = \mathrm{tr}\langle S\,\mathrm{d^p}A\rangle + \langle S\mathscr{C}^{-1}\mathrm{d^r}S\rangle, \tag{6.74}$$

which is connected to $\bar{t}$ and $\mathrm{d^p}\bar{e}$ for any measure by an equation that is analogous to equation (6.72). (Note that the total work, to second order, in a differential cycle of overall load on the polycrystalline element equals the addition of eq. (6.74) and one-half of eq. (6.57). That is,

$$\mathrm{d^r}\bar{w} \equiv \mathrm{tr}[(\bar{S} + \tfrac{1}{2}\mathrm{d}\bar{S})\mathrm{d^p}\bar{A}] = \mathrm{tr}\langle (S + \tfrac{1}{2}\mathrm{d}S)\mathrm{d^p}A\rangle \\ + \langle (S + \tfrac{1}{2}\mathrm{d^r}S)\mathscr{C}^{-1}\mathrm{d^r}S\rangle,$$

which is equivalent to the combination of eqs. $(4.12)_1$, $(4.13)_1$, and $(4.14)_1$ in Hill (1984).) When the basic definition of $\mathrm{d^p}\bar{A}$ (eq. $(6.29)_2$) and the work identity $\mathrm{tr}(\bar{S}\,\mathrm{d}\bar{A}) = \mathrm{tr}\langle S\,\mathrm{d}A\rangle$ (from eqs. (6.12) and (6.14)) are substituted into equation (6.74), there follows the scalar equality

$$\bar{S}\bar{\mathscr{C}}^{-1}\,\mathrm{d}\bar{S} = \langle S\mathscr{C}^{-1}\,\mathrm{d}S\rangle - \langle S\mathscr{C}^{-1}\,\mathrm{d^r}S\rangle. \tag{6.75}$$

Consider the integrals of equations (6.74) and (6.75) over the history of the deformation. In the context of infinitesimal-strain theory, $\int\langle S\mathscr{C}^{-1}\,\mathrm{d^r}S\rangle$ reduces to $\frac{1}{2}\langle\sigma^{\mathrm{r}}\mathscr{M}_0\sigma^{\mathrm{r}}\rangle > 0$, the residual (latent) energy at zero macrostress $\bar{\sigma}$. Consequently, the integrals of the infinitesimal-strain counterparts of equations (6.74) and (6.75) have the following interpretations: (i) Mechanical energy dissipated as heat in the polycrystalline element (i.e., internal plastic work $\int\langle\sigma\,\mathrm{d^p}\varepsilon\rangle$, the reduced form of $\int\langle S\,\mathrm{d^p}A\rangle$) is less than apparent (macroscopic) plastic work $\int\bar{\sigma}\,\mathrm{d^p}\bar{\varepsilon}$ for all paths; and (ii) internal strain energy $\frac{1}{2}\langle\sigma\mathscr{M}_0\sigma\rangle$ (the reduced form of $\int\langle S\mathscr{C}^{-1}\,\mathrm{d}S\rangle$) is necessarily greater than the mechanically recoverable work $\frac{1}{2}\bar{\sigma}\bar{\mathscr{M}}_0\bar{\sigma}$ under imagined elastic unloading (see Hill 1971; Havner, Singh & Varadarajan 1974). At finite strain, however, the interpretations are less clear. The integral $\int\langle S\mathscr{C}^{-1}\,\mathrm{d^r}S\rangle$, evaluated in loading from an unstressed state (with $\mathscr{C}$ continuously changing), does not equal the internal residual energy after (presumed) elastic unloading to zero macrostress, which energy is approximately $\frac{1}{2}\langle\sigma^{\mathrm{r}}\mathscr{M}\sigma^{\mathrm{r}}\rangle$, with $\mathscr{M}$ referred to the current state. Moreover, $\int\langle S\,\mathrm{d^p}A\rangle$ is only approximately equal to the internal plastic work $\int\langle\mathrm{d}w_{\mathrm{p}}\rangle$ during loading. The possible determination of more precise analytical connections among the various macroscopic and internal (microscopic) energies at finite strain remains an open line of investigation.

With regard to experimental studies of this subject, the first notable investigation of (mechanical) energy remaining in a polycrystalline metal after a loading cycle is the classic work of Farren & Taylor (1925) on finitely extended metal bars. They determined that the latent energy averaged 7 percent of net work in aluminum and 9 percent in copper, with mean deviations of 0.6 percent and 0.8 percent, respectively (see their table 3). Taylor & Quinney (1934) and Quinney & Taylor (1937) found the latent energy in pre-twisted copper rods to range between 7.5 and 9 percent, to large strains. Subsequent measurements of latent energy in copper by Williams (1961) and Wolfenden (1971) were in agreement with these results at small to moderate strains, as were the cyclic small-strain and temperature measurements of Dillon (1968), indicating the latent energy in aluminum is 6 percent of the macroscopic plastic work.

Latent energies during cyclic loading of discrete polycrystalline models of aluminum and copper *at small strain* were calculated by Havner et al. (1974) and Havner & Singh (1977) using Taylor hardening.[5] They determined the average values to be 7 and 8.6 percent, respectively, with a mean deviation of approximately $\frac{1}{2}$ percent. These results obviously are in very good agreement with experiment; but the success may have been somewhat fortuitous as the magnitude of the discretization errors was unknown, and most of the cited experimental values are at much higher levels of strain than can be appropriately studied within the limits of an infinitesimal-strain model. Also, a portion of the actual latent energy that is recoverable upon heating is a submicroscopic lattice-strain energy associated with newly created dislocations, as first suggested by Taylor (1935); and the modeling of that energy is outside the scope of both Havner & Singh (1977) and this monograph.

To close this section and chapter, we make use of the plastic-potential equations $(3.52)_2$, $(6.20)_2$, (6.35), and $(6.41)_2$ and the basic definitions of $\mathrm{d^p}e$, $\mathrm{d^p}A$, $\mathrm{d^p}\bar{A}$, and $\mathrm{d^p}\bar{e}$ to write a final set of five identities for the invariant differential work per unit volume of the polycrystalline macroelement (Havner 1986, eqs. (3.40)):

$$\begin{aligned}
\mathrm{d}\bar{w} &= \bar{S}\bar{\mathscr{C}}^{-1}\,\mathrm{d}\bar{S} + \bar{S}\frac{\partial(\mathrm{d}\bar{w}_\mathrm{p})}{\partial\bar{S}^\mathrm{T}} = \langle S\mathscr{C}^{-1}\,\mathrm{d}S\rangle + \left\langle S\frac{\partial(\mathrm{d}w_\mathrm{p})}{\partial S^\mathrm{T}}\right\rangle \\
&= \bar{t}\bar{\mathscr{M}}\,\mathrm{d}\bar{t} + \bar{t}\frac{\partial(\mathrm{d}\bar{w}_\mathrm{p})}{\partial\bar{t}} = \langle t\mathscr{M}\,\mathrm{d}t\rangle + \left\langle t\frac{\partial(\mathrm{d}w_\mathrm{p})}{\partial t}\right\rangle \\
&= \langle\rho_0\,\mathrm{d}\phi^*\rangle + \langle\mathrm{d}w_\mathrm{p}\rangle, \quad \mathrm{d}\bar{w}_\mathrm{p} = \langle\mathrm{d}w_\mathrm{p}\rangle = \langle\textstyle\sum(\tilde{\tau}\,\mathrm{d}\tilde{\gamma})_j\rangle.
\end{aligned} \tag{6.76}$$

Each of the 10 terms is different from every other term, and only the last two $\langle\rho_0\,\mathrm{d}\phi^*\rangle$ and $\langle\mathrm{d}w_\mathrm{p}\rangle$ are separately invariant under change in strain measure.

[5] I am not aware of any theoretical calculations of latent energy in a polycrystalline metal that have been made for finite strain.

7

APPROXIMATE POLYCRYSTAL MODELS

Turning from the rigorous theoretical analysis of Chapter 6 to the subject of (and literature on) the calculation of approximate polycrystalline aggregate models at finite strain, one can identify three prominent themes: the prediction of (i) macroscopic axial-stress–strain curves, (ii) macroscopic yield loci, and (iii) the evolution of textures (that is, the development of preferred crystal orientations in initially statistically isotropic aggregates). The topic of polycrystal calculations is vast and complex, warranting a monograph on its own (and by other hands). In this closing chapter of the present work I primarily shall review selected papers (acknowledging others) from among those contributions that are particularly significant or noteworthy in the more than 50 years' history of the subject.

7.1 The Classic Theories of Taylor, Bishop, and Hill

Near the beginning of G. I. Taylor's (1938a) May Lecture to the Institute of Metals is the following splendid sentence. "I must begin by making the confession that I am not a metallurgist; I may say, however, that I have had the advantage of help from, and collaboration with, members of your Institute, whose names are a sure guarantee that the metals I have used were all right, even if my theories about them are all wrong." More than anything else this statement reflects Taylor's irrepressible humor, for of course his theories were not "all wrong." Rather, that lecture, containing an insightful physical postulate known as the *principle of minimum shears* and introducing the "Taylor model" of aggregate behavior, together with a companion paper (Taylor 1938b) has influenced the subsequent literature of crystalline aggregate modeling and calculation to an extent matched only by the papers of Bishop & Hill (1951a, b).

Lattice strains are disregarded in Taylor's analysis, whence the components on the lattice axes of strain rate D and plastic spin Ω in f.c.c. crystals are (from Table 1 and the defining equations)

$$
\begin{aligned}
\sqrt{6}d_{11} &= (\overbrace{0, 1, -1}^{a}, \overbrace{0, 1, -1}^{b}, \overbrace{0, 1, -1}^{c}, \overbrace{0, 1, -1}^{d})\dot{\gamma}, \\
\sqrt{6}d_{22} &= (-1, 0, 1, -1, 0, 1, -1, 0, 1, -1, 0, 1)\dot{\gamma}, \\
\sqrt{6}d_{33} &= (1, -1, 0, 1, -1, 0, 1, -1, 0, 1, -1, 0)\dot{\gamma}, \\
2\sqrt{6}d_{23} &= (0, -1, 1, 0, 1, -1, 0, -1, 1, 0, 1, -1)\dot{\gamma}, \\
2\sqrt{6}d_{13} &= (1, 0, -1, -1, 0, 1, -1, 0, 1, 1, 0, -1)\dot{\gamma}, \\
2\sqrt{6}d_{12} &= (-1, 1, 0, -1, 1, 0, 1, -1, 0, 1, -1, 0)\dot{\gamma}
\end{aligned}
\tag{7.1}
$$

(with $\operatorname{tr} D = 0$ as required), and

$$
\begin{aligned}
2\sqrt{6}\Omega_{32} &= (\overbrace{2, -1, -1}^{a}, \overbrace{-2, 1, 1}^{b}, \overbrace{2, -1, -1}^{c}, \overbrace{-2, 1, 1}^{d})\dot{\gamma}, \\
2\sqrt{6}\Omega_{13} &= (-1, 2, -1, 1, -2, 1, 1, -2, 1, -1, 2, -1)\dot{\gamma}, \\
2\sqrt{6}\Omega_{21} &= (-1, -1, 2, -1, -1, 2, 1, 1, -2, 1, 1, -2)\dot{\gamma},
\end{aligned}
\tag{7.2}
$$

wherein $\dot{\gamma}$ is the vector of the 12 possible slip rates (counting both senses as one) in the order of systems $a1$ through $d3$. These equations were first given in full in Taylor (1938b, eqs. (3) and (7)).[1] In his model, every crystal within a polycrystalline aggregage subjected to a macroscopically uniform deformation is considered to sustain the same strain as the bulk material and the same rigid-body rotation relative to the reference frame, thus ensuring that the grains continue to fit together after deformation (which, as Taylor remarked, earlier aggregate models had not done).

Let ϕ, θ, and ψ denote the Euler angles of the lattice axes of an arbitrarily oriented crystal relative to the specimen frame (the mnemonic L of Chapter 2 no longer needed). From equations (2.29), the orthogonal

[1] In Taylor's equations (7), all signs are opposite those in equations (7.2) here. Evidently he defined incremental rotation of material relative to lattice as *clockwise* positive when viewed looking along a lattice axis toward the origin. Equivalently, his plastic-spin tensor would be expressed by $\operatorname{skw}\sum(\mathbf{n}\otimes\mathbf{b})_j\dot{\gamma}_j$, the transpose of equation (4.10) for Ω.

transformation Q defining the direction cosines of the crystal axes with respect to the specimen axes is

$$Q = \begin{bmatrix} \cos\phi\cos\theta\cos\psi - \sin\phi\sin\psi, & \sin\phi\cos\theta\cos\psi + \cos\phi\sin\psi, & -\sin\theta\cos\psi \\ -\cos\phi\cos\theta\sin\psi - \sin\phi\cos\psi, & -\sin\phi\cos\theta\sin\psi + \cos\phi\cos\psi, & \sin\theta\sin\psi \\ \cos\phi\sin\theta, & \sin\phi\sin\theta, & \cos\theta \end{bmatrix} \tag{7.3}$$

($Q_{ij} = \mathbf{e}_i \cdot \mathbf{e}^s_j$, $i = 1, 2, 3$, $j = x, y, z$). Because the (macroscopic) strain rate $\bar{D}$ in matrix array on the specimen frame is taken to be uniform throughout the aggregate in Taylor's model, the matrix D of components d_{ij} on local crystal axes (eqs. (7.1)) is $D = Q\bar{D}Q^T$. Since the d_{ij} vary widely over a full range of crystal orientations, no special symmetries (as in Chapter 5) are to be expected for an arbitrary crystal. Consequently, at least five slip systems must be active in each grain to satisfy equations (7.1) for the five independent components of strain rate. Drawing on an analogy with the mechanics of frictional systems, Taylor (1938a) assumed there would be *only* five active systems.

There are ${}_{12}C_5 = 792$ combinations[2] of 12 things taken 5 at a time, hence Taylor began with this number of possibilities. However, as he briefly explained, this total can be greatly reduced, although not so far as he first thought (see Taylor (1956) for his subsequent revision). The actual number of independent sets, established by Bishop & Hill (1951b), can be deduced as follows.

Clearly, as noted by Taylor, there cannot be three kinematically independent slip rates in the same plane. Equivalently, for the 6×12 matrix of coefficients in equations (7.1) (or 5×12 if any one of the interdependent first three rows is removed), $\sum a_i = 0$, $\sum b_i = 0$, $\sum c_i = 0$, and $\sum d_i = 0$ in each row. For any one of these sets of three, the number of possible combinations with the remaining nine slip systems is ${}_9C_2 = 36$. Thus, a total of 144 dependent sets can be eliminated. Of the remaining 648 combinations, 324 occur as sets of two slip rates on each of two planes with one on a third. The last 324 combinations are then sets of two slip rates on one plane and one on each of the other three planes. Among this latter group (none of which Taylor originally thought geometrically admissible), there are 24 dependent sets involving combinations (from eqs. (7.1)) $a_1 - b_1 + c_1 - d_1 = 0$, $a_2 - b_2 - c_2 + d_2 = 0$, and $a_3 + b_3 - c_3 - d_3 = 0$. An additional 12 dependent combinations are obtained by adding (or

[2] The standard definition is ${}_nC_k = \dfrac{n!}{(n-k)!\,k!}$.

subtracting) $\sum a_i$, $\sum b_i$, and so forth, one at a time, to each of the preceding expressions. A representative one of these is $a_1 - b_1 + c_1 + d_2 + d_3 = 0$.

There are 132 dependent sets (33 each) given by $a_1 + b_2 + d_3 = 0$, $a_2 + b_1 + c_3 = 0$, $a_3 + c_2 + d_1 = 0$, and $b_3 + c_1 + d_2 = 0$ (each group of 36 reduced by three cases already counted consisting of three systems in one plane). An additional 84 combinations are obtained by subtracting $\sum a_i$, and so forth, again one at a time, from these equations (but not, of course, $\sum c_i$ from the first equation, etc.). A representative group of seven dependent sets is $a_1 + b_2 - d_1 - d_2 = 0$ (the combination with d_3 already having been counted). Finally, 12 more dependent sets are found from these groups by subtracting the appropriate one of $\sum a_i$, and so forth from each of the 12 (group) equations. An example is $-a_1 + b_1 + b_3 + d_1 + d_2 = 0$. Thus, there are a total of 408 dependent sets: 144 with three slip systems in the same plane, 108 with two systems in each of two planes and one in a third, and 156 with two systems in one plane and one in each of the other three (consistent with the conclusions of Bishop & Hill (1951b, p. 1302)). Consequently, there remain 384 independent sets of five slip rates, each corresponding to a nonsingular 5×5 submatrix obtained for that set from equations (7.1) by eliminating one (d_{33}, say) of the interdependent axial strain rates. With **d** designating the reduced strain-rate vector $(d_{11}, d_{22}, 2d_{23}, 2d_{13}, 2d_{12})^{\mathrm{T}}$ on lattice axes, we have

$$\dot{\gamma}_\alpha = N_\alpha^{-1}\mathbf{d}, \quad \alpha = 1, \ldots, 384, \tag{7.4}$$

for every independent set, where $\dot{\gamma}_\alpha$ is the vector of the five slip rates and N_α is the corresponding 5×5 matrix.

Taylor further argued in his May Lecture, again by analogy with the mechanics of frictional systems, that each crystal will slip on that combination of systems that makes the (incremental) work least. Then, from his assumption of equal (positive) critical strength τ_c in every slip system, he deduced his famous principle of minimum shears (incremental slips). His most concise statement of this is in Taylor (1938b, p. 229), "Of all the possible combinations of the 12 shears which could produce an assigned strain, only that combination is operative for which the sum of the absolute values of shears is least."[3] One therefore has

$$\mathrm{tr}(\bar{\sigma}\bar{D}) = \langle \dot{w} \rangle = \langle \textstyle\sum(\tau\dot{\gamma})_j \rangle = \langle \tau_c(\min \textstyle\sum |\dot{\gamma}_j|) \rangle, \tag{7.5}$$

[3] All arguments and equations in Taylor (1938a, b) actually are expressed in terms of "strain" and "shears," assumed small, and so must be strictly interpreted as applying incrementally with the current state as reference, or as relating strain rates and slip rates as done here.

with $\min \sum |\dot{\gamma}_j|$ in a given crystal determined from among the 384 kinematically independent sets of five slip rates calculated from equation (7.4). (The absolute values must be specified in the sum because $\tau_j = -\tau_c$ in a system slipping in the opposite sense to that defined as positive in Table 1.)

In effect, Taylor adopted an average critical strength $\bar{\tau}_c$ in the aggregate that is definable from

$$d\bar{w} = \langle dw \rangle = \langle \tau_c (\min \sum |d\gamma_j|) \rangle = \bar{\tau}_c \langle \min \sum |d\gamma_j| \rangle. \tag{7.6}$$

(At the outset, τ_c is taken to be uniform throughout, whence $\bar{\tau}_c = \tau_c$.) He further restricted consideration to axisymmetric deformation of a bar pulled in tension, for which case $d\bar{w} = \bar{\sigma}_a \, d\bar{e}_L$, $\bar{d}_{xx} = (\bar{e}_L)^{\cdot}$, and $\bar{d}_{yy} = \bar{d}_{zz} = -\frac{1}{2}(\bar{e}_L)^{\cdot}$, with $\bar{\sigma}_a$ the macroscopic axial stress, $\bar{e}_L$ the macroscpic, logarithmic axial strain, and all other components zero on the specimen axes. Thus we have (from eq. (7.6))

$$\frac{\bar{\sigma}_a}{\bar{\tau}_c} = \bar{M} = \frac{\langle \min \sum |d\gamma_j| \rangle}{d\bar{e}_L}. \tag{7.7}$$

The ratio $\bar{M}$ so defined is generally known as the "Taylor factor."

For axial loading and assumed uniform axisymmetric deformation of f.c.c. polycrystals, it is only necessary to select a set of specimen axis orientations relative to the lattice that span a single spherical triangle within a standard stereographic projection. Taylor (1938a) chose 44 orientations distributed uniformly over the triangle $[101][100][11\bar{1}]$ ($d\bar{3}$ of Fig. 1.6) and obtained $\bar{M} = 3.10$, which is readily calculated from the 44 $\min \sum |d\gamma_j|$ values shown in his figure 13. However, as already noted, Taylor investigated only the 216 independent sets involving double slip in each of two planes and did not consider the 168 geometrically admissible combinations with double slip in only one plane. As it turned out, this omission made relatively little difference in the final averaged value of $\bar{M}$, for Bishop & Hill (1951b) subsequently obtained $\bar{M} = 3.06$ from calculations of their fully equivalent aggregate model.

The macroscopic stress–strain curve $\bar{\sigma}_a(\bar{e}_L)$ may be expressed (from eqs. (7.7))

$$\bar{\sigma}_a = \bar{M} \bar{\tau}_c \left(\int \bar{M} \, d\bar{e}_L \right), \tag{7.8}$$

in which (with Taylor hardening) $\bar{\tau}_c$ is the same function of its argument as is τ of γ in single slip (the Taylor–Schmid law $\tau = f(\gamma)$ discussed in Chapter 1). For the aluminum crystals of 99.6 percent purity investigated by Taylor in his tension and compression experiments with C. F. Elam and W. S. Farren, this function is very nearly parabolic (see Fig. 1.4 here, Taylor's (1938a) fig. 7, and fig. 4.68 in Bell (1973)), whence the macroscopic stress–strain curve is approximately given by

$$\bar{\sigma}_a \approx A_0 \bar{M}\left(\int \bar{M}\, d\bar{e}_L\right)^{1/2},$$

with $A_0 = 38$MPa (from Bell's fig. 4.68).

As remarked by Bishop (1955), $\bar{M}$ in equation (7.8) (his eq. (ii)) is a function of $\bar{e}_L$. Within each grain the lattice rotates at a rate $\omega = -\Omega$ calculated from equations (7.2) (W being zero relative to the specimen frame in axisymmetric deformation), thereby changing the initially uniform distribution of orientations. Consequently, $\bar{M}$ as determined by the volume average in equation (7.7) changes with the development of texture at finite strain. Taylor (1938a), however, took $\bar{M}$ to be constant, whence his calculated macroscopic curve may be expressed $\bar{\sigma} = \bar{M} f(\bar{M}\bar{e}_L)$, with $\bar{M} \approx 3.1$.[4] He then compared his approximate theoretical curve with data of Dr. Elam for a polycrystalline aluminum bar of the same purity as that for which the single-slip hardening curve $\tau = f(\gamma)$ was determined. Her data were for *nominal* stress and strain, to $\bar{e}_N \approx 0.31$ ($\bar{e}_L \approx 0.27$), and were not transformed. Consequently, the value of the comparison is questionable at the larger strains, both because $\bar{M}$ will have changed and because the theoretical curve and the experimental points are for different measures. There is another inconsistency, of course, in the assumption of uniform strain, as subsequently will be discussed. The agreement nonetheless was good, as may be seen in Taylor's figure 14, whether or not in part fortuitous.

Taylor also calculated the lattice spin of each grain (from $\omega = -\Omega$ and eqs. (7.2)) and plotted the incremental rotation of the specimen axis relative to the lattice axes for each of his 44 initial orientations. The directions and relative amounts of these rotations are shown in figure 15 of Taylor (1938a), reproduced here as Fig. 7.1. From his calculations, Taylor concluded that the aggregate would "tend to attain a state in which the crystal axes of grains have either a (111) or a (100) axis... in the direction of extension" (pp. 323–4), a typical texture in finitely stretched f.c.c. polycrystalline bars.

[4] The magnitude of Taylor's calculations for only one $\bar{M}$ (done "with the help of Mallock's equation-solving machine") were remarkable for their time. That he did not go beyond this initial value certainly is understandable.

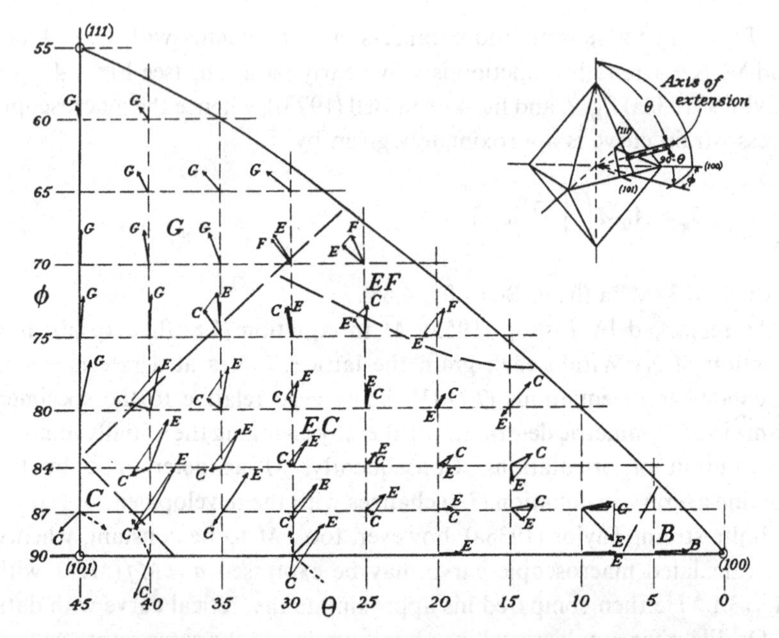

Fig. 7.1. Theoretical rotation of specimen axis relative to lattice axes for differently oriented grains in a polycrystalline bar extended 2.37% (from Taylor (1938a, fig. 13), as presented in *The Scientific Papers of Sir Geoffrey Ingram Taylor*, Vol.I).

Because different combinations of incremental slips gave the same min $\sum|d\gamma_j|$ for many of Taylor's orientations, a range of directions emanates from these initial axis positions in Fig. 7.1. Additionally, Taylor missed some of the minimum-slip combinations by not having considered the cases of double slip in only one plane, as already discussed; and an even greater degree of nonuniqueness of initial rotations is evident in figure 5 of Bishop (1954) for the opposite case of axial compression, as calculated from the equivalent aggregate model of Bishop & Hill (1951b). (Also see Taylor's note following the version of Taylor (1938b) included in Volume I of his collected works.) The matter of nonuniqueness will be further discussed later in this chapter.

Taylor did not specifically consider the stress state in a grain, but this can be determined (to within a pure pressure) as follows. Let $\mathbf{t}$ now represent the stress vector conjugate to the reduced strain-rate vector $\mathbf{d} = (d_{11}, d_{22}, 2d_{23}, 2d_{13}, 2d_{12})^{\mathrm{T}}$, whence $\mathbf{t} = (\sigma_{11} - \sigma_{33}, \sigma_{22} - \sigma_{33}, \sigma_{23}, \sigma_{13}, \sigma_{12})^{\mathrm{T}}$. For a set α of five geometrically independent slip rates, it follows from equation (7.4) and

$$\dot{w} = \mathbf{t} \cdot \mathbf{d} = \sum (\tau \dot{\gamma})_j \qquad (7.9)$$

that

$$\mathbf{t} = N_\alpha^{-\mathrm{T}} \boldsymbol{\tau}_\alpha, \tag{7.10}$$

where $\boldsymbol{\tau}_\alpha$ is the vector of the five critical shear stresses τ_j in set α, with $\tau_j = \pm\tau_c$ for Taylor hardening (the sign of course dependent on whether the jth system is slipping in the positive or negative sense as defined in Table 1). Bishop (1953) proved that a stress state determined from equations (7.4), (7.9), and (7.10) by minimizing $\dot{w}$ for a given $\mathbf{d}$ will not exceed the critical strength in any other system. Bishop & Hill (1951a) had proved the converse of this theorem, namely, that the rate of stress work in a "physically possible" set of slip rates is less than that in a set that is only geometrically admissible. (A physically possible set for prescribed D corresponds through equations (7.4) and (7.10) with a stress state that satisfies the crystal yield condition. That is, $\mathrm{tr}(N_j\sigma) = \tau_j^c$ in each active system and $\mathrm{tr}(N_j\sigma) \leqslant \tau_j^c$ in all other systems, with opposing directions of slip here distinguished by different j. For an arbitrary geometrically admissible set, the yield condition typically will be violated in one or more systems.) This theorem of Bishop and Hill's is the dual to their famous *principle of maximum plastic work* in a crystal grain. Together with the theorem proved by Bishop (1953) it establishes the validity of Taylor's postulate of minimum shears for isotropic hardening as well as his more general hypothesis of minimum work for a given strain increment.

The diverse stress states among the grains obtained by minimizing $\dot{w}$ for the same $\bar{D}$ in each crystal will not be in equilibrium, as these stresses will violate continuity of tractions across grain boundaries. Henceforth let this nonequilibrium set of stresses be denoted σ_w, and let σ and D denote the actual stress and strain-rate fields in the aggregate. We further define the aggregate average

$$\bar{\sigma}_w = \langle \sigma_w \rangle \tag{7.11}$$

to be the macroscopic stress (to within a pure pressure) corresponding to the Taylor model, which definition is consistent with equation (7.7) for axial loading. Consider now the difference between aggregate work rates of stresses σ_w and σ in a uniform strain rate $\bar{D}$:

$$\langle \mathrm{tr}[(\sigma_w - \sigma)\bar{D}] \rangle = \mathrm{tr}[(\bar{\sigma}_w - \langle \sigma \rangle)\bar{D}].$$

For an aggregate the hardening of whose grains is both known in the current state and uniform within each grain,

$$\mathrm{tr}[(\sigma_w - \sigma)\bar{D}] \geqslant 0 \tag{7.12}$$

locally from Bishop and Hill's principle of maximum plastic work (1951a, eq. (19)) because σ, the actual stress, does not violate the current yield condition whereas σ_w satisfies that condition for the prescribed $\bar{D}$ by definition. Thus

$$\mathrm{tr}(\langle\sigma\rangle\bar{D}) \leqslant \mathrm{tr}(\bar{\sigma}_w\bar{D}). \tag{7.13}$$

This inequality is equivalent to the right-hand side of Bishop and Hill's (1951a) equation (19) when $\langle\sigma\rangle$ is taken to equal the actual macroscopic aggregate stress, as is consistent with their original averaging theorem based upon a unit cube defined in the current state. Consequently, $\bar{\sigma}_w$ is an *upper bound* to the stress state at the current yield point in macrostress space for a given $\bar{D}$.

Bishop & Hill (1951b) argue that the incremental work done by $\bar{\sigma}_w$ in $\bar{D}$ very nearly equals the actual incremental work in the aggregate, that is,

$$\mathrm{tr}(\bar{\sigma}_w\bar{D}) \approx \mathrm{tr}\langle\sigma D\rangle \equiv \mathrm{tr}(\bar{\sigma}\bar{D}), \tag{7.14}$$

so the endpoint of stress vector $\bar{\sigma}_w$ lies on or very near the hyperplane in macrostress space that is orthogonal to $\bar{D}$ and tangent to the yield surface at point $\bar{\sigma}$. This idea is the basis for their well-known method of calculation of the yield locus.

Let d_H denote the perpendicular distance to the yield hyperplane corresponding to $\bar{D}$. Then, from equations (7.11) and (7.14) we have (Bishop & Hill 1951b, eq. (8))

$$d_H = \frac{\mathrm{tr}(\langle\sigma_w\rangle\bar{D})}{\|\bar{D}\|}, \quad \|\bar{D}\| = (\mathrm{tr}\,\bar{D}^2)^{1/2}, \tag{7.15}$$

and the macroscopic yield surface is the envelope of planes so defined from the complete range of unit strain rates $\bar{D}/\|\bar{D}\|$ (subject to $\mathrm{tr}\,\bar{D} = 0$). It is not necessary, however, to go through the lengthy process defined by equations (7.4), (7.9), and (7.10) to calculate σ_w within each grain. Rather, the great result proved by Bishop and Hill (through their theorems) is that the stress which *minimizes* $\dot{w}$ among all geometrically admissible slip rates within a crystal for a given $\bar{D}$ also *maximizes* $\dot{w}$ among all stress states that do not violate the crystal yield condition.[5] Moreover, in their second paper they established that for isotropic hardening (or a virgin crystal) it is only necessary to investigate 56 particular stress states

[5] A proof of the equivalence of the Taylor and Bishop and Hill models presented in the language of linear programming and optimization theory may be found in Chin & Mammel (1969).

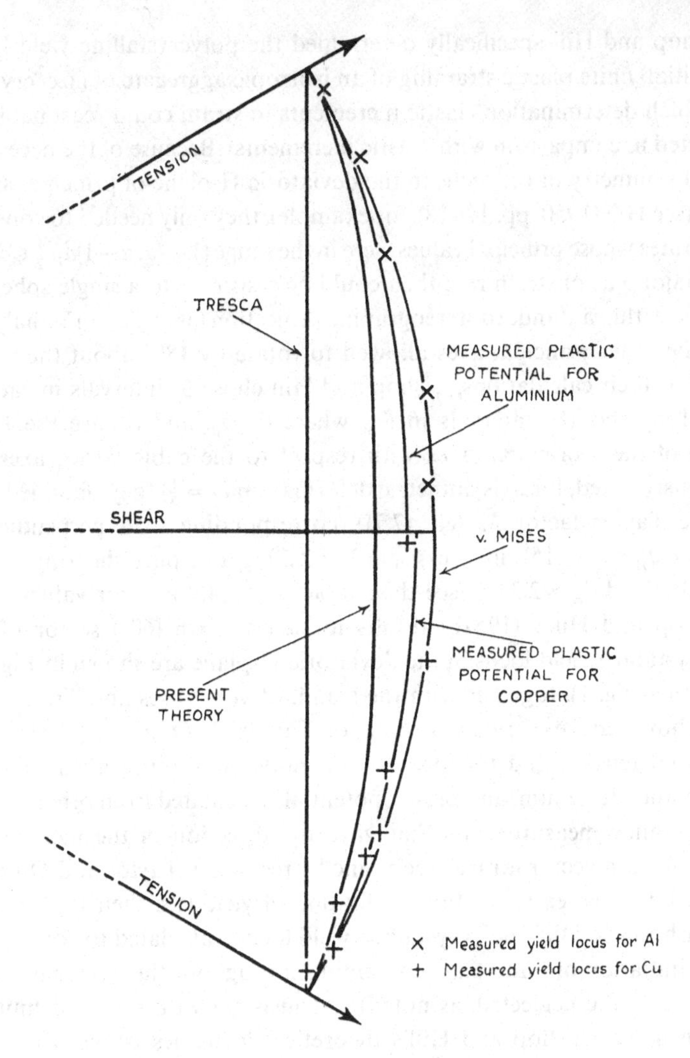

Fig. 7.2 Comparison of Bishop and Hill's theoretical polycrystalline yield locus for f.c.c. metals ("present theory") with experimentally determined plastic potentials and yield loci for aluminum and copper (data from Taylor & Quinney 1931) and with the Tresca and von Mises criteria (from Bishop & Hill 1951b, fig. 1).

corresponding to the vertices of the polyhedral crystal yield surface. These 56 states are defined in terms of the uniform critical strength τ_c in Bishop & Hill (1951b, pp. 1302–3). Thirty-two of them correspond to a sixfold vertex (that is, the critical shear stress is attained in six different slip systems) and the remaining 24 to an eightfold vertex.

Bishop and Hill specifically determined the polycrystalline yield locus for (initial) finite plastic straining of an isotropic aggregate of f.c.c. crystals (for which determination elastic increments in strain could reasonably be neglected in comparison with plastic increments). Because of the necessary sixfold symmetry of the locus in the deviatoric Π-plane of principal stress space (see Hill (1950, pp. 17–18), for example), they only needed to consider strain rates whose principal values were in the range $(1, -\alpha, \alpha-1)d_I, \frac{1}{2} \leqslant \alpha \leqslant 1$. The major axis of strain rate then could be restricted to a single spherical triangle within a standard stereographic projection (as in Taylor's analysis), with the other principal axes allowed to rotate by 180° about the major axis. For their calculations, Bishop and Hill chose 5° intervals in each of θ_p and ϕ_p and 18° intervals in ψ_p, where θ_p, ϕ_p, and ψ_p are the Euler angles of these principal axes with respect to the cubic lattice axes. As previously noted, for axisymmetric deformation ($\alpha = \frac{1}{2}$) they obtained 3.06 for the Taylor factor $\bar{M}$ (eq. (7.7)), corresponding to a perpendicular distance d_H (eq. (7.15)) in stress space of $2.50\,\tau_c$. For pure shearing ($\alpha = 1$) they obtained $d_H = 2.34\,\tau_c$ (see their table, p. 1304, for other values).

Bishop and Hill's (1951b) results for a one-sixth (60°) sector of the polycrystalline yield locus in the deviatoric Π-plane are shown in Fig. 7.2 (from their fig. 1), together with the standard von Mises and Tresca loci. Also shown are experimental data of Taylor & Quinney (1931) from combined tension and torsion tests of moderately thin-walled tubes of copper and aluminum, and plastic potentials calculated from other Taylor and Quinney measurements that gave the direction of the incremental (plastic) strain vector for each combined stress state. Taylor and Quinney adopted a "back-extrapolation" definition of yield (see their figs. 6a, b, c); and Bishop and Hill's polycrystalline yield locus, calculated to correspond to the initiation of finite plastic straining throughout the aggregate (with elastic straining neglected, as noted), is consistent with such a definition. It is seen that Bishop and Hill's theoretical locus lies between the two classical loci and is approximately equidistant with the von Mises locus from the measured plastic potentials for aluminum and copper.

7.2 Applications of Taylor–Bishop–Hill Models

Linear programming techniques to systematize calculation of the Taylor model by computer were used by Chin & Mammel (1967) for axisymmetric deformation of b.c.c. crystalline aggregates. They determined contour maps of constant M-factors for the four cases of separate

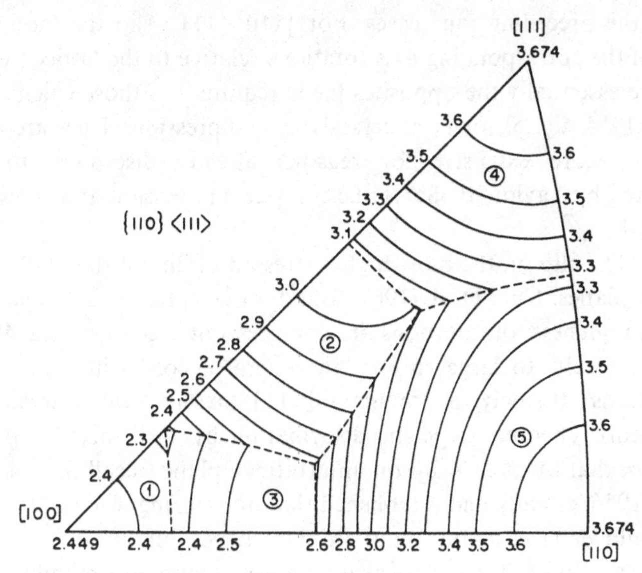

Fig. 7.3. Contours of constant M-factors for $\{110\}\langle 111\rangle$ or $\{111\}\langle 110\rangle$ slip according to the Taylor aggregate model of axisymmetric axial deformation. Dashed boundaries delineate regions within which a specific set of slip systems is active (from Chin & Mammel (1967, fig. 2); these regions were first displayed in Bishop (1953, fig. 5)). Reprinted with permission from *Transactions of the Metallurgical Society*, Vol. 239, pp. 1400–5, 1967, a publication of The Minerals, Metals & Materials Society, Warrendale, PA 15086.

and combined $\langle 111\rangle$ slip on $\{110\}$, $\{112\}$, and $\{123\}$ planes, with $M = \min \sum |\gamma'_j|$ (the prime signifying differentiation with respect to logarithmic compressive strain $\bar{e}_L$). The first of these cases is, of course, also that of $\{111\}\langle 110\rangle$ slip in f.c.c. crystals, as an interchange of vectors $\mathbf{b}_j$ and $\mathbf{n}_j$ does not affect tensor $N_j = \text{sym}(\mathbf{b} \otimes \mathbf{n})_j$, hence does not affect equations (7.1) and only changes the signs in equations (7.2). (Of course, for b.c.c. crystals a, b, c, and d designate slip directions rather than slip planes in these equations.) Correspondingly, the $\bar{M}$ value that Chin and Mammel calculated for $\{110\}\langle 111\rangle$ slip ($\bar{M} = 3.067$), based upon a spherically uniform distribution of crystal orientations, is consistent with Bishop & Hill's (1951b) and Taylor's (1938a) values (3.06 and 3.10, respectively) calculated for $\{111\}\langle 110\rangle$ slip without benefit of computers. Chin and Mammel's map of these contour lines for $\{110\}\langle 111\rangle$ (or $\{111\}\langle 110\rangle$) slip is shown in Fig. 7.3 (from their fig. 2).

Chin, Mammel & Dolan (1967) determined the initial lattice rotations of a uniform distribution of 52 b.c.c. crystallites in axial compression for each of the preceding four cases. For {110}⟨111⟩ slip the (nonunique) ranges of the corresponding axis rotations relative to the lattice (see their fig. 1) are essentially the opposites (as is required) of those calculated by Bishop (1954, fig. 5) for f.c.c. crystals in compression. They are similar (although more extensive for reasons already discussed) to those determined by Taylor (1938a) for f.c.c. crystals in tension and shown here in Fig. 7.1.

For ⟨111⟩ slip on the most highly stressed of any of the {110}, {112}, or {123} planes, Chin et al. (1967) found an essentially negligible degree of nonuniqueness of rotations in compression (see their fig. 4), with projected results to large strains "in general accord with experimental observations" (namely, a dominant [111] texture with a weak [100] component). Theoretical plastic deformation by such mixed slip differs little from that by ⟨111⟩ slip on an arbitrary plane (recall Sect. 4.2); and Taylor (1956) already had established that minimizing $\dot{w}$ for ⟨111⟩ pencil glide would lead to a unique set of slip rates, hence unique lattice rotations. Thus, the results of Chin et al. of nearly unique rotations from minimization of $\dot{w}$ over 12 different slip planes were consistent with Taylor's analysis yielding a unique prediction for pencil glide in b.c.c. crystals.

Chin (1969) addressed the nonuniqueness of incremental slips and lattice rotations in f.c.c. crystals by distinguishing between high and low "stacking-faulty energy" metals. He selected from among the admissible sets of slip rates that minimize $\dot{w}$ by favoring (i) collinear slip-system pairs (i.e., cross slip) for high-stacking-fault-energy metals (e.g., aluminum) and (ii) coplanar slip-system pairs for those of low stacking-fault energy (e.g., copper and silver). These ideas were based in part upon data from various of the latent-hardening experiments on f.c.c. crystals reviewed in Chapter 4. Although Chin did not carry out calculations beyond the first increment in deformation, his procedures would tend to produce less of the [100] texture component (in comparison with [111]) in both cases (see his figs. 7 and 10). Based upon the experimental investigations of English & Chin (1965), the [100] component actually increases with a decrease in stacking-fault energy over a wide range that includes a number of pure f.c.c. metals (see their fig. 2).

Apparently the first published work in which a Taylor–Bishop–Hill aggregate analysis was carried out in successive increments to large strains is that of Kallend & Davies (1972). They adopted plane strain to approximate the deformation state in cold rolling of metals, and assumed Taylor hardening in each of the 406 crystallites of their specific model. At each deformation step (a 5 percent logarithmic strain increment) they determined the stress state using Bishop and Hill's principle of maximum plastic work, continuing the calculations to a thickness reduction of 80 percent (a logarithmic compressive strain of 1.61). To resolve the nonuniqueness of slip rates among the six or eight critical systems at each yield-locus vertex, Kallend and Davies also distinguished between high- and low-stacking-fault-energy metals and selected from among the physically possible slip systems on the basis of related concepts from dislocation theory. The principal interest of these investigators was in the evolution of texture; and they determined the crystal grains' orientation distribution at each of 40 percent and 80 percent reductions in thickness. They found that the differences between the differently predicted textures "based on the probable effects of both high and low stacking-fault energy... were slight, and both were consistent with the experimentally determined texture of cold-rolled copper" (Kallend & Davies 1972, p. 488).

Several other ideas have been advanced to deal with the nonuniqueness in Taylor–Bishop–Hill analyses of f.c.c. crystals. Gil-Sevillano, Van Houtte & Aernoudt (1975) chose the average of all admissible rotations in their calculations of texture evolution in simple shear (corresponding to torsional deformation of cylindrical bars or tubes). They also made calculations for slip on a combination of $\{110\}$ and $\{112\}$ planes in b.c.c. crystals. They state that up to shear strains of about 4, "a good agreement with experimentally measured pole figures is found" (p. 367). Other ways of resolving the nonuniqueness are mentioned in Gil-Sevillano, Van Houtte & Aernoudt (1982), which work (in chapters 5 and 6) contains an extensive and comprehensive review of f.c.c. and b.c.c. aggregate models and texture predictions through 1978.

Tomé et al. (1984) also selected the average of all admissible rotations in their early-stage calculations of an 800-grain f.c.c. aggregate model for each of axial tension, compression, and torsional deformation, using strain increments of 0.025. They subsequently changed from a Taylor model to a "relaxed-constraints model," however. (Relaxed-constraints models have been calculated by a number of investigators, prominent among them H. Mecking and U.F. Kocks, but they will not be considered here.) For extensive experimental studies of torsional deformation textures in copper, aluminum, and α-iron over a wide range of temperatures see Montheillet, Cohen & Jonas (1984).

7.3 Analysis of Uniqueness

The apparent nonuniqueness of slip rates in the models of Taylor and Bishop and Hill can be resolved within an arbitrary crystal grain in a manner compatible with hardening theory as follows. Let σ now represent the stress state determined for a given $D = Q\bar{D}Q^{\mathrm{T}}$ and the current yield surface by Bishop and Hill's maximum-plastic-work analysis, which stress typically is unique (to within a pure pressure) for an arbitrary strain rate (Bishop 1953, p. 63). Consider the consistency conditions in the n critical systems, which may be expressed (from eqs. $(3.58)_1$ and (3.75)–(3.76), written with the current state as reference)

$$\dot{f}_k = \sum_j g_{kj}\dot{\gamma}_j - \mathrm{tr}(\Lambda_k D) \geqslant 0, \quad \dot{\gamma}_k \geqslant 0, \quad \dot{f}_k\dot{\gamma}_k = 0, \tag{7.16}$$

where

$$\begin{aligned} g_{kj} &= N_k \cdot \mathscr{L}_0 \cdot N_j + H_{kj} - \mathrm{tr}(N_j \alpha_k), \\ \Lambda_k &= N_k \cdot \mathscr{L}_0 - \alpha_k, \quad \alpha_k = 2\mathrm{sym}(\sigma\Omega_k) \end{aligned} \tag{7.17}$$

from equation (5.125) and equations (3.49) and (3.35), evaluated for the logarithmic measure in the current state. For infinitesimal lattice strains, $\mathscr{L}_0$ may be taken as the standard tensor of second-order elastic moduli c_{11}, c_{12}, and c_{44} in cubic crystals. As proved by Hill & Rice (1972) and shown here in Section 3.6, equations (7.16) have a unique solution for the given D if the parameters g_{kj} are positive definite.

Let $\Delta\dot{\gamma}_k$ denote the difference between two sets of slip rates (among six or eight critical systems, say) that produce the same D but a plastic-spin difference $\Delta\Omega$. There follow $\sum N_j \Delta\dot{\gamma}_j = 0$ and (from eq. $(7.17)_1$)

$$\sum\sum g_{kj}\Delta\dot{\gamma}_k\Delta\dot{\gamma}_j = \sum\sum H_{kj}\Delta\dot{\gamma}_k\Delta\dot{\gamma}_j, \quad j, k = 1, \ldots, n. \tag{7.18}$$

For Taylor hardening this reduces to $H(\sum\Delta\dot{\gamma}_j)^2$, which is zero because both sets of slip rates correspond to the same $\min\sum\dot{\gamma}_j$ (whence $\sum\Delta\dot{\gamma}_j \equiv 0$). The matrix g_{kj} is therefore only positive semidefinite. The nonuniqueness can be eliminated and the g parameters made positive definite by adding a small amount of self-hardening $\eta \ll H$ to the Taylor moduli, with the resultant quadratic form (from eq. (7.18))

$$H(\sum\Delta\dot{\gamma}_j)^2 + \eta\sum(\Delta\dot{\gamma}_j)^2 > 0.$$

For the simple theory, if we restore the very small term $\alpha_k \cdot \mathscr{M}_0 \sum\alpha_j\dot{\gamma}_j$ that was neglected following equation (4.6) in Section 4.4.1 (or directly apply eq. (2.7) in Havner (1981)), we have

$$\sum\sum g_{kj}\Delta\dot{\gamma}_k\Delta\dot{\gamma}_j = h(\sum\Delta\dot{\gamma}_j)^2 + 4(\sigma\Delta\Omega)\cdot\mathscr{M}_0\cdot(\sigma\Delta\Omega),$$

which will be positive unless both $\sum\Delta\dot{\gamma}_j$ and $\sigma\Delta\Omega$ (tensor product) are zero. As this may happen for certain stress states and slip-system combinations (see Havner 1981, sect. 2), the same small self-hardening η may be added to the original simple theory (see Fuh & Havner 1986, 1989) to once again guarantee positive definiteness of matrix g_{kj}. (This procedure may be applied equally well to the P.A.N. and two-parameter hardening rules.)

Consider the plane-strain compression of an f.c.c. crystal in the family of orientations having a [100] lattice direction initially coincident with the direction of maximum extension (for example, the longitudinal axis Z in a channel die). As briefly discussed in Section 5.7, there are eight critical systems $a2$, $a\bar{3}$, $b2$, $b\bar{3}$, $c2$, $c3$, $d2$, and $d\bar{3}$ (here numbered 1–8 as before) under the biaxial stress state $-\sigma_{xx} = -\sigma_{yy} = \sqrt{6}\tau_0$ at the initiation of finite plastic straining, with Y the direction of full lateral constraint $d_{yy} = 0$. Fuh & Havner (1989, sect. 2.4) carried out a rigorous analysis of this family of orientations by applying the consistency conditions (eqs. (7.16)) to the modified simple theory

$$H_{kj} = h - \mathrm{tr}(N_k \alpha_j) + \eta \delta_{kj}, \quad h \gg \eta > 0. \tag{7.19}$$

They included a compressive volumetric strain rate ε_V of the lattice, so

$$d_{11} \equiv d_{zz} = 1 - \varepsilon_V, \quad d_{22} = -\cos^2\theta, \quad d_{33} = -\sin^2\theta,$$
$$d_{23} = -\tfrac{1}{2}\sin 2\theta, \qquad d_{13} = d_{12} = 0$$

with respect to the compressive logarithmic strain e_L. Here, as in Section 5.7, θ is the counterclockwise orientation of the compression direction X from [010] on a [100] stereographic projection. (ε_V could have been disregarded without affecting the uniqueness analysis.) For these eight critical systems in an arbitrary orientation, Fuh and Havner obtained the unique solution

$$\gamma_1' = \gamma_5' = \frac{\sqrt{6}}{4}(3 - 2\varepsilon_V)/(3 + 2\bar{h}) - \frac{\sqrt{6}}{4}\cos 2\theta \approx \frac{\sqrt{6}}{2}\sin^2\theta,$$

$$\gamma_2' = \gamma_6' = \frac{\sqrt{6}}{4}(\sin 2\theta + \cos 2\theta), \quad \gamma_3' = \gamma_7' = 0,$$

$$\gamma_4' = \gamma_8' = \frac{\sqrt{6}}{4}(3 - 2\varepsilon_V)/(3 + 2\bar{h}) \tag{7.20}$$

$$- \frac{\sqrt{6}}{4}\sin 2\theta \approx \frac{\sqrt{6}}{2}(1 - \sin 2\theta),$$

$$\bar{h} = 6h/(c_{11} - c_{12}),$$

in which terms involving $\eta/(c_{11} - c_{12})$ have been neglected for simplicity. The expressions on the far right-hand sides in equations (7.20) represent the rigid/plastic limit (as $\bar{h} \to 0$, $\varepsilon_V \to 0$). This is precisely the *minimum-plastic-spin solution* defined by point F of Fig. 5.20 and discussed in Section 5.7 in relation to channel die experiments of Driver et al. (1983).

From the requirement of nonnegative slip rates, equations $(7.20)_{1,4}$ provide the following limits on θ for their validity:

$$\arccos\{(3 - 2\varepsilon_V)/(3 + 2\bar{h})\} \leqslant 2\theta \leqslant \arcsin\{(3 - 2\varepsilon_V)/(3 + 2\bar{h})\}. \tag{7.21}$$

For all θ in the very narrow range between 0 and the lower limit, systems 1 and 5 as well as 3 and 7 are inactive, and the resulting unique solution gives $a_1 = \theta' = \frac{1}{2}\sin 2\theta, a_2 = a_3 = 0$ (plastic-spin components), which converges to the minimum-plastic-spin solution of lattice stability as $\theta \to 0$ ($X = [010]$). For θ in the narrow range between the upper limit in equation (7.21) and 45°, only systems 1, 2, 5, and 6 are active, with the unique result $a_1 = \theta' = \frac{1}{2}\cos 2\theta, a_2 = a_3 = 0$. This also is the minimum-spin, stable-lattice solution in the limit $\theta = 45°$ ($X = [011]$).[6] (See Fuh & Havner (1989, pp. 206–9) for details.)

7.4 A Survey of Other Models

Apparently the first (compatible) finite-strain polycrystal model satisfying intergrain equilibrium, at least approximately (through discretization of the field equations), was that proposed in Havner (1973a). The model was analyzed as regards existence and uniqueness of solution in terms of a quadratic programming formulation, but never was calculated. However, the original *infinitesimal*-strain version (Havner 1971) was computed for aluminum and copper aggregates by Havner & Varadarajan (1973), Havner et al. (1974), and Havner & Singh (1977), with emphasis on initial and subsequent macroscopic yield loci, elastoplastic macrostress and strain curves (including unloading/reloading), and latent strain energy (see Sect. 6.4 here). Theoretical aspects of convergence of

[6] The minimization of plastic spin may itself prove to be a useful means of resolving the nonuniqueness of slip rates in aggregate analysis. This speculation is based upon the theory's highly satisfactory predictions of typical experimental behavior of f.c.c. single crystals in the three families of multiple-slip configurations analyzed by Fuh & Havner (1989). However, such aggregate calculations have not yet been performed.

the discretization were investigated by Havner & Patel (1976). Noteworthy earlier examples of polycrystal calculations at infinitesimal strain that account for both compatibility of deformations and intergrain equilibrium may be found in Lin & Ito (1965, 1966) and Hutchinson (1970) (see also Lin 1971). Hutchinson's calculations incorporate cubic elastic anisotropy (as do those of Havner et al. (1974)) and are based upon the "self-consistent" method of Hill (1965). The analysis and calculations of Lin and Ito, however (following Lin, Uchiyama & Martin (1961), Lin & Tung (1962), and Lin (1964)), require that the crystal elasticity be represented as isotropic. (As none of the foregoing models consider lattice rotation (with the exception of Havner (1973a)), they, of course, have nothing to offer regarding texture prediction.)

A self-consistent model at finite strain that essentially falls within the general aggregate analysis of Chapter 6 and approximately satisfies equilibrium, deformation compatibility, and Hill's (1972) averaging theorem is presented in Iwakuma & Nemat-Nasser (1984). Space and the complexity of their model and calculations preclude a detailed review here, but a brief description follows.

Iwakuma and Nemat-Nasser define a local, elastoplastic tensor modulus $\mathscr{F}$ (their eq. (2.16)) that connects the nominal stress rate $\dot{S}$ and the velocity gradient $\Gamma = \dot{A}$ (with the current state as reference) and involves summations "over all active systems." If every critical system is active, then $\mathscr{F}$ is essentially the tensor $\mathscr{C}_{\mathrm{p}}$ of fully plastic response defined in equation (3.108) here (with minor differences related in part to the approximation of infinitesimal lattice straining). Otherwise, it is not clear how the active set would be selected a priori from among the critical systems, for neither the local strain rate D nor the lattice-corotational rate $\hat{\sigma}$ of Cauchy stress is known in advance in the nonuniform field problem that these authors define. (Their critical slip-system inequalities are expresssed in terms of $\hat{\sigma}$, based upon Hill (1966). See the unnumbered inequality involving $\hat{\sigma} \equiv \mathscr{D}^*\sigma/\mathscr{D}t$ that follows eq. (3.56) here and encompasses their eqs. $(2.9)_{2,3}$.)

Given $\mathscr{F}$, a macroscopic tensor $\bar{\mathscr{F}}$ is defined from $\mathrm{d}\bar{S} = \bar{\mathscr{F}}\,\mathrm{d}\bar{A}$, with $\mathrm{d}\bar{S} = \langle \mathrm{d}S \rangle$ and $\mathrm{d}\bar{A} = \langle \mathrm{d}A \rangle$ (or $\bar{\Gamma} = \langle \Gamma \rangle$) consistent with equations (6.14) and the averaging theorem. Iwakuma and Nemat-Nasser further define a "concentration tensor" $\mathscr{A}_{\mathrm{p}}$ by $\mathrm{d}A = \mathscr{A}_{\mathrm{p}}\,\mathrm{d}\bar{A}$. (This obviously is distinct from Hill's (1984) influence tensor $\mathscr{A}$, eqs. (6.24)–(6.25) here, which is expressed in terms of purely elastic incremental response of the aggregate.) There follows $\bar{\mathscr{F}} = \langle \mathscr{F}\mathscr{A}_{\mathrm{p}} \rangle$ (their eq. (3.8)), analogous to equation $(6.26)_2$ for the tensor of macroscopic (nominal) elastic moduli. Thus, in order to calculate incremental macroscopic response, they must determine $\mathscr{A}_p$ at each stage of the deformation. To accomplish this, they extend Hill's (1965) self-consistent method to finite strain and consider the problem of a single ellipsoid crystal $\mathscr{E}$, of nominal moduli $\mathscr{F}$, embedded in an unbounded macroscopic medium (of unknown moduli $\bar{\mathscr{F}}$) subjected to the overall macroscopic stress rate $(\bar{S})^{\cdot}$. It is required that

$$\dot{S}(\mathbf{x}) = \begin{cases} \mathscr{F}\dot{A} \text{ within } \mathscr{E}; \\ \bar{\mathscr{F}}\dot{A} \text{ outside } \mathscr{E}, \end{cases} \tag{7.22}$$

with $\operatorname{Div} \dot{S} = \mathbf{0}$ everywhere, $\mathbf{n}_0 \dot{S}$ (nominal traction rate) continuous across $\partial\mathscr{E}$, and $\dot{S}(\mathbf{x}) \to (\bar{S})^{\cdot}$ as $|\mathbf{x}| \to \infty$ (with origin at the center of $\mathscr{E}$). After a moderate development, they are able to express $\mathscr{A}_p$ as a nonlinear function of $\mathscr{F}$ and the unknown $\bar{\mathscr{F}}$ that involves the Green function for the macroscopic material (see Iwakuma & Nemat-Nasser 1984, pp. 93–5). The averaging in the nonlinear calculation

$$\bar{\mathscr{F}} = \langle \mathscr{F}\mathscr{A}_p(\mathscr{F}, \bar{\mathscr{F}}) \rangle \tag{7.23}$$

is then to be performed over a representative set of crystal orientations.

For a two-dimensional crystal model originally proposed by Asaro (1979), Iwakuma and Nemat-Nasser achieve an analytic solution for the Green function (see their appendix B), with the concentration tensor expressed as

$$\mathscr{A}_p = [\mathscr{I} + \mathscr{J}(\mathscr{F} - \bar{\mathscr{F}})]^{-1} \tag{7.24}$$

(eq. (B9)), where $\mathscr{I}$ is the fourth-order identity tensor and the components of tensor $\mathscr{J}$ are determined from the Green function (see their eqs. (B7)). Among two alternative ways of normalizing the concentration tensor (to ensure that $\langle \mathscr{A}_p \rangle = \mathscr{I}$), Iwakuma and Nemat-Nasser obtained greater numerical stability in calculations of uniaxial stressing when they replaced $\mathscr{A}_p$ by $\mathscr{A}_p \langle \mathscr{A}_p \rangle^{-1}$, a method introduced by Walpole (1969) for an analogous tensor in a linearly elastic polycrystal model. Unfortunately, and apparently related to the relatively small number of 46 crystal orientations that they used, "localization" of deformation (i.e., shear bands) became possible and the iterative scheme for the evaluation of $\bar{\mathscr{F}}$ broke down at rather modest strains. The greatest deformation achieved is a tensile strain of approximately 7.5 percent (see their fig. 8b) corresponding to linear Taylor hardening, with $\tau_0 = 2.5 \times 10^{-3} c_{44}$, $H_0 = 10^{-2} c_{44} = 4\tau_0$, and the crystal elasticity taken to be isotropic.

Hutchinson (1976) introduced a simple constitutive equation for *steady creep* of crystals and applied it in small-strain analyses of f.c.c. and ionic crystalline aggregates. This equation may be expressed

$$\dot{\gamma}_k = \dot{\gamma}_0 \frac{\tau_k}{\tau_k^{\mathrm{R}}} \left| \frac{\tau_k}{\tau_k^{\mathrm{R}}} \right|^{n-1}, \quad \dot{\gamma}_0 > 0, \tag{7.25}$$

where τ_k^{R} is a positive "reference stress" (which Hutchinson took to be constant and the same in all systems for f.c.c. crystals), $\dot{\gamma}_0$ is a "convenient reference creep rate," and $1/n$ is the positive rate-sensitivity parameter (n need not be an integer). Although Hutchinson's simple model was intended for crystalline materials at relatively high homologous temperatures, it has been adopted by many other investigators for finite-strain polycrystal analyses at ordinary temperatures (see Asaro & Needleman 1985; Nemat-Nasser & Obata 1986; Molinari, Canova, & Ahzi 1987; Toth, Gilormini & Jonas 1988; Wenk et al. 1989; Harren & Asaro 1989; Harren et al. 1989; and Neale, Toth & Jonas 1990, among others). A rate-dependent plasticity theory is thereby defined (with the τ_k^{R} no longer necessarily constant) that eliminates the common nonuniqueness problem of rate-independent theory by declaring that *every* slip-system is active in the sense of its current resolved shear stress τ_k, however small (unless zero). Such continuous viscoplastic flow at all levels of stress may be realistic for metals at elevated temperatures (as in hot rolling, for example); but for ordinary temperatures at the human scale, it seems to me that models of this form are to a certain extent computational expedients. However, it must be acknowledged that there have been a number of generally successful texture predictions using the rate-dependent form, equation (7.25). (On the other hand, were this form essentially correct for the plastic deformation of metal crystals, tens of thousands of large structures worldwide, to say nothing of tens of millions of automobiles, aircraft, ships, and trains, would long since have collapsed under their own weight!)

The papers cited in the preceding paragraph, and other similar finite-strain, rate-dependent polycrystal analyses, fall into three categories: Taylor models (that is, uniform aggregate strain), self-consistent models, and (two-dimensional) finite-element models. There also is a distinction based upon whether the reference stresses τ_k^{R} are taken as constant (pure creep) or deformation dependent (strain hardening). Asaro & Needleman (1985), following Peirce, Asaro, & Needleman (1983), adopted a hardening rule for the reference stresses that is essentially the two-parameter rule for critical strengths of equation (4.29) here, and investigated both $H_2/H_1 = 1.0$ (Taylor hardening) and 1.4, with $H_1(\gamma) = H_0 \operatorname{sech}^2\{H_0\gamma/(\tau_s - \tau_0)\}$, where $\gamma = \int \sum |d\gamma_k|$. In their calculations the rate-sensitivity parameter $1/n$ (or m) equals 0.005, $\tau_s = 1.8\tau_0$ (the "saturation strength"), and $H_0 = 8.9\tau_0$, which values lead to essentially zero hardening of the slip systems after $\gamma \approx 0.1$. They use an "extended Taylor" polycrystal model (including lattice elasticity) of 196 f.c.c. grains, with either the uniform, incremental deformation gradient $d\bar{A}_{ji}$ given or the corresponding macroscopic nominal stress $\bar{s}_{ij} = \langle s_{ij} \rangle$ specified to be zero for each of the nine tensor components. Asaro and Needleman present extensive results for "constant effective plastic strain" yield surfaces (at various levels of prestrain) and the evolution of textures for uniaxial and plane-strain tensile and compressive loading. Their paper is a very substantial work and should be consulted by anyone interested in rate-dependent polycrystal calculations.

In contrast to Asaro and Needleman's model, Toth et al. (1988) and Neale et al. (1990) completely disregard crystal hardening (as well as lattice elasticity) and adopt a pure creep law in which the τ_k^{R} equal a constant τ_0; whence equation (7.25) can be rearranged to give (with $m = 1/n$)

$$\tau_k = \tau_0 \frac{\dot{\gamma}_k}{\dot{\gamma}_0} \left| \frac{\dot{\gamma}}{\dot{\gamma}_0} \right|^{m-1} \tag{7.26}$$

(also see Toth, Neale & Jonas (1989) and Toth, Jonas, Gilormini & Bacroix (1990)). Toth et al. (1988) and Neale et al. (1990) are concerned with the development of textures and the evolution of axial stress to very large deformations in fixed-end torsion, using the Taylor model of uniform aggregate straining (although Neale et al. also include calculations using the relaxed-constraints model). The former work introduces "orientation stability maps" to characterize the evolving textures to a macroscopic shear $\gamma = 12$, with $m = 0.125$. In the latter work, texture predictions for an 800-grain model are compared with experiments of Montheillet et al. (1984) for copper and aluminum, at $\gamma = 8$, for two different m values. The best correlation for copper was obtained using $m = 0.125$ (the higher value) and the Taylor model (see Neale et al., fig. 7).

The paper by Nemat-Nasser & Obata (1986) contains a parallel development to that of Iwakuma & Nemat-Nasser (1984) reviewed earlier in this section, the principal difference being that here the rate-dependent equation (7.25) is adopted, entailing a further modification of Hill's (1965) self-consistent method. The finite-strain calculations again are based upon a two-dimensional, double-slip model. Thirty-one distinct orientations are considered, with isotropic lattice elasticity and linear Taylor hardening of the reference stresses. Nemat-Nasser and Obata present stress–strain curves for different strain rates and rate-sensitivity parameters and find that "rate-dependent crystallographic slip produces a more stable overall behaviour of the polycrystals" (p. 370) in comparison with the rate-independent calculations of Iwakuma and Nemat-Nasser.

Molinari et al. (1987) also use a self-consistent, rate-dependent model and Taylor hardening of the (equal) reference stresses in each grain, although they neglect lattice elasticity and their computational methodology is different from that of the preceding paper (see sects. 3–5 of Molinari et al.). These authors consider an f.c.c. crystalline aggregate of 100 grains, but make the simplifying assumption that the *macroscopic* modulus of the equivalent homogeneous medium is isotropic. They present texture predictions corresponding to each of uniaxial tension, uniaxial compression, plane-strain compression, and torsion for $m = 0.05$ and a von Mises equivalent strain (i.e., $\int \sqrt{\frac{2}{3}} \| \bar{D} \| \, dt$) of 1.0.

Wenk et al. (1989) give an interesting geological application of the preceding (approximate) self-consistent model. An aggregate of simple cubic crystals of halite (NaCl) is investigated, based upon dominant $\{110\}\langle 1\bar{1}0\rangle$ slip and secondary slip on $\{100\}$ and $\{111\}$ planes. Each of zero hardening and linear Taylor hardening of the reference stresses is investigated, with $H_0 = 100\,\text{MPa}$ for the latter in all systems. However, the initial values τ_k^0 are taken to be 6.6 times as great in the secondary systems as in the primary systems (Wenk et al. 1989, p. 2020), "representative of experimental data at room temperature and a strain rate of $10^{-4}\,\text{s}^{-1}$." A model of 200 grains is considered, with $m = 1/9$. A viscoplastic Taylor aggregate model also is calculated. Extensive results for macroscopic stress–strain curves, overall texture evolution, and loading-axis trajectories relative to lattice axes of selected grains are presented for each of uniaxial tension and compression. The differing predictions of the two models are contrasted, but the authors state that they "are not in a position to give general recommendations about applicability" (p. 2018) because of the lack of adequate experiments.

In Harren et al. (1989), an extensive and comprehensive work dealing with rate-dependent finite shear of f.c.c. aggregates, the nonlinear hardening of the reference stresses adopted by Asaro & Needleman (1985) is modified to include an asymptotic constant slope at infinite slip (analogous to $\mathscr{L}_T$ in eq. (5.76) here), which the authors label h_s. Extended Taylor models of 300 and 489 grains are investigated for a variety of boundary conditions, including fully constrained torsion, torsion with end constraint, and unconstrained torsion. Macroscopic stress–strain curves for both Taylor and two-parameter hardening ($H_2/H_1 = 1.4$), zero and nonzero h_s, and a range of m factors from 0.005 to 0.5 are calculated and compared. The texture predictions displayed by Harren et al. in their figures 5, 6, 7, 10, 12, and 21, which include comparisons with experiments of Williams (1962) on copper plate, are based upon $h_s = 0$, Taylor hardening, and $m = 0.005$. These include results of calculations for reverse straining from a macroscopic shear γ of 3.36 to -1.16 (see their fig. 21). This major work also includes an extensive review and comparison of phenomenological macroscopic theories of plasticity.

Harren & Asaro (1989) take account of the nonuniform deformation within crystalline aggregates by considering a two-dimensional, finite-element model consisting of 27 grains subdivided into 8960 uniform-strain triangular regions (representing one-fourth of the macroscopic unit cell). They use the same rate-sensitivity and hardening parameters as were chosen by Asaro & Needleman (1985), but adopt isotropy for the (nonlinear) elasticity of their two-dimensional crystals. Uniaxial tension and compression and simple shear are investigated, and Harren and Asaro show the development of diffuse shear bands in tension and shear by displaying deformed finite-element meshes at several strain levels (see their figs. 5 and 10). They also make comparisons of macroscopic stress–strain curves and (two-dimensional) texture development with corresponding results of a rate-dependent Taylor model. For further studies of the same two-dimensional model, including a display of deformed finite-element meshes in compression, see McHugh et al. (1989).

The various viscoplastic, finite-strain aggregate calculations reviewed in this section, and similar ones in the literature, are computationally impressive (although which approximate polycrystal model is superior appears to be an open question). However, one hopes that there may soon evolve a theory of rate-dependent crystalline slip in metals that would leave such great structural landmarks as the Eifel Tower, Empire State Building, and Golden Gate Bridge still standing.

APPENDIX: THE GENERAL THEORY OF WORK-CONJUGATE STRESS AND STRAIN

A.1 Spin Tensors of the Principal Directions of Stretch

Any pair of stress and strain measures t, e which satisfy

$$\dot{w} \equiv \mathrm{tr}(\sigma D)/\rho = t\dot{e}/\rho_0, \tag{A1}$$

with $t\dot{e}$ representing the scalar product, are called *work-conjugate* (or simply conjugate) stress and strain measures. As is well known, contravariant Kirchhoff stress with Green strain and covariant Kirchhoff stress with Almansi strain are two such conjugate pairs. In this and the following sections we develop the general equations of Hill (1970) (see also Hill 1978) for the stress measure t conjugate to the family of symmetric strain measures

$$e = \sum e_\alpha \mathbf{u}_\alpha \otimes \mathbf{u}_\alpha, \quad e_\alpha = f(\lambda_\alpha), \quad \alpha = 1, 2, 3, \tag{A2}$$

in spectral representation on the Lagrangian (reference) axes of principal stretch. Here the $\mathbf{u}_\alpha$ are unit vectors and $f(\lambda)$ is an arbitrary, smooth monotone function satisfying $f(1) = 0$, $f'(1) = 1$ (hence all strains are zero and all rates of strain equal the Eulerian strain rate D in the reference state). Because the reference directions of principal stretch at a point X are ordinarily not fixed in the material it is first necessary to establish an equation for the spin tensor of the Lagrangian triad $\mathbf{u}_\alpha$, $\alpha = 1, 2, 3$. Remarkably, considering that the basic continuum kinematics used in this monograph date from the eighteenth and nineteenth centuries, this pure kinematic result was not found until 1969 (Hill 1970). For completeness, we also derive the Biot–Hill equation for the relative spin of the Eulerian (current) principal-axis directions $\mathbf{v}_\alpha$ (unit vectors).

Let $Q_{\mathrm{L}}, Q_{\mathrm{E}}$ denote the orthogonal transformations that give the orientations of the Lagrangian and Eulerian triads relative to a reference frame $\mathscr{F}$ (i.e., $Q^{\mathrm{L}}_{\alpha j} = \mathbf{u}_\alpha \cdot \mathbf{e}_j$, $Q^{\mathrm{E}}_{\alpha j} = \mathbf{v}_\alpha \cdot \mathbf{e}_j$, the $\mathbf{e}_j$ being orthonormal basis vectors). Then

$$R_{\mathrm{L}} = Q^{\mathrm{T}}_{\mathrm{L}} = \sum \mathbf{u}_\alpha \otimes \mathbf{e}_\alpha, \quad R_{\mathrm{E}} = Q^{\mathrm{T}}_{\mathrm{E}} = \sum \mathbf{v}_\alpha \otimes \mathbf{e}_\alpha \tag{A3}$$

are the rotations of the respective triads from the reference frame, and we have

$$R_{\mathrm{E}} = R R_{\mathrm{L}}, \quad Q_{\mathrm{E}} = Q_{\mathrm{L}} Q_{\mathrm{R}}, \tag{A4}$$

with R the relative rotation of the Eulerian and Lagrangian principal axes and $Q_{\mathrm{R}} = R^{\mathrm{T}}$ their relative orientation (i.e., $Q^{\mathrm{R}}_{\alpha\beta} = \mathbf{v}_\alpha \cdot \mathbf{u}_\beta$). Further, let Ω_{E}, Ω_{L}, and Ω_{R} denote the respective spin tensors of these principal triads relative to frame $\mathscr{F}$ and to each other. As $\Omega = -Q^{\mathrm{T}}\dot{Q}$ is the relative spin of any pair of frames of relative orientation Q (with components on the axes of the base frame), we have

$$\Omega_{\mathrm{L}} = -Q^{\mathrm{T}}_{\mathrm{L}}\dot{Q}_{\mathrm{L}}, \quad \Omega_{\mathrm{E}} = -Q^{\mathrm{T}}_{\mathrm{E}}\dot{Q}_{\mathrm{E}}, \quad \Omega_{\mathrm{R}} = -Q^{\mathrm{T}}_{\mathrm{R}}\dot{Q}_{\mathrm{R}} \tag{A5}$$

from which, with equations (A3),

$$\dot{R}_{\mathrm{L}} = \Omega_{\mathrm{L}} R_{\mathrm{L}}, \quad \dot{R}_{\mathrm{E}} = \Omega_{\mathrm{E}} R_{\mathrm{E}}, \quad \dot{R} = \Omega_{\mathrm{R}} R. \tag{A6}$$

Upon differentiation of equation $(\mathrm{A4})_1$ and substitution of equations (A6), we obtain

$$R^{\mathrm{T}}(\Omega_{\mathrm{E}} - \Omega_{\mathrm{R}})R = \Omega_{\mathrm{L}}, \tag{A7}$$

giving the components of $\Omega_{\mathrm{E}} - \Omega_{\mathrm{R}}$ on the Eulerian axes when the components of Ω_{L} are expressed on the Lagrangian axes (Hill 1978, eq. (1.39)).

As $R_{\mathrm{L}}(X, \theta)$ (θ here representing time) defines the directions of the (embedded) orthogonal triad of differential line elements in B_0 at X that are again momentarily orthogonal in $B(\theta)$, R_{L} is independent of the rotational part of the motion of an infinitesimal neighborhood of X from θ_0 to θ. Therefore R_{L} is an intrinsically objective tensor field, transforming according to $R^*_{\mathrm{L}} = Q_0 R_{\mathrm{L}} Q^{\mathrm{T}}_0$ under change in observer frame (with Q_0 independent of time). Consequently, so is $\dot{R}_{\mathrm{L}}$, whence Ω_{L} is intrinsically objective from equation $(\mathrm{A6})_1$ (or eq. $(\mathrm{A5})_1$, as equivalent statements can be made about Q_{L}). In contrast, R_{E}, R, Ω_{E}, and Ω_{R} are affected by rigid-body spin and are not objective. Hill's fundamental kinematic result, which we now develop, is an equation connecting the components on the Lagrangian axes of the objective spin tensor Ω_{L} with the off-diagonal components on the Eulerian axes of the (objective) Eulerian strain rate D.

From $\mathbf{u}_\alpha = R_L \mathbf{e}_\alpha$ (by virtue of eq. $(A3)_1$) and equation $(A6)_1$,

$$\dot{\mathbf{u}}_\alpha = \Omega_L \mathbf{u}_\alpha = \sum_\beta \Omega^L_{\beta\alpha} \mathbf{u}_\beta, \tag{A8}$$

where the $\Omega^L_{\alpha\beta}$ are components on the Lagrangian triad (i.e., the components of the spin of the Lagrangian strain ellipsoid about its own axes, which spin is zero if D is principal on the Eulerian axes). As the (right) stretch tensor Λ (from the polar decomposition $A = R\Lambda$) may be expressed $\Lambda = \sum \lambda_\alpha \mathbf{u}_\alpha \otimes \mathbf{u}_\alpha$ in spectral representation, we have

$$\dot{\Lambda} = \sum (\dot{\lambda}_\alpha \mathbf{u}_\alpha \otimes \mathbf{u}_\alpha + \lambda_\alpha \dot{\mathbf{u}}_\alpha \otimes \mathbf{u}_\alpha + \lambda_\alpha \mathbf{u}_\alpha \otimes \dot{\mathbf{u}}_\alpha).$$

Upon substituting equation (A8) and simplifying, we obtain

$$\dot{\Lambda} = \sum\sum \{\dot{\lambda}_\alpha \delta_{\alpha\beta} + (\lambda_\alpha - \lambda_\beta)\Omega^L_{\beta\alpha}\} \mathbf{u}_\alpha \otimes \mathbf{u}_\beta. \tag{A9}$$

Furthermore, from

$$\dot{A}A^{-1} = R\dot{\Lambda}\Lambda^{-1}R^T + \Omega_R \tag{A10}$$

(a consequence of $A = R\Lambda$ with eq. $(A6)_3$) and the standard equation (1.8), one has

$$D^R \equiv R^T D R = \sum\sum d_{\alpha\beta} \mathbf{u}_\alpha \otimes \mathbf{u}_\beta = \operatorname{sym}(\dot{\Lambda}\Lambda^{-1}), \tag{A11}$$

with $d_{\alpha\beta}$ the components of strain rate D on the Eulerian axes. Hence, from equation (A9) and the spectral representation of Λ^{-1},

$$d_{\alpha\beta} = \tfrac{1}{2}(\lambda_\alpha^{-1} + \lambda_\beta^{-1})\dot{\Lambda}_{\alpha\beta}, \tag{A12}$$

with

$$\dot{\Lambda}_{\alpha\beta} = \dot{\lambda}_\alpha \delta_{\alpha\beta} + (\lambda_\alpha - \lambda_\beta)\Omega^L_{\beta\alpha}. \tag{A13}$$

Thus, for $\alpha = \beta$,

$$d_{\alpha\alpha} = \dot{\lambda}_\alpha / \lambda_\alpha = (\ln \lambda_\alpha)^{\cdot} \tag{A14}$$

which, as is well known, holds in every material direction, not necessarily principal. For $\alpha \neq \beta$,

$$(\lambda_\alpha^2 - \lambda_\beta^2)\Omega^L_{\beta\alpha} = 2\lambda_\alpha\lambda_\beta \mathrm{d}_{\alpha\beta}, \tag{A15}$$

so for distinct principal stretches

$$\Omega^L_{\beta\alpha} = \frac{2\lambda_\alpha\lambda_\beta}{\lambda_\alpha^2 - \lambda_\beta^2} d_{\alpha\beta}, \tag{A16}$$

which is *Hill's fundamental kinematic equation* (Hill 1970, eq. (15)). (For equal stretches but distinct stretch rates (at the outset in simple shearing, for example) the components of the Lagrangian spin tensor may be determined from

$$\Omega^{L}_{\beta\alpha} = \frac{\lambda_\alpha}{\dot{\lambda}_\alpha - \dot{\lambda}_\beta} d_{\alpha\beta},$$

obtained by differentiation of eq. (A15) followed by substitution of $\lambda_\alpha = \lambda_\beta$, $d_{\alpha\beta} = 0$.)

To develop the *Biot–Hill equation*[1] for $\Omega_E - W$ (the spin of the Eulerian axes relative to a material corotational frame) we need only an equation for $W - \Omega_R$, as we already have $\Omega_E - \Omega_R$ expressed in terms of Ω_L in equation (A7). From $D + W = \dot{A}A^{-1}$ and equations (A10)–(A11), the required equation is

$$R^{T}(W - \Omega_R)R = \mathrm{skw}(\dot{\Lambda}\Lambda^{-1}). \tag{A17}$$

Thus, with equation (A7),

$$R^{T}(\Omega_E - W)R = \Omega_L - \mathrm{skw}(\dot{\Lambda}\Lambda^{-1}). \tag{A18}$$

When we substitute the representations for $\dot{\Lambda}$, Λ^{-1}, and Ω_L on the Lagrangian triad, the left-hand sides of equations (A17) and (A18) give the components of the skew tensors $W - \Omega_R$ and $\Omega_E - W$ on the Eulerian triad (as both tensors are rotated by R from the right-hand-side reference frame). Thus

$$W_{\alpha\beta} - \Omega^{R}_{\alpha\beta} = \tfrac{1}{2}(\lambda_\beta^{-1} - \lambda_\alpha^{-1})\dot{\Lambda}_{\alpha\beta} = \frac{\lambda_\alpha - \lambda_\beta}{\lambda_\alpha + \lambda_\beta} d_{\alpha\beta} \tag{A19}$$

from equation (A12), or for $\alpha \neq \beta$ (Hill 1978, eq. (1.48))

$$W_{\alpha\beta} - \Omega^{R}_{\alpha\beta} = \frac{(\lambda_\alpha - \lambda_\beta)^2}{2\lambda_\alpha\lambda_\beta}\Omega^{L}_{\beta\alpha}. \tag{A20}$$

Finally, from equations (A18) and (A13), for $\alpha \neq \beta$,

$$\Omega^{E}_{\alpha\beta} - W_{\alpha\beta} = \frac{\lambda_\alpha^2 + \lambda_\beta^2}{2\lambda_\alpha\lambda_\beta}\Omega^{L}_{\alpha\beta}, \tag{A21}$$

whence for distinct principal stretches (from eq. (A16))

$$\Omega^{E}_{\alpha\beta} - W_{\alpha\beta} = \frac{\lambda_\alpha^2 + \lambda_\beta^2}{\lambda_\alpha^2 - \lambda_\beta^2} d_{\alpha\beta}, \tag{A22}$$

the Biot–Hill equation for the spin of the Eulerian strain ellipsoid about

[1] Here the derivation in Hill (1978, section 5) is generally followed.

its own axes, relative to a material corotational frame momentarily coincident with the Eulerian triad (Biot 1965; Hill 1970).

A.2 Hill's General Family of Conjugate Stress and Strain Measures

We now turn to our main objective, determination of the general family of stress measures t conjugate to arbitrary symmetric strain measures e. First, by analogy with equation (A9) (or by differentiating eq. (A2), using eq. (A8)) we have

$$\dot{e} = \sum\sum \{\dot{e}_\alpha \delta_{\alpha\beta} + (e_\alpha - e_\beta)\Omega^{\mathrm{L}}_{\beta\alpha}\}\mathbf{u}_\alpha \otimes \mathbf{u}_\beta, \quad \dot{e}_\alpha = f'(\lambda_\alpha)\dot{\lambda}_\alpha. \tag{A23}$$

Consequently, for $\alpha = \beta$, from equation (A14),

$$\dot{e}_{\alpha\alpha} = \lambda_\alpha f'(\lambda_\alpha) d_{\alpha\alpha}, \tag{A24}$$

and for $\alpha \neq \beta$, $\lambda_\alpha \neq \lambda_\beta$, from (Hill's) equation (A16),

$$\dot{e}_{\alpha\beta} = 2\lambda_\alpha\lambda_\beta \frac{e_\alpha - e_\beta}{\lambda_\alpha^2 - \lambda_\beta^2} d_{\alpha\beta}, \tag{A25}$$

with $\dot{e}_{\alpha\beta} = 0$ if $\lambda_\alpha = \lambda_\beta$.

Let $t_{\alpha\beta}$ denote the components of t on the embedded Lagrangian axes and $\tau_{\alpha\beta}$ designate the components of Kirchhoff stress $\tau = (\rho_0/\rho)\sigma$ on the Eulerian axes. Stress and strain tensors t, e are necessarily coaxial only in isotropic elastic materials. Thus for the general case we must write

$$t = \sum\sum t_{\alpha\beta}\mathbf{u}_\alpha \otimes \mathbf{u}_\beta, \quad \tau = \sum\sum \tau_{\alpha\beta}\mathbf{v}_\alpha \otimes \mathbf{v}_\beta, \tag{A26}$$

and (from eq. (A1))

$$\rho_0 \dot{w} = \sum\sum t_{\alpha\beta}\dot{e}_{\alpha\beta} = \sum\sum \tau_{\alpha\beta} d_{\alpha\beta}. \tag{A27}$$

Then, upon substituting equations (A24) and (A25) into equation (A27) and separately comparing axial and shearing strain-rate terms, we obtain *Hill's general equations for conjugate stress measures* (Hill 1970, eqs. (24))

$$t_{\alpha\alpha} = \frac{\tau_{\alpha\alpha}}{\lambda_\alpha f'(\lambda_\alpha)}, \quad \alpha = 1, 2, 3, \tag{A28}$$

and for $\alpha \neq \beta$

$$t_{\alpha\beta} = \begin{cases} \dfrac{\lambda_\alpha^2 - \lambda_\beta^2}{e_\alpha - e_\beta} \dfrac{\tau_{\alpha\beta}}{2\lambda_\alpha\lambda_\beta}, & \lambda_\alpha \neq \lambda_\beta; \\[2ex] \dfrac{\tau_{\alpha\beta}}{\lambda_\alpha f'(\lambda_\alpha)}, & \lambda_\alpha = \lambda_\beta, \end{cases} \tag{A29}$$

the last relation following from L'Hôpital's rule (by substituting $\dot{e}_\alpha - \dot{e}_\beta = f'(\lambda_\alpha)(\dot{\lambda}_\alpha - \dot{\lambda}_\beta)$ when $\lambda_\alpha = \lambda_\beta$). Mathematically the $t_{\alpha\beta}$ are tensor components of t on the Lagrangian axes in B_0. However, because these axes are embedded and principal, the $t_{\alpha\beta}$ also may be interpreted as weighted stresses in $B(\theta)$ that are on the same orthogonal planes and directions as the $\tau_{\alpha\beta}$ (corresponding to Eulerian axes $\mathbf{v}_\alpha$), with equations (A28)–(A29) giving the respective weighting factors for the normal and shearing components. Were Kirchhoff stress τ coaxial with left stretch V, every tensor t would be coaxial with right stretch Λ, hence with the general family of strain measures e.

The contravariant and covariant Kirchhoff stress tensors T and T_c readily follow from Hill's general equations by substituting the Green and Almansi measures, with $f'_G = \lambda$ and $f'_A = \lambda^{-3}$. Thus

$$T = \sum\sum \lambda_\alpha^{-1}\lambda_\beta^{-1}\tau_{\alpha\beta}\mathbf{u}_\alpha \otimes \mathbf{u}_\beta, \quad T_c = \sum\sum \lambda_\alpha\lambda_\beta\tau_{\alpha\beta}\mathbf{u}_\alpha \otimes \mathbf{u}_\beta. \tag{A30}$$

Moreover, the stress conjugate to nominal strain

$$E_N = \Lambda - I \tag{A31}$$

is found to be (Hill 1978, eq. (1.78))

$$T_N = \tfrac{1}{2}\sum\sum(\lambda_\alpha^{-1} + \lambda_\beta^{-1})\tau_{\alpha\beta}\mathbf{u}_\alpha \otimes \mathbf{u}_\beta. \tag{A32}$$

This can be expressed in terms of the unsymmetric nominal stress S as follows. Let S^R denote the matrix array of rotated nominal stress components $S^R_{\alpha\beta} = S_{\alpha\gamma}R_{\gamma\beta}$ on the Eulerian planes and directions of principal stretch. Hence in dyadic representation on the Lagrangian triad (as we rotate only the components of nominal traction vectors and not their planes) we have

$$S^R = SR = \sum\sum \lambda_\alpha^{-1}\tau_{\alpha\beta}\mathbf{u}_\alpha \otimes \mathbf{u}_\beta. \tag{A33}$$

Upon comparison with equation (A32) it is evident that

$$T_N = \mathrm{sym}(SR). \tag{A34}$$

Hill (1968) has called this symmetric nominal stress tensor the *Biot stress* (see Biot 1965, pp. 58–62).

With regard to the stress T_L conjugate to the logarithmic measure of strain, this cannot be expressed in a simple way in direct tensor notation as can T, T_c, and T_N. However, on the principal axes of stretch, from equations (A28)–(A29) (with $f'_L = \lambda^{-1}$), we have for $\alpha = \beta$

$$t^L_{\alpha\alpha} = \tau_{\alpha\alpha}, \tag{A35}$$

and for $\alpha \neq \beta$

$$t^{L}_{\alpha\beta} = \begin{cases} \dfrac{\lambda_\alpha^2 - \lambda_\beta^2}{\ln(\lambda_\alpha/\lambda_\beta)} \dfrac{\tau_{\alpha\beta}}{2\lambda_\alpha\lambda_\beta}, & \lambda_\alpha \neq \lambda_\beta; \\ \tau_{\alpha\beta}, & \lambda_\alpha = \lambda_\beta. \end{cases} \tag{A36}$$

Although equation (A35) was long known for the case of principal axes fixed in the material, equation (A36) was unavailable prior to Hill's development of the general equations (A28)–(A29). Last, it is worth emphasizing that since all work-conjugate stress and strain measures t, e are defined on the reference directions of principal stretch at a material point, they are intrinsically objective tensor fields; hence so are their time rates of change to all orders.

A.3 An Invariant Bilinear Form

For all strain and stress measures in the general family of equations (A2) and (A28)–(A29), the differential work per unit mass $\mathrm{d}w = t\,\mathrm{d}e/\rho_0$ is a fundamental scalar invariant in the deepest sense in mechanics of being *independent of choice of strain measure and reference state* as well as invariant under change of observer frame. We now introduce another work-related scalar quantity (due to Hill (1972)) that is a fundamental invariant in the same sense and enables the qualitative comparison of different possible responses of a material from its current state.

Let $\mathrm{d}_1 t, \mathrm{d}_1 e$ signify related incremental changes at a material point X (i.e., jointly possible changes for that material), in a real or virtual motion between times θ and $\theta + \mathrm{d}\theta$, and $\mathrm{d}_2 t, \mathrm{d}_2 e$ denote related changes in another possible motion during the interval $\mathrm{d}\theta$, with each differential motion dependent upon incremental boundary conditions on $\partial B(\theta)$. Also, let $\dot{t}_1, \dot{e}_1$ and $\dot{t}_2, \dot{e}_2$ denote the respective material derivatives in the two motions. Although $\dot{t}\dot{e}$ (scalar product) in any deformation from the current state is intrinsically invariant under observer transformations, it is not invariant under change in measure and so takes on a different value for each strain and conjugate stress chosen. However, we shall prove that, just as for work-rate $\dot{w}(X, \theta)$,

$$(\mathrm{d}_1 t\,\mathrm{d}_2 e - \mathrm{d}_2 t\,\mathrm{d}_1 e)/\rho_0 \text{ is invariant in value} \tag{A37}$$

under change in both strain measure and reference state. This expression is *Hill's invariant bilinear form* (Hill 1972), and we subsequently shall present a virtual-work interpretation of this basic scalar invariant. (We note in passing that in nonlinear elasticity the value of the bilinear invariant is zero, which result is identically the reciprocal theorem on incremental responses from the current state, generalizing for all measures the classical linear elasticity theorem at infinitesimal strain.)

For any strain measure e, the Cartesian components of $\dot{e}$ in frame $\mathscr{F}$ are

$$\dot{e}_{ij} = \sum_{\alpha}\sum_{\beta} Q^{L}_{\alpha i} Q^{L}_{\beta j} \dot{e}_{\alpha\beta}. \tag{A38}$$

On the Lagrangian principal axes the strain-rate components of e are connected to those of some other strain measure e^* (from eqs. (A24)–(A25)) by

$$\dot{e}_{\alpha\alpha} = \frac{f'(\lambda_\alpha)}{f^{*\prime}(\lambda_\alpha)}\dot{e}^*_{\alpha\alpha}, \quad \dot{e}_{\alpha\beta} = \frac{e_\alpha - e_\beta}{e^*_\alpha - e^*_\beta}\dot{e}^*_{\alpha\beta} \tag{A39}$$

(for the same reference state), the latter equation obviously requiring $\lambda_\alpha \neq \lambda_\beta$. (Explicit relations also can be written for different reference states.) Then, for any two particular strain measures, the components $\dot{e}_{ij}$ and $\dot{e}^*_{ij}$ can be related through equations (A38) and (A39). For simplicity, we express this connection in the general case by

$$\dot{e} = \frac{\partial e}{\partial e^*}\dot{e}^*, \tag{A40}$$

or

$$\dot{e}_{ij} = \frac{\partial e_{ij}}{\partial e^*_{kl}}\dot{e}^*_{kl};$$

whence, from the fundamental invariance of $t\dot{e}/\rho_0$, we have (Hill 1968, 1972)

$$(t/\rho_0)^* = (t/\rho_0)\frac{\partial e}{\partial e^*}, \tag{A41}$$

or

$$(t_{kl}/\rho_0)^* = (t_{ij}/\rho_0)\frac{\partial e_{ij}}{\partial e^*_{kl}}.$$

(Observe that in the simple, direct system of notation used in eqs. (A40) and (A41) the denominator of a tensor derivative always forms a scalar product with the following tensor, and the numerator with the preceding tensor.) Upon taking the d_1-differential of $(t/\rho_0)^*$ and forming the scalar product with a d_2-differential of e^*, one obtains

$$d_1 t^* \, d_2 e^* / \rho_0^* = d_1 t \, d_2 e / \rho_0 + (t/\rho_0) \frac{\partial^2 e}{\partial e^* \partial e^*} \cdot (d_1 e^* \otimes d_2 e^*), \tag{A42}$$

or in component form

$$d_1 t_{kl}^* \, d_2 e_{kl}^* / \rho_0^* = d_1 t_{ij} \, d_2 e_{ij} / \rho_0 + (t_{ij}/\rho_0) \frac{\partial^2 e_{ij}}{\partial e_{mn}^* \partial e_{kl}^*} d_1 e_{kl}^* \, d_2 e_{mn}^*,$$

so $\dot{t}\dot{e}/\rho_0$ (in contrast to work rate $t\dot{e}/\rho_0$) is not invariant under change in measure and reference state as already remarked. However, the invariance of Hill's bilinear form (A37) readily follows by interchanging the d_1- and d_2-differentials and subtracting the resulting expression from equation (A42), using

$$\frac{\partial^2 e_{ij}}{\partial e_{mn}^* \partial e_{kl}^*} = \frac{\partial^2 e_{ij}}{\partial e_{kl}^* \partial e_{mn}^*}$$

from the smoothness of the relationship between strain measures for smooth monotone functions $f(\lambda)$. Thus (Hill 1972)

$$(d_1 t^* \, d_2 e^* - d_2 t^* \, d_1 e^*)/\rho_0^* = (d_1 t \, d_2 e - d_2 t \, d_1 e)/\rho_0, \tag{A43}$$

or

$$(\dot{t}_1^* \dot{e}_2^* - \dot{t}_2^* \dot{e}_1^*)/\rho_0^* = (\dot{t}_1 \dot{e}_2 - \dot{t}_2 \dot{e}_1)/\rho_0, \tag{A44}$$

and the statement of invariance is proved.

For subsequent reference we develop a similar equation to equation (A44) involving the conjugate variables S and A, which are neither symmetric nor objective tensors. From $S = TA^{\mathrm{T}}$ and $\dot{A} = \Gamma A$ one obtains the standard result

$$\dot{S} = \dot{T} A^{\mathrm{T}} + S \Gamma^{\mathrm{T}}; \tag{A45}$$

whence forming the scalar product of a d_1-differential of nominal stress S with a d_2-differential of deformation gradient A, we find

$$\mathrm{tr}(\dot{S}_1 \dot{A}_2) = \mathrm{tr}(\dot{T}_1 \dot{E}_2) + \mathrm{tr}(\tau \Gamma_1^{\mathrm{T}} \Gamma_2), \tag{A46}$$

where we have used $\dot{E} = \mathrm{sym}(A^{\mathrm{T}} \dot{A})$, the symmetry of $\dot{T}$, $\dot{A}_2 = \Gamma_2 A$, and $AS = \tau$. Then, upon interchanging the differentials and noting that $\mathrm{tr}(\tau \Gamma_1^{\mathrm{T}} \Gamma_2) = \mathrm{tr}(\tau \Gamma_2^{\mathrm{T}} \Gamma_1)$ from the symmetry of Kirchhoff stress τ, we obtain

$$\mathrm{tr}(\dot{S}_1 \dot{A}_2 - \dot{S}_2 \dot{A}_1) = \mathrm{tr}(\dot{T}_1 \dot{E}_2 - \dot{T}_2 \dot{E}_1). \tag{A47}$$

As T, E belong to the general family of conjugate stress and strain measures of equations (A2) and (A28)–(A29) for which invariance was proved, $\mathrm{tr}(\mathrm{d}_1 S\,\mathrm{d}_2 \mathrm{A} - \mathrm{d}_2 S\,\mathrm{d}_1 A)/\rho_0$ equals the value of Hill's fundamental bilinear invariant.

Although the proof of invariance of the bilinear form (A37) is simple, as seen, a more physical interpretation provides insight, and this has been given by Hill (1972, 1978) in terms of the second-order virtual work done around a very small cycle of strain (from the current state) by an associated but virtual cycle of stress. Consider first the work done per unit reference volume along a path from a state designated by e_A, t_A to a nearby state e_B, t_B with $\delta e = e_B - e_A, \delta t = t_B - t_A$, and $\|\delta e\| \ll 1$. From a Taylor series expansion in a timelike parameter θ and subsequent integration one obtains

$$\int_{e_A}^{e_B} t\,\mathrm{d}e = \int_0^\theta \{(te')_A + (te')'_A\theta + \cdots\}\,\mathrm{d}\theta$$

$$= (t_A + \tfrac{1}{2}t'_A\theta)(e'_A\theta + \tfrac{1}{2}e''_A\theta^2) + O(\theta^3).$$

But

$$t_A + \tfrac{1}{2}t'_A\theta = \tfrac{1}{2}(t_B + t_A) + O(\theta^3), \quad e'_A\theta + \tfrac{1}{2}e''_A\theta^2 = e_B - e_A + O(\theta^3)$$

from Taylor expansions of t and e about state A. Thus

$$\int_{e_A}^{e_B} t\,\mathrm{d}e = \tfrac{1}{2}(t_B + t_A)(e_B - e_A) + O(\theta^3), \tag{A48}$$

with $\theta = O(\|\delta e\|)$. Moreover, the incremental work per unit mass from current state e, t, correct to second order in δe, is given by

$$\delta w \approx (t + \tfrac{1}{2}\delta t)\delta e/\rho_0. \tag{A49}$$

Now consider two states e_1, t_1 and e_2, t_2 near to and physically attainable from e, t (but not necessarily attainable from one another), and imagine a small incremental strain cycle $e \to e_1 \to e_2 \to e$. The initial stress state is likely not recoverable in an actual strain cycle (save for purely elastic response throughout), but we may define the virtual work that would be done in the strain cycle by an imagined cycle of stress $t \to t_1 \to t_2 \to t$ (with only the first leg necessarily real). Thus, the virtual work per unit reference volume is

$$\oint t\,\mathrm{d}e = \int_e^{e_1} t\,\mathrm{d}e + \int_{e_1}^{e_2} t\,\mathrm{d}e + \int_{e_2}^{e} t\,\mathrm{d}e,$$

and upon substituting equation (A48) for each integral and simplifying, we have to second order in incremental strain

$$\oint t\,\mathrm{d}e \approx \tfrac{1}{2}\{(t_1 - t)(e_2 - e) - (t_2 - t)(e_1 - e)\}.$$

Therefore, with $\delta_1 t = t_1 - t, \delta_1 e = e_1 - e$, and so forth

$$\oint t\,\mathrm{d}e/\rho_0 \approx \tfrac{1}{2}(\delta_1 t \delta_2 e - \delta_2 t \delta_1 e)/\rho_0 \qquad \text{(A50)}$$

(Hill 1972, eq. (2.18)), which gives a virtual work interpretation, in an incremental strain cycle, of Hill's invariant bilinear form $(\mathrm{d}_1 t\,\mathrm{d}_2 e - \mathrm{d}_2 t\,\mathrm{d}_1 e)/\rho_0$.

A.4 Measure-Dependent Transformations

From equation (A42), the measure-dependent scalar product of conjugate stress and strain rates transforms according to

$$\dot{t}^*\dot{e}^* = \dot{t}\dot{e} + t\frac{\partial^2 e}{\partial e^* \partial e^*}\cdot(\dot{e}^* \otimes \dot{e}^*) \qquad \text{(A51)}$$

at fixed reference state. As an example of this scalar transformation, let $e = E$ (Green strain) and $e^* = E_{\mathrm{N}}$ (nominal strain). Then, from $E = \frac{1}{2}(\Lambda^2 - I)$ and $E_{\mathrm{N}} = \Lambda - I$, we obtain (treating each component of E and Λ separately)

$$\frac{\partial^2 e_{ij}}{\partial e^*_{kl}\partial e^*_{mn}} = \frac{\partial^2 E_{ij}}{\partial \Lambda_{kl}\partial \Lambda_{mn}} = \frac{1}{2}(\delta_{ik}\delta_{lm}\delta_{jn} + \delta_{im}\delta_{kn}\delta_{jl}),$$

which is symmetric with respect to interchange of kl and mn as required. Upon substitution into equation (A51), with $t = T$ and $t^* = T_{\mathrm{N}}$, we obtain

$$\mathrm{tr}(\dot{T}_{\mathrm{N}}\dot{E}_{\mathrm{N}}) = \mathrm{tr}(\dot{T}\dot{E}) + \mathrm{tr}(T\dot{\Lambda}^2), \qquad \text{(A52)}$$

or in Cartesian-components form (as $\dot{E}_{\mathrm{N}} = \dot{\Lambda}$)

$$\dot{t}^{\mathrm{N}}_{ij}\dot{\Lambda}_{ij} = \dot{T}_{ij}\dot{E}_{ij} + T_{ij}\dot{\Lambda}_{ik}\dot{\Lambda}_{kj},$$

for the transformation of $\dot{t}\dot{e}$ between the nominal and Green measures of strain.

The transformation (A51) may be given in a much more explicit form when the principal axes of stretch, hence of strain rate, are fixed in the material. Moreover, development of the equation for that case enables a direct illustration of change in sign of $\dot{t}\dot{e}$ with change in measure. From equation $(A39)_1$ with a common reference state,

$$\frac{\partial e_\alpha}{\partial e^*_\alpha} = \frac{f'(\lambda_\alpha)}{f^{*\prime}(\lambda_\alpha)},$$

and upon differentiation and substitution in equation (A51) we find, in spectral representation on the Lagrangian axes (as all $\dot{e}_{\alpha\beta} = 0$ for $\Omega_L = 0$),

$$\sum \dot{t}^*_{\alpha\alpha}\dot{e}^*_\alpha = \sum \dot{t}_{\alpha\alpha}\dot{e}_\alpha + \sum t_{\alpha\alpha}\left\{ f''(\lambda_\alpha) - \frac{f'(\lambda_\alpha)}{f^{*\prime}(\lambda_\alpha)} f^{*\prime\prime}(\lambda_\alpha) \right\} \dot{\lambda}^2_\alpha. \tag{A53}$$

Let the base measure be logarithmic strain; whence $t_{\alpha\alpha} = \tau_{\alpha\alpha}$ (not necessarily principal) and equation (A53) becomes

$$\sum \dot{t}_{\alpha\alpha}\dot{e}_\alpha = \sum \dot{t}^L_{\alpha\alpha}\dot{\lambda}_\alpha/\lambda_\alpha - \sum \tau_{\alpha\alpha}\left\{ 1 + \lambda_\alpha \frac{f''(\lambda_\alpha)}{f'(\lambda_\alpha)} \right\} (\dot{\lambda}_\alpha/\lambda_\alpha)^2, \tag{A54}$$

in which we have deleted the asterisk as it is no longer needed. For illustration consider two measures, Green strain and nominal strain. For the first, with $f'(\lambda_\alpha) = \lambda_\alpha$ and $t_{\alpha\alpha} = T_{\alpha\alpha}$ (contravariant Kirchhoff stress),

$$\sum \dot{T}_{\alpha\alpha}\lambda_\alpha\dot{\lambda}_\alpha = \sum \dot{t}^L_{\alpha\alpha}\dot{\lambda}_\alpha/\lambda_\alpha - 2\sum \tau_{\alpha\alpha}(\dot{\lambda}_\alpha/\lambda_\alpha)^2. \tag{A55}$$

For the second, with $f'(\lambda_\alpha) = 1$ and $t_{\alpha\alpha} = t^N_{\alpha\alpha}$,

$$\sum \dot{t}^N_{\alpha\alpha}\dot{\lambda}_\alpha = \sum \dot{t}^L_{\alpha\alpha}\dot{\lambda}_\alpha/\lambda_\alpha - \sum \tau_{\alpha\alpha}(\dot{\lambda}_\alpha/\lambda_\alpha)^2, \tag{A56}$$

whence

$$\sum \dot{t}^N_{\alpha\alpha}\dot{\lambda}_\alpha - \sum \dot{T}_{\alpha\alpha}\lambda_\alpha\dot{\lambda}_\alpha = \sum \tau_{\alpha\alpha}(\dot{\lambda}_\alpha/\lambda_\alpha)^2,$$

which also is directly obtainable from equation (A52). Now consider a material response (characteristic in metals) wherein

$$\mathrm{tr}(\dot{T}_L\dot{E}_L) = \sum \dot{t}^L_{\alpha\alpha}\dot{\lambda}_\alpha/\lambda_\alpha > 0. \tag{A57}$$

Then, if dominant normal stress components are tensile on the fixed principal axes of stretch, it is clear from equations (A55) and (A56) that, as the deformation progresses, first $(\dot{t}\dot{e})_G$ (i.e., in the Green measure) and then $(\dot{t}\dot{e})_N$ may become negative while $(\dot{t}\dot{e})_L$ remains positive from the stipulated response, equation (A57). The conjectural inequality $(\dot{t}\dot{e})_L > 0$ without restriction to fixed principal axes is due to Hill (1968, part II).

Another explicit equation for a measure-dependent transformation which is useful is that for objective stress rate $\dot{t}$ evaluated with the current state as reference. This equation was first given by Hill (1968) and may be derived as follows. We begin with a power series expansion of $f(\lambda)$ about the reference state in the logarithmic measure (using $f(1)=0, f'(1)=1$):

$$f(\lambda)=\ln\lambda+\tfrac{1}{2}\{1+f''(1)\}(\ln\lambda)^2 + \tfrac{1}{6}\{1+3f''(1)+f'''(1)\}(\ln\lambda)^3+\cdots, \tag{A58}$$

which converges in some neighborhood of $\lambda=1$ for all measures. Let

$$m=\tfrac{1}{2}\{1+f''(1)\}. \tag{A59}$$

Then $e_\alpha=\ln\lambda_\alpha+m(\ln\lambda_\alpha)^2+\cdots$, and as all measures in the general family (A2) are coaxial we can write

$$e=E_{\mathrm{L}}+mE_{\mathrm{L}}^2+\cdots, \tag{A60}$$

or

$$e_{ij}=e_{ij}^{\mathrm{L}}+me_{in}^{\mathrm{L}}e_{nj}^{\mathrm{L}}+O(e_{\mathrm{L}}^3),$$

with $e_{\mathrm{L}}=\|E_{\mathrm{L}}\|$. Thus

$$\frac{\partial e_{ij}}{\partial e_{kl}^{\mathrm{L}}}=\delta_{ik}\delta_{jl}+m(\delta_{ik}e_{lj}+e_{ik}\delta_{jl})+O(e_{\mathrm{L}}^2),$$

and upon substituting this result into equation (A41) (with $t^*=T_{\mathrm{L}}, e^*=E_{\mathrm{L}}$, and $\rho_0^*=\rho_0$) we obtain

$$t_{ij}^{\mathrm{L}}=t_{ij}+m(t_{ik}e_{kj}^{\mathrm{L}}+e_{ik}^{\mathrm{L}}t_{kj})+O(e_{\mathrm{L}}^2);$$

whence after differentiating

$$\dot{t}_{ij}^{\mathrm{L}}=\dot{t}_{ij}+m(t_{ik}\dot{e}_{kj}^{\mathrm{L}}+\dot{e}_{ik}^{\mathrm{L}}t_{kj})+O(e_{\mathrm{L}}).$$

Now take the current state as reference so that, momentarily, all strains are zero, all strain rates equal D, and all stresses equal Cauchy stress σ. Therefore, *exactly* (Hill 1968, part I, eq. (5))

$$\dot{T}_{\mathrm{L}}^{(0)}=\dot{t}^{(0)}+m(\sigma D+D\sigma), \tag{A61}$$

with (0) signifying evaluation in the reference state.

We may determine $\dot{T}_{\mathrm{L}}^{(0)}$ from equation (A61) by choosing a measure for which the conjugate stress rate is already known. Consider the Green measure $f(\lambda)=\frac{1}{2}(\lambda^2-1)$, for which $m=1$ and the conjugate stress is T. From the standard equations $\tau=ATA^{\mathrm{T}}$ and $\dot{A}=(D+W)A$ one obtains (with θ again representing time)

$$\frac{\mathscr{D}\tau}{\mathscr{D}\theta}-(\tau D+D\tau)=A\dot{T}A^{\mathrm{T}}, \tag{A62}$$

where

$$\frac{\mathscr{D}\tau}{\mathscr{D}\theta} = \dot{\tau} + \tau W - W\tau \tag{A63}$$

is the material corotational (or Jaumann–Zaremba) derivative of Kirchhoff stress τ. Consequently, $\dot{T}^{(0)}$ equals the left-hand side of equation (A62); and upon substitution into equation (A61) we evidently have (as $\tau = \sigma$ momentarily)

$$\dot{T}_{\mathrm{L}}^{(0)} = \frac{\mathscr{D}\tau}{\mathscr{D}\theta}. \tag{A64}$$

Thus, the stress rate conjugate to logarithmic strain, evaluated with the current state as reference, equals the material corotational derivative of Kirchhoff stress. Finally, from equations (A61) and (A64), the conjugate stress rate for any measure in the reference state is (Hill 1968, p. 231)

$$\dot{t}^{(0)} = \frac{\mathscr{D}\tau}{\mathscr{D}\theta} - m(\sigma D + D\sigma). \tag{A65}$$

Consider now the transformation of $(\dot{t}\dot{e})^{(0)}$ with change in measure. From equation (A65) there immediately follows

$$(\dot{t}\dot{e})^{(0)} = \mathrm{tr}\left(\frac{\mathscr{D}\tau}{\mathscr{D}\theta} D\right) - 2m\,\mathrm{tr}(\sigma D^2). \tag{A66}$$

We may compare this result with equation (A54) evaluated in the reference state since all $\dot{e}_{\alpha\beta} = 0$ there whether or not the principal axes are fixed in the material. (That is, only the condition $\dot{e}_{\alpha\beta} = 0$ rather than $\Omega_{\mathrm{L}} = 0$ is required for the validity of eq. (A54). $\Omega_{\mathrm{L}} = 0$ of course guarantees $\dot{e}_{\alpha\beta} = 0$, but the converse does not hold in the reference state.) Thus, setting all $\lambda_\alpha = 1$, $\dot{e}_\alpha = \dot{\lambda}_\alpha$, and $\tau_{\alpha\alpha} = \sigma_{\alpha\alpha}$ in equation (A54), we obtain

$$\sum \dot{t}_{\alpha\alpha}^{(0)} \dot{\lambda}_\alpha = \sum \dot{t}_{\alpha\alpha}^{\mathrm{L}(0)} \dot{\lambda}_\alpha - 2m \sum \sigma_{\alpha\alpha} \dot{\lambda}_\alpha^2. \tag{A67}$$

Choosing the Green measure, one has

$$\dot{t}_{\alpha\alpha}^{(0)} = \dot{T}_{\alpha\alpha}^{(0)} = \left(\frac{\mathscr{D}\tau}{\mathscr{D}\theta}\right)_{\alpha\alpha} - 2\sigma_{\alpha\alpha}\dot{\lambda}_\alpha$$

from equation (A62) as then, momentarily, $A = I$, $d_{\alpha\alpha} = \dot{\lambda}_\alpha$, and $d_{\alpha\beta} = 0$. Thus, from equation (A67) (with $m = 1$)

$$\sum \left\{ \dot{t}_{\alpha\alpha}^{\mathrm{L}(0)} - \left(\frac{\mathscr{D}\tau}{\mathscr{D}\theta}\right)_{\alpha\alpha} \right\} \dot{\lambda}_\alpha = 0,$$

which is consistent with equation (A64). Finally, for any measure with the current state as reference,

$$\sum t_{\alpha\alpha}^{(0)}\dot{\lambda}_\alpha = \sum\left(\frac{\mathscr{D}\tau}{\mathscr{D}\theta}\right)_{\alpha\alpha}\dot{\lambda}_\alpha - 2m\sum\sigma_{\alpha\alpha}\dot{\lambda}_\alpha, \tag{A68}$$

which obviously is equation (A66) expressed in component form on the principal axes of stretch, thereby confirming the necessary connection between equations (A54) and (A66).

REFERENCES

Ambrosi, P., Göttler, E., & Schwink, C. (1974). On the dislocation arrangement and flow stress of [111]-copper single crystals deformed in tension. *Scripta Met.* 8: 1093–8.

Ambrosi, P. & Schwink, C. (1978). Slip line length of copper single crystals oriented along [100] and [111]. *Scripta Met.* 12: 303–8.

Asaro, R. J. (1979). Geometrical effects in the inhomogeneous deformation of ductile single crystals. *Acta Met.* 27: 445–53.

Asaro, R. J. & Needleman, A. (1985). Texture development and strain hardening in rate dependent polycrystals. *Acta Met.* 33: 923–53.

Barrett, C. S., Ansel, G. & Mehl, R. F. (1937). Slip, twinning and cleavage in iron and silicon ferrite. *Trans. Am. Soc, Metals* 25: 702–33.

Barsch, G. R. & Chang, Z. P. (1967). Adiabatic, isothermal, and intermediate pressure derivatives of the elastic constants for cubic symmetry – II. Numerical results for 25 materials. *Phys. Status Solidi* 19: 139–51.

Basinski, S. J & Basinski, Z. S. (1979). Plastic deformation and work hardening. In *Dislocations in Solids,* Vol. 4, *Dislocations in Metallurgy*, ed. F. R. N. Nabarro, pp. 261–362. Amsterdam: North-Holland.

Basinski, Z. S. & Jackson, P. J. (1965a). The effect of extraneous deformation on strain hardening in Cu single crystals. *Appl. Phys. Lett.* 6: 148–50.

Basinski, Z. S. & Jackson, P. J. (1965b). The instability of the work-hardened state – I. Slip in extraneously deformed crystals. *Phys. Status Solidi* 9: 805–23.

Bell, J. F. (1964). A generalized large deformation behaviour for face-centred cubic solids – high purity copper. *Phil. Mag.* ser. 8, 10: 107–26.

Bell, J. F. (1965). Generalized large deformation behaviour for face-centred cubic solids: Nickel, aluminium, gold, silver and lead. *Phil. Mag.* ser. 8, 11: 1135–56.

Bell, J. F. (1973). The experimental foundations of solid mechanics. In *Handbuch der Physik*, Vol. VIa/1, ed. C Truesdell, pp. 1–813. Berlin: Springer.

Bell, J. F. & Green, R. E., Jr. (1967). An experimental study of the double slip deformation hypothesis in face-centred cubic single crystals. *Phil Mag.* ser. 8, 15: 469–76.

Biot, M. A. (1965). *Mechanics of Incremental Deformation.* New York: Wiley.

Bishop, J. F. W. (1953). A theoretical examination of the plastic deformation of crystals by glide. *Phil. Mag.* ser. 7, 44: 51–64.

Bishop, J. F. W. (1954). A theory of the tensile and compressive textures of face-centred cubic metals. *J. Mech. Phys. Solids* 3: 130–42.

Bishop, J. F. W. (1955). On the contribution of crystallographic fibring to hardening under uniaxial straining conditions. *J. Mech. Phys. Solids.* 3: 259–66.

Bishop, J. F. W. & Hill, R. (1951a). A theory of the plastic distortion of a polycrystalline aggregate under combined stresses. *Phil. Mag.* ser. 7, 42: 414–427.

Bishop, J. F. W. & Hill, R. (1951b). A theoretical derivation of the plastic properties of a polycrystalline face-centred metal. *Phil. Mag.* ser. 7, 42: 1298–307.

Bowen, D. K. & Christian, J. W. (1965). The calculation of shear stress and shear strain for double glide in tension and compression. *Phil. Mag.* ser. 8, 12: 369–78.

Brugger, K. (1964). Thermodynamic definition of higher order elastic constants. *Phys. Rev. A*133: 1611–12.

Chen, N. K. & Maddin, R. (1954). Slip planes and the energy of dislocations in a body-centered cubic structure. *Acta Met.* 2: 49–51.

Chidambarrao, D. (1988). Comparative analyses and assessment of different hardening rules in channel die compression of f.c.c. crystals. Ph.D. thesis, North Carolina State University.

Chidambarrao, D. & Havner, K. S. (1988a). Finite deformation analysis of f.c.c. crystals in $(110)(1\bar{1}\bar{2})(\bar{1}1\bar{1})$ channel die compression. *Int. J. Plasticity* 4: 1–27.

Chidambarrao, D. & Havner, K. S. (1988b). On finite deformation of f.c.c. crystals in (110) channel die compression. *J. Mech. Phys. Solids* 36: 285–315.

Chin, G. Y. (1969). Tension and compression textures. In *Textures in Research and Practice*, ed. J. Grewen and G. Wassermann, pp. 51–80. Berlin: Springer.

Chin, G. Y. & Mammel, W. L. (1967). Computer solutions of the Taylor analysis for axisymmetric flow. *Trans. Met. Soc. AIME* 239: 1400–5.

Chin, G. Y. & Mammel, W. L. (1969). Generalization and equivalence of the minimum work (Taylor) and maximum work (Bishop–Hill) principles for crystal plasticity. *Trans. Met. Soc. AIME* 245: 1211–14.

Chin, G. Y., Mammel, W. L. & Dolan, M. T. (1967). Taylor's theory of texture for axisymmetric flow in body-centered cubic metals. *Trans. Met. Soc. AIME* 239: 1854–5.

Chin, G. Y., Nesbitt, E. A. & Williams, A. J. (1966). Anisotropy of strength in single crystals under plane strain compression. *Acta Met.* 14: 467–76.

Chin, G. Y., Thurston, R. N. & Nesbitt, E. A. (1966). Finite plastic deformation due to crystallographic slip. *Trans. Met. Soc. AIME* 236: 69–76.

Dillon, O. W. (1968). Plastic deformation waves and heat generated near the yield point of annealed aluminum. In *Mechanical Behavior of Materials under Dynamic Loads*, ed. U. S. Lindholm, pp. 21–60. Berlin: Springer.

Driver, J. H. & Skalli, A. (1982). L'essai de compression plane de monocristaux encastres: methode d'etude du comportement d'un cristal soumis a une deformation plastique imposee. *Revue Phys. Appl.* 17:447–51.
Driver, J. H., Skalli, A. & Wintenberger, M. (1983). Etude theorique et experimentale de la deformation plastique de monocristaux d'aluminium. Premiere partie: compression plane pure. *Mem. Et. Sci. Rev. Met.* 80:241–50.
Driver, J. H., Skalli, A. & Wintenberger, M. (1984). A theoretical and experimental study of the plastic deformation of f.c.c. crystals in plane strain compression. *Phil. Mag.* A49: 505–24.
Drucker, D. C. (1950). Some implications of work hardening and ideal plasticity. *Q. Appl. Math.* 7:411–8.
Drucker, D. C. (1951). A more fundamental approach to plastic stress–strain relations. In *Proc. 1st U.S. Natl. Cong. Appl. Mech.*, ed. E. Sternberg, pp. 487–91. New York: American Society of Mechanical Engineers.
Elam, C. F. (1926). Tensile tests of large gold, silver and copper crystals. *Proc. R. Soc. Lond.* A112: 289–94.
Elam, C. F. (1927a). Tensile tests on alloy crystals. Part II. Solid solution alloys of copper and zinc. *Proc. R. Soc. Lond.* A115: 148–69.
Elam, C. F. (1927b). Tensile tests on alloy crystals. Part IV. A copper alloy containing five percent aluminium. *Proc. R. Soc. Lond.* A116: 694–702.
Elam, C. F. (1935). *Distortion of Metal Crystals.* Oxford: Oxford University Press.
Elam, C. F. (1936). The distortion of β-brass and iron crystals. *Proc. R. Soc. Lond.* A153: 273–301.
English, A. T. & Chin, G. Y. (1965). On the variation of wire texture with stacking fault energy in f.c.c. metals and alloys. *Acta Met.* 13: 1013–16.
Fahrenhorst, W. & Schmid, E. (1932). Ueber die plastiche Dehnung von α-Eisenkristallen. *Z. Phys.* 78: 383–94.
Farren, W. S. & Taylor, G. I. (1925). The heat developed during plastic extension of metals. *Proc. R. Soc. Lond.* A107: 422–51.
Ferriss, D. P., Rose, R. M. & Wulff, J. (1962). Deformation of tantalum single crystals. *Trans. AIME* 224: 975–81.
Franciosi, P., Berveiller, M. & Zaoui, A. (1980). Latent hardening in copper and aluminium single crystals. *Acta Met.* 28: 273–83.
Franciosi, P. & Zaoui, A. (1982a). Multislip in f.c.c. crystals, a theoretical approach compared with experimental data. *Acta Met.* 30: 1627–37.
Franciosi, P. & Zaoui, A. (1982b). Multislip tests on copper crystals: a junctions hardening effect. *Acta Met.* 30: 2141–51.
Fuh, S. & Havner, K. S. (1986). On uniqueness of multiple-slip solutions in constrained and unconstrained f.c.c. crystal deformation problems. *Int. J. Plasticity* 2: 329–45.
Fuh, S. & Havner, K. S. (1989). A theory of minimum plastic spin in crystal mechanics. *Proc. R. Soc. Lond.* A422: 193–239.
Gil-Sevillano, J., Van Houtte, P. & Aernoudt, E. (1975). Deutung der Schertexturen mit Hilfe der Taylor-Analyse. *Z. Metallk.* 66: 367–73.
Gil-Sevillano, J., Van Houtte, P. & Aernoudt, E. (1982). Large strain work hardening and textures. In *Progress in Materials Science*, Vol. 25, ed.

J. W. Christian, P. Haasen, and T. B. Massalski, pp. 69–412. Oxford: Pergamon Press.

Göttler, E. (1973). Versetzungsstruktur und Verfestigung von [100]-Kupfereinkristallen – I. Versetzungsanordnung und Zellstruktur zugverformter Kristalle. *Phil Mag.* ser. 8, 28: 1057–76.

Green, G. (1839). On the laws of reflexion and refraction of light at the common surface of two non-crystallized media. *Trans. Camb. Phil. Soc.* 7: 1–24.

Haasen, P. & Lawson, A. W., Jr. (1958). Der Einfluss hydrostatischen Druckes auf die Zugverformung von Einkristallen. *Z. Metallk.* 49: 280–91.

Harren, S. V. & Asaro, R. J. (1989). Nonuniform deformations in polycrystals and aspects of the validity of the Taylor theory. *J. Mech. Phys. Solids* 37: 191–232.

Harren, S. V., Lowe, T. C., Asaro, R. J. & Needleman, A. (1989). Analysis of large-strain shear in rate-dependent face-centred cubic polycrystals: correlation of micro- and macromechanics. *Phil. Trans. R. Soc. Lond.* *A*328: 443–500.

Havner, K. S. (1971). A discrete model for the prediction of subsequent yield surfaces in polycrystalline plasticity. *Int. J. Solids Struct.* 7: 719–30.

Havner, K. S. (1973a). An analytical model of large deformation effects in crystalline aggregates. In *Foundations of Plasticity*, ed. A. Sawczuk, pp. 93–106. Leyden: Noordhoff.

Havner, K. S. (1973b). On the mechanics of crystalline solids. *J. Mech. Phys. Solids* 21: 383–94.

Havner, K. S. (1974). Aspects of theoretical plasticity at finite deformation and large pressure. *Z. angew. Math. Phys.* 25: 765–81.

Havner, K. S. (1977). On uniqueness criteria and minimum principles for crystalline solids at finite strain. *Acta Mech.* 28: 139–51.

Havner, K. S. (1978). On unifying concepts in plasticity theory and related matters in numerical analysis. *Nucl. Engr. Design* 46: 187–201.

Havner, K. S. (1979). The kinematics of double slip with application to cubic crystals in the compression test. *J. Mech. Phys. Solids* 27: 415–29.

Havner, K. S. (1981). A theoretical analysis of finitely deforming f.c.c. crystals in the sixfold symmetry position. *Proc. R. Soc. Lond.* *A*378: 329–49.

Havner, K. S. (1982a). The theory of finite plastic deformation of crystalline solids. In *Mechanics of Solids, The Rodney Hill 60th Anniversary Volume*, ed. H. G. Hopkins and M. J. Sewell, pp. 265–302. Oxford: Pergamon Press.

Havner, K. S. (1982b). Minimum plastic work selects the highest symmetry deformation in axially loaded f.c.c. crystals. *Mech. Mater.* 1: 97–111.

Havner, K. S. (1984). First- and second-order analysis of axially loaded crystals in *n*-fold symmetry. *Phil. Trans. R. Soc. Lond.* *A*311: 469–93.

Havner, K. S. (1985). Comparisons of crystal hardening laws in multiple slip. *Int. J. Plasticity* 1: 111–24.

Havner, K. S. (1986). Fundamental considerations in micromechanical modeling of polycrystalline metals at finite strain. In *Large Deformations of Solids: Physical Basis and Mathematical Modelling*, ed. J. Gittus, J. Zarka, and S. Nemat-Nasser, pp. 243–65. Amsterdam: Elsevier.

Havner, K. S. (1987a). On the continuum mechanics of crystal slip. In *Continuum Models of Discrete Systems*, ed. A. J. M. Spencer, pp. 47–59. Rotterdam: Balkema.

Havner, K. S. (1987b). On the kinematics of bi- and tri-crystals in channel die compression. In *Abstracts, 6th Engng. Mech. Spec. Conf.*, p. 158. New York: American Society of Civil Engineers.

Havner, K. S. & Baker, G. S. (1979). Theoretical latent hardening in crystals – II. Bcc crystals in tension and compression. *J. Mech. Phys. Solids* 27: 285–314.

Havner, K. S., Baker, G. S. & Vause, R. F. (1979). Theoretical latent hardening in crystals – I. General equations for tension and compression with application to f.c.c. crystals in tension. *J. Mech. Phys. Solids* 27: 33–50.

Havner, K. S. & Chidambarrao, D. (1987). Analysis of a family of unstable lattice orientations in (110) channel die compression. *Acta Mech.* 69: 243–69.

Havner, K. S. & Patel, H. P. (1976). On convergence of the finite-element method for a class of elastic-plastic solids. *Q. Appl. Math.* 34: 59–68.

Havner, K. S. & Salpekar, S. A. (1983). Theoretical latent hardening of crystals in double-slip – II. F.c.c. crystals slipping on distinct planes. *J. Mech. Phys. Solids* 31: 231–50.

Havner, K. S. & Shalaby, A. H. (1977). A simple mathematical theory of finite distortional latent hardening in single crystals. *Proc. R. Soc. Lond. A* 358: 47–70.

Havner, K. S. & Shalaby, A. H. (1978). Further investigation of a new hardening law in crystal plasticity. *J. Appl. Mech.* 45: 500–6.

Havner, K. S. & Singh, C. (1977). Application of a discrete polycrystal model to the analysis of cyclic straining in copper. *Int. J. Solids Struct.* 13: 395–407.

Havner, K. S., Singh, C. & Varadarajan, R. (1974). Plastic deformation and latent strain energy in a polycrystalline aluminum model. *Int. J. Solids Struct.* 10: 853–62.

Havner, K. S. & Sue, P. L. (1985). Theoretical analysis of the channel die compression test – II. First- and second-order analysis of orientation $[110][00\bar{1}][\bar{1}10]$ in f.c.c. crystals. *J. Mech. Phys. Solids* 33: 285–313.

Havner, K. S. & Varadarajan, R. (1973). A quantitative study of a crystalline aggregate model. *Int. J. Solids Struct.* 9: 379–94.

Heidenreich, R. D. (1949). Structure of slip bands and cold-worked metal. In *Cold Working of Metals*, pp. 57–64. Cleveland, OH: American Society for Metals.

Hill, R. (1950). *The Mathematical Theory of Plasticity*. Oxford: Oxford University Press.

Hill, R. (1958). A general theory of uniqueness and stability in elastic-plastic solids. *J. Mech. Phys. Solids* 6: 236–49.

Hill, R. (1965). Continuum micro-mechanics of elastoplastic polycrystals. *J. Mech. Phys. Solids* 13: 89–101.

Hill, R. (1966). Generalized constitutive relations for incremental deformation of metal crystals by multislip. *J. Mech. Phys. Solids* 14: 95–102.

Hill, R. (1968). On constitutive inequalities for simple materials – I, II. *J. Mech. Phys. Solids* 16: 229–42, 315–22.

Hill, R. (1970). Constitutive inequalities for isotropic elastic solids under finite strain. *Proc. R. Soc. Lond.* *A*314:457–72.

Hill, R. (1971). On macroscopic measures of plastic work and deformation in micro-heterogeneous media. *J. Appl. Math. Mech.* 35: 11–17. (English version of *Prikl. Mat. Mekh.* 35: 31–9.)

Hill, R. (1972). On constitutive macro-variables for heterogeneous solids at finite strain. *Proc. R. Soc. Lond.* *A*326: 131–47.

Hill, R. (1978). Aspects of invariance in solid mechanics. In *Advances in Applied Mechanics*, Vol. 18, ed. C.-S. Yih, pp. 1–75. New York: Academic.

Hill, R. (1984). On macroscopic effects of heterogeneity in elastoplastic media at finite strain. *Math. Proc. Camb. Phil. Soc.* 95: 481–94.

Hill, R. (1985). On the micro-to-macro transition in constitutive analyses of elastoplastic response at finite strain. *Math. Proc. Camb. Phil. Soc.* 98: 579–90.

Hill, R. & Havner, K. S. (1982). Perspectives in the mechanics of elastoplastic crystals. *J. Mech. Phys. Solids* 30: 5–22.

Hill, R. & Rice, J. R. (1972). Constitutive analysis of elastic-plastic crystals at arbitrary strain. *J. Mech. Phys. Solids* 20: 401–13.

Hill, R. & Rice, J. R. (1973). Elastic potentials and the structure of inelastic constitutive laws. *SIAM J. Appl. Math.* 25: 448–61.

Hirth, J. P. & Lothe, J. (1982). *Theory of Dislocations*, 2nd ed. New York: Wiley.

Hutchinson, J. W. (1970). Elastic-plastic behaviour of polycrystalline metals and composites. *Proc. R. Soc. Lond.* *A*319: 247–72.

Hutchinson, J. W. (1976). Bounds and self-consistent estimates for creep of polycrystalline materials. *Proc. R. Soc. Lond.* *A*348: 101–27.

Ilyushin, A. A. (1961). On the postulate of plasticity. *J. Appl. Math. Mech.* 25: 746–52. (Translation of *Prikl. Mat. Mekh.* 25: 503–7.)

Iwakuma, T. & Nemat-Nasser, S. (1984). Finite elastic-plastic deformation of polycrystalline metals. *Proc. R. Soc. Lond.* *A*394: 87–119.

Jackson, P. J. & Basinski, Z. S. (1967). Latent hardening and the flow stress in copper single crystals. *Can. J. Phys.* 45: 707–35.

Kallend, J. S. & Davies, G. J. (1972). A simulation of texture development in f.c.c. metals. *Phil. Mag.* ser. 8, 25: 471–90.

Karnop, R. & Sachs, G. (1927). Das verhalten von Aluminiumkristallen bei Zugversuchen. II. Experimenteller Teil. *Z. Phys.* 41: 116–39.

Keh, A. S. (1965). Work hardening and deformation sub-structure in iron single crystals deformed in tension at 298 °K. *Phil. Mag.* ser. 8, 12: 9–30.

Khedro, T. (1989). Comparative investigations of a general latent hardening theory for axially loaded f.c.c. crystals in tension & compression using the Green & Almansi strain measures. M.S. thesis, North Carolina State University.

Khedro, T. & Havner, K. S. (1991). Investigation of a one-parameter family of hardening rules in single slip in f.c.c. crystals. *Int. J. Plasticity* 7, 477–503.

Kocks, U. F. (1964). Latent hardening and secondary slip in aluminum and silver. *Trans. Met. Soc. AIME* 230: 1160–7.

Kocks, U. F. & Brown, T. J. (1966). Latent hardening in aluminum. *Acta Met.* 14: 87–98.

Kocks, U. F. & Chandra, H. (1982). Slip geometry in partially constrained deformation. *Acta Met.* 30:695–709.

Kocks, U. F., Nakada, Y. & Ramaswami, B. (1964). Compression testing of fcc crystals. *Trans. Met. Soc. AIME* 230:1005–9.

Kratochvil, J. (1971). Finite-strain theory of crystalline elastic–inelastic materials. *J. Appl. Phys.* 42:1104–8.

Krisement, O., de Vries, G. F. & Bell, F. (1964). Zur latenten Verfestigung in Kupfer-Einkristallen. *Phys. Status Solidi* 6:73–81.

Lee, E. H. (1969). Elastic-plastic deformation at finite strains. *J. Appl. Mech.* 36:1–6.

Lee, E. H. & Liu, D. T. (1967). Finite-strain elastic-plastic theory with application to plane-wave analysis. *J. Appl. Phys.* 38:19–27.

Lin, T. H. (1964). Slip and stress fields of a polycrystalline aggregate at different stages of loading. *J. Mech. Phys. Solids* 12:391–408.

Lin, T. H. (1971). Physical theory of plasticity. In *Advances in Applied Mechanics*, Vol. 11, ed. C.-S. Yih, pp. 255–311. New York: Academic.

Lin, T. H. & Ito, M. (1965). Theoretical plastic distortion of a polycrystalline aggregate under combined and reversed stresses. *J. Mech. Phys. Solids* 13:103–15.

Lin, T. H. & Ito, M. (1966). Theoretical plastic stress–strain relationship of a polycrystal and the comparisons with the von Mises and the Tresca plasticity theories. *Int. J. Engng. Sci.* 4:543–61.

Lin, T. H. & Tung, T. K. (1962). The stress field produced by localized plastic slip at a free surface. *J. Appl. Mech.* 29:523–32.

Lin, T. H., Uchiyama, S. & Martin, D. (1961). Stress field in metals at initial stage of plastic deformation. *J. Mech. Phys. Solids* 9:200–9.

McHugh, P. E., Varias, A. G., Asaro, R. J. & Shih, C. F. (1989). Computational modeling of microstructures. *Future Generation Computer Systems* 5:295–318.

McLean, D. (1962). *Mechanical Properties of Metals*. New York: Wiley.

Mair, W. M. & Pugh, H. L. D. (1964). Effect of pre-strain on yield surfaces in copper. *J. Mech. Engng. Sci.* 6:150–63.

Mandel, J. (1966). Contribution theorique a l'étude de l'ecrouissage et des lois de l'ecoulement plastique. In *Proc. 11th Intl. Cong. Appl. Mech.* (Munich 1964), ed. H. Görtler, pp. 502–9. Berlin: Springer.

Mark, H., Polanyi, M. & Schmid, E. (1923). Vorgänge bei der Dehnung von Zinkkristallen. I, II, III. *Z. Phys.* 12:58–77, 78–110, 111–16.

Molinari, A., Canova, G. R. & Ahzi, S. (1987). A self consistent approach of the large deformation polycrystal viscoplasticity. *Acta Met.* 35:2983–94.

Montheillet, F., Cohen, M. & Jonas, J. J. (1984). Axial stresses and texture development during the torsion testing of Al, Cu and α-Fe. *Acta Met.* 32:2077–89.

Nakada, Y. (1970). Unpublished experiments. Bell Telephone Laboratories, Allentown PA.

Nakada, Y. & Keh. A. S. (1966). Latent hardening in iron single crystals. *Acta Met.* 14:961–73.

Nakada, Y., Kocks, U. F. & Chalmers, B. (1964). Plastic deformation of oriented gold crystals. *Trans. Met. Soc. AIME* 230:607–9.

Neale, K. W., Toth, L. S. & Jonas, J. J. (1990). Large strain shear and torsion of rate-sensitive fcc polycrystals. *Int. J. Plasticity* 6: 45–61.

Nemat-Nasser, S. & Obata, M. (1986). Rate-dependent, finite elasto-plastic deformation of polycrystals. *Proc. R. Soc. Lond.* A407: 343–75.

Newton, I. (1687). *Mathematical Principles of Natural Philosophy*. English translation of third edition by A. Motte (1729) as revised by F. Cajori. Berkeley: University of California Press, 1934.

Nye, J. F. (1957). *Physical Properties of Crystals*. Oxford: Oxford University Press.

Opinsky, A. J. & Smoluchowski, R. (1951a). The crystallographic aspect of slip in body-centered cubic single crystals. I. Theoretical considerations. *J. Appl. Phys.* 22: 1380–4.

Opinsky, A. J. & Smoluchowski, R. (1951b). The crystallographic aspect of slip in body-centered cubic single crystals. II. Interpretation of experiments. *J. Appl. Phys.* 22: 1488–92.

Peirce, D., Asaro, R. J. & Needleman, A. (1982). An analysis of nonuniform and localized deformation in ductile single crystals. *Acta Met.* 30: 1087–119.

Peirce, D., Asaro, R. J. & Needleman, A. (1983). Material rate dependence and localized deformation in crystalline solids. *Acta Met.* 31: 1951–76.

Piercy, G. R., Cahn, R. W. & Cottrell, A. H. (1955). A study of primary and conjugate slip in crystals of alpha-brass. *Acta Met.* 3: 331–8.

Quinney, H. & Taylor, G. I. (1937). The emission of the latent energy due to previous cold working when a metal is heated. *Proc. R. Soc. Lond.* A163: 157–81.

Ramaswami, B., Kocks, U. F. & Chalmers, B. (1965). Latent hardening in silver and an Ag–Au alloy. *Trans. Met. Soc. AIME* 233: 927–31.

Rice, J. R. (1971). Inelastic constitutive relations for solids: an internal-variable theory and its application to metal plasticity. *J. Mech. Phys. Solids* 19: 433–55.

Rice, J. R. (1975). Continuum mechanics and thermodynamics of plasticity in relation to microscale deformation mechanisms. In *Constitutive Equations in Plasticity*, ed. A. S. Argon, pp. 23–79. Cambridge, MA: MIT Press.

Ronay, M. (1968). Second-order elongation of metal tubes in cyclic torsion. *Int. J. Solids Struct.* 4: 509–16.

Schmid, E. (1924). Neuere Untersuchungen an Metallkristallen. In *Proc. 1st Intl. Cong. Appl. Mech.*, ed. C. B. Biezeno and J. M. Burgers, pp. 343–53. Delft: Waltman.

Schmid, E. (1926). Ueber die Schubverfestigung von Einkristallen bei plasticher Deformation. *Z. Phys.* 40: 54–74.

Schmid, E. & Boas, W. (1935). *Kristallplastizität*. Berlin: Springer. English ed., *Plasticity of Crystals*. London: Hughes, 1950 (reissue, Chapman and Hall, 1958).

Sestak, B. & Libovicky, T. (1963). Transition to crystallographic slip on Fe–3% Si single crystals at 78° K. *Acta Met.* 11: 1190–1.

Sewell, M. J. (1972). A survey of plastic buckling. In *Stability*, ed. H. Leipholz, pp. 85–197. Waterloo: University of Waterloo Press.

Sewell, M. J. (1974). On application of saddle-shaped and convex generating functionals. In *Physical Structure in Systems Theory*, ed. J. J. van Dixhoorn and F. J. Evans, pp. 219–45. New York: Academic.

Sewell, M. J. (1987). *Maximum and Minimum Principles*. Cambridge: Cambridge University Press.

Shalaby, A. H. & Havner, K. S. (1978). A general kinematical analysis of double slip. *J. Mech. Phys.. Solids* 26: 79–92.

Skalli, A. (1984). Etude theorique et experimentale de la deformation plastique en compression plane de cristaux d'aluminium. State doctoral thesis. School of Mines of Saint-Etienne.

Skalli, A., Driver, J. H. & Wintenberger, M. (1983). Etude theorique et experimentale de la deformation plastique de monocristaux d'aluminium. Deuxieme partie: compression plane partiellement imposée. *Mem. Et. Sci. Rev. Met.* 80: 293–302.

Sue, P. L. & Havner, K. S. (1984). Theoretical analysis of the channel die compression test – I. General considerations and finite deformation of f.c.c. crystals in stable lattice orientations. *J. Mech. Phys. Solids* 32: 417–42.

Taylor, G. I. (1925). Notes on the 'Navier effect.' Aeronautical Research Committee. (Reprinted in *The Scientific Papers of Sir Geoffrey Ingram Taylor*, ed. G. K. Batchelor, Vol. I, *Mechanics of Solids*, pp. 130–2. Cambridge: Cambridge University Press, 1958.)

Taylor, G. I. (1926). The distortion of single crystals of metals. In *Proc. 2nd Intl. Cong. Appl. Mech.*, ed. E. Meissner, pp. 46–52. Zurich: Orell Füssli.

Taylor, G. I. (1927a). The distortion of crystals of aluminium under compression – Part II. Distortion by double slipping and changes in orientation of crystal axes during compression. *Proc. R. Soc. Lond.* *A*116: 16–38.

Taylor, G. I. (1927b). The distortion of crystals of aluminium under compression – Part III. Measurements of stress. *Proc. R. Soc. Lond.* *A*116: 39–60.

Taylor, G. I. (1928). The deformation of crystals of β-brass. *Proc. R. Soc. Lond.* *A*118: 1–24.

Taylor, G. I. (1935). Lattice distortion and latent heat of cold work in copper. Aeronautical Research Committee. (Reprinted in *The Scientific Papers of Sir Geoffrey Ingram Taylor*, ed. G. K. Batchelor, Vol. I, *Mechanics of Solids*, pp. 399–401. Cambridge: Cambridge University Press, 1958.)

Taylor, G. I. (1938a). Plastic strain in metals. *J. Inst. Metals* 62: 307–24.

Taylor, G. I. (1938b). Analysis of plastic strain in a cubic crystal. In *Stephen Timoshenko 60th Anniversary Volume*, ed. J. M. Lessels, pp. 218–24. New York: Macmillan. (Reprinted with added note in *The Scientific Papers of Sir Geoffrey Ingram Taylor*, ed. G. K. Batchelor, Vol. I, *Mechanics of Solids*, pp. 439–45. Cambridge: Cambridge University Press, 1958.)

Taylor, G. I. (1956). Strains in crystalline aggregates. In *Proc. Colloq. Deform. Flow Solids* (Madrid 1955), ed. R. Grammel, pp. 3–12. Berlin: Springer.

Taylor, G. I. & Elam, C. F. (1923). The distortion of an aluminium crystal during a tensile test. *Proc. R. Soc. Lond. A*102: 643–67.

Taylor, G. I. & Elam, C. F. (1925). The plastic extension and fracture of aluminium crystals. *Proc. R. Soc. Lond. A*108: 28–51.

Taylor, G. I. & Elam, C. F. (1926). The distortion of iron crystals. *Proc. R. Soc. Lond. A*112: 337–61.

Taylor, G. I. & Farren, W. S. (1926). The distortion of crystals of aluminium under compression – Part I. *Proc. R. Soc. Lond. A*111: 529–51.

Taylor, G. I. & Quinney, H. (1931). The plastic distortion of metals. *Phil. Trans. R. Soc. Lond. A*230: 323–62.

Taylor, G. I. & Quinney, H. (1934). The latent energy remaining in a metal after cold working. *Proc. R. Soc. Lond. A*143: 307–26.

Tomé, C., Canova, G. R., Kocks, U. F., Christodoulou, N. & Jonas, J. J. (1984). The relation between macroscopic and microscopic strain hardening in f.c.c. polycrystals. *Acta Met.* 32: 1637 – 53.

Toth, L. S., Gilormini, P. & Jonas, J. J. (1988). Effect of rate sensitivity on the stability of torsion textures. *Acta Met.* 36: 3077–91.

Toth, L. S., Jonas, J. J., Gilormini, P. & Bacroix, B. (1990). Length changes during free end torsion: a rate sensitive analysis. *Int. J. Plasticity* 6: 83–108.

Toth, L. S., Jonas, J. J. & Neale, K. W. (1990). Comparison of the minimum plastic spin and rate sensitive slip theories for loading of symmetrical crystal orientations. *Proc. R. Soc. Lond. A*427: 201–19.

Toth, L. S., Neale, K. W. & Jonas, J. J. (1989). Stress response and persistence characteristics of the ideal orientations of shear textures. *Acta Met.* 37: 2197–210.

Vause, R. F. & Havner, K. S. (1979). Theoretical latent hardening in crystals – III. F.c.c. crystals in compression. *J. Mech. Phys. Solids* 27: 393–414.

v. Göler & Sachs, G. (1927). Das Verhalten von Aluminiumkristallen bei Zugversuchen. I. Geometrische Grundlagen. *Z. Phys.* 41: 103–15.

von Mises, R. (1928). Mechanik der plastischen Formanderung von Kristallen. *Z. angew. Math. Mech.* 8: 161–85.

Vorbrugg, W., Goetting, H. C. & Schwink, C. (1971). Work-hardening and surface investigations on copper single crystals oriented for multiple glide. *Phys. Status Solidi B*46: 257–64.

Walpole, L. J. (1969). On the overall elastic moduli of composite materials. *J. Mech. Phys. Solids* 17: 235–51.

Wenk, H. R., Canova, G. R., Molinari, A. & Mecking, H. (1989). Texture development in halite: comparison of Taylor model and self-consistent theory. *Acta Met.* 37: 2017–29.

Wever, F. & Schmid, W. E. (1930). Texturen kaltverformter Metalle. *Z. Metallk.* 22: 133–40.

Williams, R. O. (1961). The stored energy in deformed copper: the effect of grain size and silver content. *Acta Met.* 9: 949–57.

Williams, R. O. (1962). Shear textures in copper, brass, aluminum, iron, and zirconium. *Trans. AIME* 224: 129–40.

Wolfenden, A. (1971). The energy stored in polycrystalline copper deformed at room temperature. *Acta Met.* 19: 1373–7.

Wonsiewicz, B. C. & Chin, G. Y. (1970a). Inhomogeneity of plastic flow in constrained deformation. *Met. Trans.* 1: 57–61.

Wonsiewicz, B. C. & Chin, G. Y. (1970b). Plane strain compression of copper, Cu 6Wt Pct Al, and Ag 4Wt Pct Sn crystals. *Met. Trans.* 1: 2715–22.

Wonsiewicz, B. C., Chin, G. Y. & Hart, R. R. (1971). Lateral constraints in plane strain compression of single crystals. *Met. Trans.* 2:2093–6.

INDEX